ANNALS *of* THE NEW YORK ACADEMY OF SCIENCES

EDITOR-IN-CHIEF
Douglas Braaten

ASSOCIATE EDITOR
Rebecca E. Cooney

PROJECT MANAGER
Steven E. Bohall

EDITORIAL ADMINISTRATOR
Daniel J. Becker

Artwork and design by Ash Ayman Shairzay

The New York Academy of Sciences
7 World Trade Center
250 Greenwich Street, 40th Floor
New York, NY 10007-2157

annals@nyas.org
www.nyas.org/annals

**The New York
Academy of Sciences**

Published by Wiley-Blackwell Publishing
On behalf of the New York Academy of Sciences

Boston, Massachusetts
2012

ANNALS *of* THE NEW YORK ACADEMY OF SCIENCES

VOLUME
1257

ISSUE

Barriers and Channels Formed by Tight Junction Proteins I

ISSUE EDITORS

Michael Fromm and Jörg-Dieter Schulzke

Charité, Campus Benjamin Franklin, Berlin, Germany

TABLE OF CONTENTS

Regulation of the tight junction and barrier function

Tight junctions in skin, lung, endothelia, and nervous tissues

Academy Membership: Connecting you to the nexus of scientific innovation

Since 1817, the Academy has carried out its mission to bring together extraordinary people working at the frontiers of discovery. Members gain recognition by joining a thriving community of over 25,000 scientists. Academy members also access unique member benefits.

 Network and exchange ideas with the leaders of academia and industry

 Broaden your knowledge across many disciplines

 Gain access to exclusive online content

 Select one free *Annals* volume each year of membership and get additional volumes for just $25

Join or renew today at **www.nyas.org**.
Or by phone at **800.843.6927** (**212.298.8640** if outside the US).

Ann. N.Y. Acad. Sci. ISSN 0077-8923

ANNALS OF THE NEW YORK ACADEMY OF SCIENCES

Issue: *Barriers and Channels Formed by Tight Junction Proteins*

Perspectives on tight junction research

Jörg-Dieter Schulzke,[1] Dorothee Günzel,[2] Lena J. John,[1] and Michael Fromm[2]

[1]Department of Gastroenterology, Infectious Diseases, and Rheumatology, Division of Nutritional Medicine, Charité, Universitätsmedizin Berlin, Germany. [2]Institute of Clinical Physiology, Charité, Universitätsmedizin Berlin, Germany

Address for correspondence: Prof. Dr. Jörg-Dieter Schulzke, Department of Gastroenterology, Infectious Diseases, and Rheumatology, Division of Nutritional Medicine, Charité, Universitätsmedizin Berlin, Campus Benjamin Franklin, Hindenburgdamm 30, 12203 Berlin, Germany. joerg.schulzke@charite.de

The tight junction connects neighboring epithelial or endothelial cells. As a general function, it seals the paracellular pathway and thus prevents back-leakage of just transported solutes and water. However, not all tight junctions are merely tight: some tight junction proteins build their own transport pathways by forming channels selective for small cations, anions, or water. Two families of tight junction proteins have been identified, claudins (27 members in mammals) and tight junction-associated MARVEL proteins ((TAMPs) occludin, tricellulin, and MarvelD3); an additional, structurally different, junction protein is junction adhesion molecule (JAM). Besides classification by genetic or molecular kinship, classification of tight junction proteins has been suggested according to permeability attributes. Recent studies describe specific *cis* and *trans* interactions and manifold physiologic regulations of claudins and TAMPs. In many inflammatory and infectious diseases they are found to be altered, for example, causing adversely increased permeability. Currently, attempts are being made to alter the paracellular barrier for therapeutic interventions or for transiently facilitating drug uptake. This overview concludes with a list of open questions and future topics in tight junction research.

Keywords: barrier; paracellular channel; claudin; occludin; tricellulin; inflammatory bowel disease

Introduction

The tight junction (TJ; zonula occludens) is a structure that allows epithelia to control passage of solutes and water through the lateral paracellular space. Without this barrier function, all absorbed ions would easily diffuse back, making vectorial net transport impossible. However, TJ proteins can also specifically allow the passage of single solutes, if functionally required. Thus, TJ permeability and selectivity is highly regulated and may range from leaky, as in the proximal segments of the nephron or the GI tract, to tighter, as in their distal segments, to more or less completely impermeable, as in the urinary bladder. This limiting property of the TJ is determined by its molecular composition. Only a few years after the first TJ protein, occludin, had been discovered by Tsukita's group in the early 1990s,[1] the claudin family of TJ proteins was identified.[2,3] So far 27 members of this family have been shown to participate in TJ assembly in mammals.[4] Most recently, the TJ proteins tricellulin and MarvelD3

were identified.[5,6] All of these share the specific feature of being tetraspanning membrane proteins. By possessing two extracellular loops, contact between each other and to TJs of neighboring cells is realized in a zipper-like organization. Two intracellular domains exist and the carboxy-terminal ends connect to scaffold proteins and, finally, to the cytoskeleton. As a result, TJs undergo rapid and pronounced regulation from cellular signaling.

This paper attempts to highlight current and future perspectives in TJ research, including a recently defined family of TJ proteins, TJ-associated MARVEL proteins (TAMPs); a classification of TJ proteins according to their barrier or channel function; specific *cis* and *trans* interaction of TJ proteins; physiologic regulations of claudins and TAMPs, altered claudin and TAMP expression in inflammatory and infectious diseases; attempts to alter the paracellular barrier for therapeutic interventions; and attempts to alter the paracellular barrier for transiently facilitating drug uptake. We close with

doi: 10.1111/j.1749-6632.2012.06485.x

a listing of open questions and future topics in TJ research.

TJ proteins as barrier or channel formers

TJ structure

TJs are located at the most apical position of the lateral intercellular space between neighboring epithelial cells.[7] In transmission electron microscopy (EM), an electron dense cytoplasmic material can be observed at the site of the TJ (kissing points). Freeze-fracture EM (FFEM) subsequently showed that kissing points are composed of intramembranous particles grouped together to form TJ strands. Strands are arranged in meshworks with various levels of complexity depending on the individual tissue.[8] Claude postulated a logarithmic correlation between strand number and electrical resistance, which was confirmed for a broad range of epithelia.[9] However, there are also tissues with similar strand numbers that greatly differ in resistance. This is due to the presence or absence of channel-forming TJ proteins (e.g., claudin-2, see Ref. 10) or reflective of the properties of the apical cell membrane, that is, the transcellular pathway.

Where three neighboring cells meet at a tricellular TJ, TJ strands expand considerably along the lateral membrane[11] and a central tube is established that is thought to be permeable to higher molecular weight solutes.

Single-span TJ proteins

To date, more than 30 different proteins have been identified in mammals that contribute to the formation of TJ strands and determine paracellular leakiness or tightness. In contrast to the four-transmembrane domain proteins, there are other strand-forming proteins with one or three transmembrane domains, for example, the JAM (junctional adhesion molecule) family with at least four members, as well as Crb3, CAR, and Bves.[12–16] These proteins are probably not directly responsible for permeability properties, but rather seem to influence strand formation processes.

Connection of strand-forming proteins to the cytoskeleton is realized via their PDZ domains, which bind to scaffold or TJ adapter proteins like ZO-1 to ZO-3, MAGI-1-3, MUPP-1, Par3, Par6, PALS1, or PATJ.[17,18] ZO-1 and ZO-3, but not ZO-2, directly bind to the actin cytoskeleton. A multitude of TJ-associated proteins, such as ZONAB, AP-1, symplekin, aPKA, Rab3b, Rab13, and WNK4, are important for regulatory influences on the TJ, including organization of cell polarity, cell proliferation/migration, and TJ assembly.[19,20]

TJ-associated Marvel proteins

Traditionally, tetraspan TJ proteins were subdivided into occludin and the claudin family; in 2005 this was subdivided further to include tricellulin. Only recently were occludin, tricellulin (also called MarvelD2), and MarvelD3[6] grouped as a new family called TAMPs.[21] TAMPs have in common that they contain a conserved four-transmembrane MARVEL (MAL [myelin and lymphocyte] and related proteins for vesicle trafficking and membrane link) domain. TAMPs and claudins share no sequence homology.[22] Splice variants of TAMPs are abundant, with four for occludin[23] and four for tricellulin,[23] and two splice variants for MarvelD3. TAMPs are found in all TJs throughout the body. Interaction with cytoskeleton proteins occurs via ZO-1 in the cases of occludin and tricellulin, whereas MarvelD3 only binds to other TAMPs.[21]

Claudin family of TJ strand proteins

It has only recently been recognized in mammals the claudin family of TJ proteins, with molecular weights between 22 and 25 kDa, should be expanded from 24 to 27 members from 24 to 27 different claudins.[4] In contrast, the pufferfish (*Takifugu rubripes*) expresses 56 different claudins.[24] Functionally, claudins are highly important strand components as is evidenced by knockout and overexpression studies, as well as by expressing claudins in mouse L-fibroblasts, which do not express endogenous TJ proteins.[2] Most claudins possess C-terminal PDZ binding motifs.[25]

PDZ binding motifs, which can bind ZO-1, were characterize as components of claudin-1–11, claudin-14–16, claudin-18, and claudin-19b. ZO-2 binding motifs were found in claudin-1–4, claudin-7–10, claudin-14, claudin-15, claudin-18, claudin-19b; and ZO-3 binding motifs were found in for claudin-8. Binding to MAGI-1, MAGI-2, or MAGI-3 could be corroborated for claudin-16 and claudin-23, and binding to MUPP-1 could be corroborated for claudin-3, claudin-8, and claudin-14.

Phylogenetic classification of claudins according to amino acid sequence homology has been performed.[4,24,26] By publishing a phylogenetic tree for the full-length sequences of claudins, a subdivision of claudins into *classic* (claudins 1–10, claudin-14, claudin-15, claudin-17, and claudin-19) and *non-classical* claudins (11–13, 16, 18, and 20–24) has been made according to their degree of sequence similarity.[22] Recently, a unifying modification of that phylogenetic tree was published considering all available data.[27] A corresponding tree was presented for the MARVEL proteins, including the TAMPs.[28]

It is by consensus that the function of the tetraspan TJ proteins is determined mainly by their two extracellular loops (ECL). The larger one, ECL1, specifies the paracellular barrier or selective channel properties. ECL2, in contrast, seems to be crucial for narrowing and holding the opposing lateral cell membranes.[22] Although phylogenetic trees provide deep insight in the molecular kinship within the claudin family and the TAMP family, they do not reflect functional characteristics of the involved TJ components.

Therefore, a classification on functional criteria was introduced recently. It subdivides the TJ proteins according to their permeability attributes into: sealing, channel forming, those with inconsistent functionality, and those with limited functional characterization (Table 1). This classification is, of course, not fixed, because, for example, as soon as they are clearly characterized, claudins of the groups 3 and 4 may change to groups 1 or 2.

TJ proteins with sealing function

On first view, TJ proteins with pronounced sealing properties are considered to be rather typical for tight than for leaky epithelia. The definition for tight and leaky is derived from electrical resistance measurements. Tight epithelia are characterized by high resistance, due to a low paracellular conductivity that is lower than transcellular conductivity. Furthermore, paracellular passage of molecules of either size is rather low.[30] In contrast, paracellular conductivity is higher than transcellular conductivity in leaky epithelia like proximal tubule and small intestine. Such epithelia show high absorptive transport rates without significant concentration gradients.[31] Sealing TJ proteins are claudin-1, claudin-3, claudin-5, claudin-11, claudin-14, and claudin-19, the function of which is indicated by experimental evidence from gene deficient animals and symptoms from gene mutations in humans.

Claudin-1. Claudin-1 knockout mice die immediately after birth from skin barrier defects. In humans, claudin-1 mutations lead to ichthyosis and sclerosing cholangitis. Further experimental evidence for the functional role as a tightening TJ molecule comes from cell experiments with low-resistance MDCK II cells, where claudin-1 overexpression increased resistance about fourfold.[32]

Claudin-3. In cell culture experiments with low-resistance MDCK cells, claudin-3 overexpression increased resistance threefold.[33] Interestingly, overexpression of claudin-3 had a special phenotype when significant amounts of claudin-2 are coexpressed (e.g., in low ohmic MDCK-cells). Then, claudin-3 overexpression appears to directly interact with the cation permeability of claudin-2, leading to an overproportional sealing effect, while water permeability of claudin-2 is not affected.[34] The mechanism is suggested to be heteromerization of claudin-3 with claudin-2.[35] Although in a low-resistance epithelium, claudin-3 exhibited definite sealing properties; however, it seems not to be completely impermeable, as in alveolar cells, which form a high-resistance epithelium, claudin-3 decreases the transepithelial resistance from 550 to 400 $\Omega \cdot cm$.[2,35]

Claudin-1 and claudin-3 appear to represent a minimum requirement of TJ formation, because only coexpression of both resulted in functional barrier formation of NIH/3T3 fibroblasts normally not possessing TJs.[36]

Claudin-5. The role of claudin-5 is less clear. Claudin-5 knockout mice suffer from skin and blood–brain barrier (BBB) defects. Claudin-5 has sealing functions in endothelial cells,[37] but does not tighten all epithelial cell line models as effectively. Whereas Caco-2 cells showed higher resistances, some MDCK cell lines were unaffected upon transfection.[38] Thus, the effect of claudin-5 may depend on the background of the other claudins, and it has been proposed that the claudin-5 sealing function requires claudin-12 as a binding partner, which is present in endothelial cells, for example.[36]

Claudin-11. Knockout mice exhibit neuronal defects, male sterility, and deafness.[39,40] Claudin-11 is present in corpus cavernosum endothelium.[41] In cell experiments, knockdown or overexpression of

Table 1. Current classification of claudins and TAMPs according to their permeability attribute[27,29]

Permeability attribute	Tight junction protein
Clearly sealing	Claudin-1, claudin3, claudin-5, claudin-11, claudin-14, claudin-19, tricellulin
Channel-forming	
Cation-selective	Claudin-2, claudin-10b, claudin-15
Anion-selective	Claudin-10a, claudin-17
Water-permeable	Claudin-2
Inconsistent function	Claudin-4, claudin-7, claudin-8, claudin-16, occludin
Unknown function	Claudin-6, claudin-9, claudin-12, claudin-13, claudin-18, claudin-20–27, MarvelD3

claudin-11 altered barrier function in endothelia and epithelia.

Claudin-14. In knockout mice claudin-14 deficiency leads to deafness. Similarly, in humans, claudin-14 mutations are characterized by deafness. Claudin-14 can seal the epithelial barrier, as indicated by overexpression in MDCK cells.[42]

Claudin-19. In mice claudin-19 deficiency causes Schwann cell defects and altered behavior,[43] as well as impaired kidney function.[44] In humans, claudin-19 mutations are associated with renal hypomagnesemia and ocular defects. Claudin-19 overexpression results in an increase in barrier function in MDCK II and LLC-PK1 cells.[45,46]

Tricellulin. In addition to the claudins and their sealing properties, tricellulin has a sealing function for TJs.[5] Tricellulin-A mutation induces inner ear deafness,[47] while overexpression of tricellulin limits macromolecule permeability through the central tube of the tricellular junction. Pronounced overexpression of tricellulin-A decreases ionic permeability by insertion into bicellular TJs.[46] Under physiological conditions, tricellulin appears to primarily regulate the central tube of the tricellular TJ.

Channel formers

A major change in paradigm occurred when it became clear that not all claudins seal the TJ more or less perfectly, but that several claudins form paracellular channels. A recent paper by Krug *et al.* explains the selectivity of claudins for cations, anions, and/or water. Thus, the channel-forming claudins are included in Table 1 but only explained briefly here.

Claudin-2. Claudin-2 was the first TJ protein identified to reduce transepithelial resistance if overexpressed.[48] This was observed to be based solely on permeability changes to small cations, which led to the conclusion that claudin-2 forms a paracellular cation channel.[49] Recently, it was demonstrated that claudin-2 is also permeable to water, which means that claudin-2 forms the paracellular counterpart of the transmembrane water channels, the aquaporins.[50]

Claudin-10b. Claudin-10 has two major splice variants, 10a and 10b, which differ in their first transmembrane region and ECL. Claudin-10b forms a cation-selective channel.[51,52]

Claudin-15. Claudin-15 is present in the intestine.[53] Overexpression in LLC-PK1 cells increased cation permeability.[54] Claudin-15–deficient mice develop a mega-intestine. Importantly, in adult mice claudin-15 is expressed in intestinal villi, where the sodium-glucose symporter SGLT1 predominates.[55] Claudin-15 and SGLT1 thus act in concert to recirculate sodium.

Claudin-10a. Claudin-10 may act as an anion pore in some cell systems and is found in the proximal tubule of the kidney.[51,52] Mutations that alter the charge of the claudin-10a pore cause a loss of this anion selectivity.[52]

Claudin-17. Recently, claudin-17 has been shown to form TJ channels with distinct anion selectivity.[56] Expression was found in two organs: marginal in brain but abundant in kidney proximal nephrons. This tubule segment exhibits substantial, though molecularly undefined, paracellular chloride reabsorption, suggesting that claudin-17 has a unique physiological function in that process.

TJ proteins with inconsistent function

In general, some claudins do not possess a clear sealing or channel-forming activity, although their functionality in transfection or knockdown experiments depends on the background of the other TJ proteins. On the one hand, this may be due to a direct interaction of these TJ proteins with specific partners, the presence or absence of which determines the functional consequences. On the other hand, this may be due to simple displacement of TJ proteins with more pronounced sealing or channel-forming function. Thus, TJ molecules with inconsistent function may be described as *sealing* in one cellular system but as *pore forming* in another. This means that definitively channel-forming TJ proteins (e.g., claudin-2) must have a very pronounced ability to increase cation permeability independently of the TJ protein background in the respective epithelium. The only limitation would be an epithelium that is already rich in cation pores like claudin-2, for example, low-resistance MDCK cells, which do not change barrier function after additional claudin-2 insertion. A more detailed definition of pore-forming and tightening TJ proteins has recently provided by Shen *et al.*[57]

Claudin-4. Claudin-4 overexpression in low-resistance MDCK cells increased resistance and decreased sodium permeability in favor of a barrier role,[53] whereas in M-1 and mIMCD3 cells, the opposite design—knockdown of claudin-4—also increased resistance indicating its role as an ion pore.[58] These surprising results in different cell models could either be the result of an interaction of claudin-4 with claudin-2 or may simply reflect a displacement of claudin-2 from the TJ in MDCK II cells. Alternatively, phosphorylation of claudin-4 may be a cofactor, because anion permeability was shown to correlate with claudin-4 phosphorylation in RCCD2 renal-collecting duct cells.[59] Finally, claudin-8 has been shown to be important for claudin-4 insertion into the TJ in M-1 and mIMCD3 cells[58] and thus could be another cofactor, although claudin-4 expression without significant coexpression of claudin-8 has been observed in the skin.[60]

Claudin-7. Knockdown of claudin-7 decreased resistance by an increase in sodium permeability in MDCK II cells.[61] Overexpression of claudin-7 in LLC-PK1 cells, on the other hand, caused an increase in resistance due to a decrease in chloride permeability, with a concomitant increase in sodium permeability.[62]

Claudin-8. Overexpression of claudin-8 resulted in decrease in mono- and divalent cation permeability, including protons and ammonium ions, and also to bicarbonate, in MDCK II cells. However, at least some of these effects seem to be due to claudin-2 displacement.[45,63]

Claudin-16. Claudin-16 (former name paracellin-1) has been identified as a protein causing familial hypomagnesemia, hypercalciuria, and nephrocalcinosis when it is mutated. Claudin-16 is localized in the thick ascending Henle's loop where magnesium is reabsorbed in the kidney. Its function and/or regulatory influences are not clear. Originally, claudin-16 was thought to be a magnesium pore, but this could not be confirmed in cell transfection experiments; an alternative explanation was that claudin-16–dependent changes in sodium permeability affect transport of magnesium or interaction with the Ca^{2+}-sensing receptor, which phosphorylates claudin-16.[64] Recently, Hou *et al.* found that claudin-16 is inserted into the TJ in response to claudin-19,[44] a result that needs further exploration. However, all attempts to explain the pathophysiology of magnesium transport and to define the role of claudin-16 have thus far been unsuccessful, called into question by experiments of other models, or are based on results obtained under different conditions that do not support the respective explanation.

Occludin. Occludin, discovered in 1993,[1] can be seen in every TJ and, consequently, is often used as a TJ marker. Even after two decades, however, the functional role of occludin is still a matter of debate. In knockout animals, occludin seems to have more regulatory functions, resulting in growth retardation and chronic atrophic gastritis, even though TJs in these animals appear morphologically and functionally normal.[65,66] Similarly, knockdown of occludin in MDCK II cells did not show barrier disturbances.[67] In contrast, overexpression of occludin increased resistance in MDCK II cells.[67] Taken together, occludin may tighten the epithelial barrier when transfected into a suitable background of TJ proteins; but when occludin is deficient, other TJ proteins can compensate (a phenomenon known as *substitutional redundancy*).

TJ proteins with unknown function

Claudin-6. Claudin-6 is transitorily expressed during development of the intestine, as a marker of the endoderm and the kidney (proximal tubule). In adults, claudin-6 is not detectable. Surprisingly, claudin-6 knockout mice do not suffer from intestinal differentiation arrest or show significant structural or functional changes, which points again toward substitutional redundancy for this TJ protein. Surprisingly, overexpression of claudin-6 leads to skin defects in mice, which might be caused by displacement of other claudins.[68] Both claudin-6 and claudin-9 overexpression increase resistance and reduce anion and cation permeability of the TJ in MDCK II cells.[69]

Claudin-9. Claudin-9, similar to claudin-6, is mainly transitorily expressed during early development in the proximal tubule;[70] its overexpression also increases resistance (see Claudin-6, above). In adult mice, claudin-9 expression is found in the cochlea, and mutation of claudin-9 causes deafness.[71,72]

Claudin-12. Claudin-12 is found in the BBB, together with claudin-3 and claudin-5, and in the urinary bladder, epidermis, hair follicles,[73] along the GI tract,[53] and in peripheral nerve cells.[74] Claudin-12 may be a barrier-forming TJ protein, for example, as part of the very tight urinary bladder, but final experimental evidence is lacking.

Claudin-13. In adult mice claudin-13 is expressed in the urinary bladder and the large intestine, whereas during development it is also expressed in the kidney.[53] Functional data are not available. Humans do not express claudin-13.

Claudin-18. Claudin-18 has two splice variants, claudin-18A1.1 is expressed in the lung and claudin-18A2.1 is expressed in stomach epithelium.[75] High claudin-18 expression has also been observed in Barrett's mucosa. Overexpression of claudin-18 in MDCK II cells tightens the TJ via reduction in sodium and proton permeability, either directly or by replacing claudin-2.[76]

Claudin-20, -22, and -23. Claudin-20, claudin-22, and claudin-23 may be involved in BBB, as they are expressed in mouse brain endothelial cells.[74]

Claudin-21, -24, and -27. The role of these claudins is unknown. After MDCK cell transfection, claudin-24 and claudin-27 are detected intracellularly and not in the TJ.

Claudin-25 and -26. Claudin-25 and claudin-26 colocalize with ZO-1 after transfection into MDCK cells.

MarvelD3. There is no published information on the permeability features of this protein.

Interactions of TJ proteins

TJ proteins bind to other TJ proteins within the same membrane-forming TJ strands (*cis* interaction), and also to TJ proteins in the plasma membrane of neighboring cells (*trans* interaction). Although a single claudin, such as claudin-1, is sufficient to form a TJ meshwork in L-fibroblasts,[2] all TJs contain many different TJ proteins under various physiological conditions. Some TJ proteins like occludin need claudins as a backbone structure for their insertion into the TJ.

Furuse *et al.* have shown through transfection of different claudins into L-fibroblasts, and cocultivation of these cells, that both claudin-1 and claudin-2 bind in *trans* to claudin-3 but not to each other.[34] This has recently been confirmed by Lim *et al.* and Piontek *et al.*[77–79] Piontek *et al.* studied BBB-relevant TJ proteins and found homophilic *trans* interactions between claudin-1, claudin-2, claudin-3, and claudin-5 but not between claudin-12, and heterophilic *trans* interactions between claudin-1 and claudin-5, claudin-1 and claudin-3, and claudin-3 and claudin-5,[78] while strong *trans* interaction from immunoprecipitations are postulated between claudin-1 and claudin-3, and claudin-3 and claudin-5, but not between claudin-1 and claudin-5.[36]

Daugherty *et al.* have further studied *trans* interactions between claudin-1, claudin-3, claudin-4, and claudin-5.[80] Despite high structural homology between caludin-3 and claudin-4, which makes them binding partners of *Clostridium perfringens* enterotoxin, these two claudins do not bind to each other in *trans*. In addition, claudin-1 and claudin-5 do not bind in *trans* to claudin-4, although both interact with claudin-3. A chimeric claudin-4 that includes either ECL1 or ECL2 of claudin-3 interacts in *trans* with claudin-1 and claudin-5. On the other hand, the amino acid sequence of the ECL is not the only determinant for *trans* binding; a chimer with a claudin-3 backbone and both claudin-4 ECLs binds to a chimera with a claudin-4 backbone and both claudin-3 ECLs. This finding points to the importance of the orientation of the transmembrane domains that determine the conformation. Investigating *cis* interactions, Van Itallie *et al.* found claudin-4 monomers, and claudin-2 homo-dimerization occurred via the second transmembrane domain (helix–helix interaction). Van Itallie *et al.* also recently detected TJ complexes of 600 kDa containing occludin, claudin-1, and claudin-2.[81] *Cis* homo-dimerization was also found by FRET analysis for occludin (via the C-terminal domain) and for claudin-5.[82] Piontek *et al.* described *cis* interactions of claudins in an order of 5/5 > 5/1 > 3/1 > 3/3 > 3/5, but not for 3/2.[78] Interestingly, despite lacking *trans* interaction, claudin-3 and claudin-4 can interact in *cis*.[80]

Furthermore, interactions between claudin-4 and claudin-8, and claudin-16 and claudin-19, have been studied by yeast two-hybrid methodology.[44,58] Claudin-4 and claudin-8 are cotransported from the Golgi to the TJ and interact in *cis* with each other. Claudin-16 and claudin-19 interact in *cis*, but not in *trans*. HOMGO-mutations of claudin-19 disturb either claudin-19 function or homo-dimerization, leading to a lack in TJ localization. Claudin-19 also binds to claudin-10 (to both isoforms) and to claudin-18, although weaker compared to claudin-16.[44]

Tricellulin shows homo- and hetero-dimerization in *cis*, namely with occludin.[83] However, no *trans* interactions were detected. Table 2 lists the various *cis* and *trans* interactions of TJ proteins, including claudins and TAMPs. This diversity may easily explain controversial results with knockdown or over-expression experiments in the literature, as background of the other TJ proteins present in a TJ is an important determinant of the functional consequences.

TJ regulation

TJs are a cell domain with high plasticity. Assembly and disassembly of TJs is a continuous process in every epithelium, either by endocytotic internalization of TJ proteins or by supply of new TJ proteins along the membrane flow from the Golgi to the TJ (but also via lateral membrane diffusion to the TJ). Interestingly, endocytosis of TJs additionally leads to internalization of the in *trans* interacting TJ proteins with the neighboring epithelial cell.[86]

Measured with FRAP, the half-life time of different TJ proteins within the TJ was as short as five minutes.[87] Regulatory inputs that modulate the velocity of these processes are abundant and highly important.

Numerous regulatory effects have been described for occludin, the first integral TJ protein identified. It is relocalized in response to its phosphorylation status.[88] PKC, PKA, MAPK, PI3K, CK1, CK2, tyrosine kinases, and RhoK are responsible for phosphorylation at different sites (serine or threonine residues). Consequently, the phosphorylation status can affect binding to claudin-1 and claudin-2, as well as to ZO-1.[89] Phosphorylation by some kinases stabilizes the TJ (such as PKCη[90]), whereas others lead to disassembly (such as c-src[91]). Several phosphatases with distinct specificity, such as PP1, PP2A, or DEP-1, can terminate the effect of these kinases and contribute to the differential regulation of TJ protein phosphorylation.[92] PP1 dephosphorylates serine and PP2A threonine; DEP-1 dephosphorylates tyrosine- and, as a consequence, resistance is increased and, for example, 10 kDa dextran permeability is reduced.[93] Tricellulin has 35% homology to occludin,[5] but tricellulin is not a CK1 or CK2 substrate.[94]

For claudins, several phosphorylation sites have been detected. Protein kinases that can phosphorylate claudins include PKC, PKA, MAPK, WNK, myosin light chain kinase (MLCK), c-Src, RhoK, and the ephrin receptor family. However, a role for regulation has only been confirmed for a few of these kinases. For example MAPK-, PKC-, and PKA-dependent phosphorylation of claudin-1 leads to insertion into the TJ;[95] dephosphorylation of claudin-1 by PP2A antagonizes this effect.[92] WNK4 and WNK4 mutations, causing hypertension due to pseudohyperaldosteronism type II, led to phosphorylation of claudin-1 in MDCK II cells and increased chloride permeability.[96] The PKA-dependent phosphorylation status of claudin-3 determines TJ disassembly and thus attenuates barrier function. Claudin-4 phosphorylation by PKC is important for TJ formation.[97] Claudin-16 is ERK- and PKA-dependently phosphorylated, which leads to claudin-16 to insertion into the TJ—regulation that is modulated by the Ca^{2+}-sensing receptor.[98] In addition to phosphorylation, palmitoylation has also been shown to modulate claudin-2, claudin-4, and claudin-14 tight-junction localization.[99]

Table 2. Confirmed *cis* and *trans* interactions between tight junction proteins

Interaction	*Trans*	*Cis*	Unspecified	Methods	Reference
cldn1–cldn1	+			*Trans*: co-enrichment in HEK cells	79
		+		*Cis*: blue native PAGE + cross-linking	67
			–	Unspecified: PFO- and SDS-PAGE	37
cldn1–cldn2	–	+		*Cis*, *trans*: co-enrichment and FFEM in L cells	35
	–			*Trans*: AFM	77
cld1–cld3	+	+		*Cis*: FRET, *trans*: co-enrichment in HEK cells	79
	+	+		*Cis*, *trans*: co-enrichment and FFEM in L cells	35
	+			*Trans*: co-enrichment in HeLa cells, co-IP	80
				Trans: co-IP	37
cld1–cld4	–			*Trans*: co-enrichment in HeLa cells, co-IP	80
cld1–cld5	+	+		*Cis*: FRET, *trans*: co-enrichment in HEK cells	79
	–			*Trans*: co-enrichment in HeLa cells, co-IP	80
	–			*Trans*: co-IP	37
cld1–cld10b	–			*Trans*: FFEM on HEK cells	84
cld1–cld15	–			*Trans*: FFEM on HEK cells	84
cld2–cld2	+			*Trans*: AFM	77
		+		*Cis*: blue native PAGE + cross linking	67
cld2–cld3	+	–		*Cis*: FRET, *trans*: co-enrichment in HEK cells	79
	+	+		*Cis*, *trans*: co-enrichment and FFEM in L cells	35
cld3–cld3	+	+		*Cis*: FRET, *trans*: co-enrichment in HEK cells	79
			–	Unspecified: yeast 2-hybrid	58
			–	Unspecified: PFO- and SDS-PAGE	37
cld3–cld4			+	Unspecified: yeast 2-hybrid	58
	-	+		*Trans*: co-enrichment in HeLa cells, co-IP	80
cld3–cld5	+	+		*Cis*: FRET, *trans*: co-enrichment in HEK cells	79
	+			*Trans*: co-enrichment in HeLa cells, co-IP	80
	+			*Trans*: permeability measurements, co-IP	37
cld3–cld7			–	Unspecified: yeast 2-hybrid	58
cld3–cld8			+	Unspecified: yeast 2-hybrid	58
cld4–cld4			–	Unspecified: yeast 2-hybrid	58
		–		*Cis*: blue native PAGE + cross linking	67
cld4–cld5	–			*Trans*: co-enrichment in HeLa cells, co-IP	80
cld4–cld7			–	Unspecified: yeast 2-hybrid	58
cld4–cld8			+	Unspecified: yeast 2-hybrid, coimmunoprec.	58
cld5–cld5	+	+		*Cis*: FRET,	79
				Trans: co-enrichment in HEK cells	
			+	Unspecified: PFO- and SDS-PAGE	37
cld7–cld7			+	Unspecified: yeast 2-hybrid	58
	+			*Cis* : blue native PAGE + cross linking	67
cld7–cld8			+	Unspecified: yeast 2-hybrid	58
cld8–cld8			+	Unspecified: yeast 2-hybrid	58
cld10a/b–cld16			–	Unspecified: yeast 2-hybrid	44
cld10a/b–cld19			+	Unspecified: yeast 2-hybrid	44
cld12–cld12	–			*Trans*: co-enrichment in HEK cells	79

Continued

Table 2. *Continued*

Interaction	*Trans*	*Cis*	Unspecified	Methods	Reference
cld16-cld16	–	–		FFEM in L fibroblasts	85
cld16-cld18			–	Unspecified: yeast 2-hybrid	44
cld16-cld19	–		+	Unspecified: yeast 2-hybrid, *trans*: co-enrichm. in L cells	44, 85
cld18-cld19			+	Unspecified: yeast 2-hybrid	44
occ-occ		+		*Cis*: FRET	82
occ-tric	–	+		*Cis*, *trans*: co-enrichment in HEK cells, co-IP	83
tric-tric	–	+		*Cis*, *trans*: co-enrichment in HEK cells, co-IP	83

Altered TJs in disease

There is an emerging understanding of the interaction of the various TJ proteins and their (dys-) regulation in states of disease, ranging from barrier breakdown during infection and inflammation to loss of cell polarity and changed migratory behavior during tumorigenesis. The following will describe this in detail, with special attention to the GI tract.

TJs—virus binding and interaction with bacterial toxins

Viral and bacterial pathogens have been shown to interfere with TJ proteins. They can be important to invasion by acting as receptors or infection cofactors. For example, claudin-1, claudin-6, and claudin-9, as well as occludin, enable hepatitis C virus (HCV) entry,[100,101] claudin-7 promotes HIV infection,[102] and claudin-1 facilitates dengue virus uptake.[103] JAM is a reovirus receptor[104] and CAR is a coxsackie virus and adenovirus receptor.[105] Claudin-3 and claudin-4 are receptors for *C. perfringens* enterotoxin (CPE-receptor-2 and CPE-receptor-1[106]) via the C-terminus of CPE binding to a helix–loop–helix structure of ECL2.[107] CagA of *Helicobacter pylori* causes translocation of ZO-1 and JAM-A to the *H. pylori* adhesion site, which leads to a loss of barrier and fence function of the TJ.[108]

TJ changes in infectious or inflammatory diseases

Infectious or inflammatory diseases can cause TJ alterations, and as a result, noxious agents such as antigens or toxic components can pass through the epithelial barrier and interfere with the immune system to perpetuate inflammation and loss of ions and water through the epithelium, which can lead to leak flux diarrhea.[109] Changes in TJ protein expression and distribution in infectious and inflammatory diseases of the intestine are presented in Table 3. Whereas infectious agents induce downregulation of TJ proteins, inflammatory diseases trigger differential alterations in TJ proteins, for example upregulation of claudin-2 while other TJ proteins are downregulated. This is explained by the action of proinflammatory cytokines specifically regulating the TJ. Claudin-2 can also be upregulated in infectious diseases when infectious agents activate the immune system to release cytokines, for example in untreated HIV infection.[128]

TJs in inflammatory bowel disease

In inflammatory bowel disease (IBD), epithelial barrier dysfunction, with altered TJ structure and function, contributes to the onset or perpetuation of disease activity (leaky gut) and to clinical symptoms (leak flux diarrhea). In the distal colon of patients with mild Crohn's disease, TJs have a reduced number of horizontally oriented strands, with reduced main meshwork depth and more frequent strand breaks. Sealing TJ proteins are downregulated, as occludin, claudin-3, claudin-5, and claudin-8 are distributed away from the TJ to the basolateral plasma membrane or into the epithelial cell.[110] On the other hand, the pore-forming TJ protein claudin-2 is upregulated. The pattern of TJ alterations in moderately inflamed distal colon of patients with ulcerative colitis is very similar, with a reduction in strand count and main meshwork depth, although no strand breaks are detected.[134] Occludin, claudin-1, and claudin-4 are downregulated and claudin-2 is upregulated (claudin-5 and claudin-8 were not studied).[135] In addition, apoptotic leaks have been shown to contribute to the barrier disturbance found in ulcerative colitis and Crohn's

Table 3. Changes in tight junction protein expression in inflammation or infection

	Cld-1	Cld-2	Cld-3	Cld-4	Cld-5	Cld-7	Cld-8	Occl	Involved cytokines	Reference
Inflammatory or immune-mediated disease										
Crohn's disease	=	↑		=	↓	=	↓	↓	TNF-α, IFN-γ	110
Ulcerative colitis	=	↑	=	↓		↓			IL-13	111
									IL-1β, IL-6, IL-8, TNF-α	112
Mouse DSS-colitis	↓	=	↓	↓	↓			↓		113
									Acute: TNF-α, IL-6, IL-17	114
Pouchitis	↓	↑	=	=	=	=		=		115
									IL-1β, IL-8	116
									IL-1β, IL-6, IL-8, TNF-α	117
Collagenous colitis	=		=	↓	=			↓		118
									IFN-γ, IL-15, TNF-α,	119
									TGF-β1	120
Atopic dermatitis	↓									121
Pleural inflammation	↓	↑	↓		↓	↓		↓		122
									IL-5, IL-6, IL-8, TNF-α	123
Celiac disease		↑	↑	=						124
									IFN-γ, IL-1β, TNF-α, IL-4, IL-6, IL-8, IL-10	125
Gliadin application	=		↓	↓				↓		126
Enterohemorrhagic E. coli		↑	Displaced					↓	TNF-α	127
HIV	↓	↑		=				=	TNF-α, IL-2, IL-4, IL-13	128
Cytokines										
TNF-α	↓	↑								129
TNF-α + IFN-γ	↓	↑		=	↓	↓		↓		130
IL-13	=	↑		=						131
TGF-β	↓	↑						↑		132
				↑						133

disease.[110,135,136] These apoptotic leaks can develop toward microerosions and even ulcers if epithelial restitution is hampered, such as in ulcerative colitis, where interleukin-13 affects epithelial repair mechanisms such as cell migration.[135] Another mechanism that interferes with TJ barrier function during intestinal inflammation is epithelial–mesenchymal transition, which has recently been shown to be important in celiac disease.[137] Epithelial-mesenchymal transition changes cell polarization within the epithelium and can increase transcytotic antigen uptake and TJ protein internalization by endocytosis. Even bacterial translocation in IBD can be enabled by endo-/transcytosis. There is also preliminary evidence that tricellulin, which restricts macromolecule permeability through the tricellular TJ, is impaired in ulcerative colitis.[46,138]

TJs and inflammatory cytokines

In IBD cytokines are important regulators for epithelial barrier dysfunction and TJ changes. In Crohn's disease, TNF-α and IFN-γ are predominant (Th1 profile), whereas in ulcerative colitis IL-13—together with TNF-α—is the key effector cytokine. Similar TJ disturbances are seen in cell and animal models in response to these cytokines. TNF-α and IL-13 upregulate claudin-2 expression via the phosphatidylinositol-3-kinase pathway;[139,140] TNF-α and IFN-γ up-regulate claudin-2 and downregulate claudin-1, claudin-5, and claudin-7 in rat colon.[130] Transforming growth factor-β (TGF-β) is a protective factor in IBD, which can increase claudin-4 expression via claudin-4 promoter activation.[133] MLCK is another TJ regulator in IBD.[141] TNF-α and IL-1β stimulate MLCK gene expression via NF-κB;[142] MLCK phosphorylation induces reorganization of perijunctional F-actin, with redistribution of TJ proteins off the TJ,[143] and this is mediated by an MLCK-induced exchange in ZO-1.[144] MLCK also induces caveolin-1–dependent endocytosis of TJ proteins such as occludin.[145] IFN-γ causes TJ redistribution via Rho/ROCK-induced macropinocytosis.[146] As the result of one of these mechanisms, TNF-α, IL-13, and IFN-γ also increase bacterial translocation.[147] Finally, TNF-α and IL-13 also induce epithelial apoptosis.[134,135]

TJs in the etiology of IBD-induced barrier dysfunction

Ulcerative colitis is associated with SNPs in PAR-3 and MAGI2, both of which regulate TJ assembly.[148] In addition, HNF4α was identified as a susceptibility locus in a genome-wide association study on ulcerative colitis.[149] HNF4α transcriptionally regulates TJs, adherens junctions, and desmosomes;[150] for example, claudin-15 is regulated via HNF4α,[151] and intestinal HNF4α deletion leads to barrier dysfunction and exacerbation of dextran sulfate sodium (DSS)-induced colitis.[152] Also, regulation of claudin-2 expression may be involved, as its promoter has Cdx, GATA-4, and HNF1α binding sites.[153]

Risk factors for Crohn's disease include mutations in the caspase-activated recruitment domain (CARD15) gene. NOD2, the product of *CARD15*, is a sensor for bacterial components[154] and *CARD15* mutations are associated with barrier dysfunction.[155] NOD2 signaling defects can lead to downregulation of defensins, which can initiate inflammatory barrier defects as the result of pathogen access to the mucosa.[156]

Gut microbiota are another important factor that influences intestinal barrier function, especially, for example, in IL-2 or IL-10 knockout mice. The spontaneous intestinal inflammation in these models is significantly attenuated under germ-free conditions.[157] The same applies to DSS-induced colitis.[158] Microbial alterations in IBD have been shown to reduce the concentration of short-chain fatty acids.[159] Butyrate, an important energy source for the intestine, has anti-inflammatory properties and decreases bacterial translocation.[160,161]

Pearl string-like (particle type) and non- continuous type TJ strands occur in IBD more often than in control in freeze-fracture EM studies. However, the significance of this phenomenon is unknown. On the one hand, it has been related to claudin-2 in the TJ, and claudin-2 is upregulated in IBD.[48] On the other hand, it seems to seems to be related more with TJ strand breaks or discontinuities, which develop due to the incompatibility of claudin-1 and claudin-2.[48] But while the latter feature may indeed represent some kind of a leak for larger molecules, it is too rare to have a major influence on ion permeability.

Transiently facilitating drug uptake

Specifically and transiently opening the paracellular barrier for purposes of enhancing drug uptake is important, as numerous drugs are neither absorbed in the intestinal nor in the respiratory tract. The idea here is to selectively open the TJ for the passage of macromolecules. This effect should be fast and reversible, not only to facilitate the passage of a drug but also to minimize unwanted translocation of noxious agents (see Ref. 162).

Therapeutic influences on TJs

Specific therapies that affect TJs are rare. In principle, therapeutic interventions targeting epithelial barrier function via the TJ, perhaps in form of a multimodal approach, seems to be an attractive concept for attenuating IBD activity as long as a causative therapy is lacking.

In Crohn's disease, zinc has been shown stabilize epithelial barrier function,[163] although the mechanism has not yet been determined (upregulation of occludin and ZO-1 upregulation has only been demonstrated in an animal model[164]). Among other things, zinc decreases TNF-α and IL-1 release, and the zinc finger protein A20 inhibits NF-kB signaling.[140]

Flavonoids can have direct effects on TJs, as shown, for example, with quercetin, which Quercetin improves barrier function in Caco-2 cells by upregulating claudin-4 expression.[165] In part, this is PKC mediated.

Berberine is a traditional remedy and demonstrates anti-inflammatory action in experimental colitis.[166] In Caco-2 cells, berberine prevents TNF-α– and INF-γ–induced barrier dysfunction by preventing upregulation of claudin-2 and redistribution of claudin-1 away from TJs.[129]

Probiotics influence pro-inflammatory cytokine expression[167] and can affect TJ protein expression and composition. In a mouse colitis model, reduction and redistribution of ZO-1, occludin, claudin-1, claudin-3, claudin-4, and claudin-5 are prevented by the probiotic VSL#3®.[113] The probiotic strain *Escherichia coli* Nissle 1917 maintains remission in ulcerative colitis[168] and enhances mucosal integrity;[169] however, the direct mechanism of action on the TJ has not been identified so far.

Unsolved questions and future topics in TJ research

Thus far this review has underscored several points that have not been resolved and questions that have not yet been answered. Therefore, the concluding section outlines topics and goals that need to be addressed in the future. Goals for future TJ research can be sorted into four categories: structure, regulation, function, and pharmacological tools.

Structure
• Identification of intramolecular factors that determine channel formation and channel selectivity. Although several amino acids have been identified that are responsible for channel selectivity,[52,170,171] their nature of the selectivity filter is only starting to emerge, and basic differences between barrier-forming and channel-forming TJ proteins are unknown.[63]

• Identification of intermolecular interactions that lead to strand formation, causing barrier or channel properties. Nearly all TJs contain several claudins that need to interact within single strands,[172] and some interactions between two different claudins have been studied.[27,79] However, how multiple interactions take place on a molecular level and how they modulate TJ function is unclear.

• Determination of the three-dimensional ultra-structure of the TJ proteins, particles, and strands. Knowledge of the three-dimensional structure of TJ units will be indispensable for answering the previous questions. A first step may be the recently reported isolation of 600 kDa multiprotein TJ particles.[81]

• Identification of further TJ proteins. Recent publications presenting, for example, three further mammalian claudins and proteins, such as MarvelD3 or LSR (lipolysis-stimulated lipoprotein receptor), indicate that we are far from knowing all players involved in TJ building.[4,6,173]

• Identification and characterization of splice variants. Splice variants exist for many members of the TAMP and claudin families and findings, such as those for claudin-10 and claudin-18, suggest that different splice variants may even have complementary function. However, in many cases, the physiological role of splice variants is still unknown.

• Fence function of the TJ. Between the two functions of TJs, *gate* and *fence*, the gate function is more prevalent. Although both functions are based on the same structure, they may be dissociated from each other under certain experimental conditions, for example, those that cause a loss of barrier but maintain fence function.[174,175] Thus, for a complete understanding of TJ function, further investigation of the fence function will be indispensable.

Regulation
• Investigation of transcriptional regulation of TJ proteins. As illustrated in Table 3, common regulatory patterns, for example, pattern apparent under inflammatory conditions, start to emerge, but are far from being understood in detail.

• Investigation of posttranslatory regulation of TJ proteins. Almost all TJ proteins possess multiple phosphorylation sites; however, in most cases, it is not yet known under which conditions, and to what effect, phosphorylation occurs.[176] Phosphorylation-specific antibodies need to be developed to allow detection/quantification of the phosphorylation status.

- Regulation of insertion/withdrawal of TJ proteins in/from TJ strands. Endocytosis has been described as being responsible for TJ protein internalization.[86] In this context, interaction between TJ proteins and caveolin emerges to be of major importance.[177] However, receptor-dependent endocytosis should also be considered, especially as a potential ligand for bacterial toxin.
- Dynamics of TJ strands. TJ strands have to be continuously remodeled to maintain the paracellular barrier, as cells move within epithelial layers.[86] Further development of techniques, such as high resolution LSM, will possibly allow such questions to be addressed to TJ strand dynamics.

Function

- Function on the cellular level. It is not yet clear to what extent other cell–cell junctions contribute to the gate function of TJs. This question is in part covered by studies performed during the development of epithelial confluence, but is less well studied thereafter. More information is needed on the interaction of para- and transcellular transport processes. So far, only a few examples are available—for example, hormone-dependent transport activation affecting the paracellular path via regulation of TJ protein expression.[178]
- Function on the tissue level. In stratified epithelia, TJ structure is very complex and thus requires refined morphological techniques for analysis. Only when this structural complexity is resolved can barrier function of these epithelia be understood.
- Function on the organismic level. Animal models with gene deficiency of TJ proteins not only provide insight into the role of the proteins but also allow study of regulatory functions of TJ proteins—for example, by investigation of compensatory up/downregulation of other TJ proteins. This may in part be achieved by reevaluation of existing knockout models. In this context, a major aspect may turn out to be the role of claudins in embryonic developmental.[179] Because TJ protein expression is dramatically changed in tumors, it is often used as a marker for patient outcome. Yet, the functional meaning of these changes for tumor formation is completely unknown. Recent studies indicate, however, that changes in claudins may induce changes in cell motility, and thus may play a role in the metastatic process.[180] Conversely, changes in cell motility may also be important in wound healing.[181]

Pharmacological tools

- Modification of TJ proteins as pharmacological or therapeutic measures. More specific therapies should be identified that target TJ proteins. Such approaches may also be used later in multimodal therapies.
- Interaction of viral/bacterial and TJ proteins. Similar to *C. perfringens* toxin, other pathogens successfully interact with TJ proteins to overcome TJ barriers. One example of such an interaction is hepatitis C virus.[100,101] And the list of viral/bacterial proteins with binding to TJ proteins is growing. All of these proteins may potentially be exploited as vehicles to direct drugs to their targets or to specifically affect TJs during therapy.

Acknowledgment

This work was supported by the Deutsche Forschungsgemeinschaft (FOR 721, SFB 852).

Conflicts of interest

The authors declare no conflicts of interest.

References

1. Furuse, M., T. Hirase, M. Itoh, *et al.* 1993. Occludin: a novel integral membrane protein localizing at tight junctions. *J. Cell Biol.* **123:** 1777–1788.
2. Furuse, M., H. Sasaki, K. Fujimoto & S. Tsukita. 1998. A single gene product, claudin-1 or -2, reconstitutes tight junction strands and recruits occludin in fibroblasts. *J. Cell Biol.* **141:** 1539–1550.
3. Morita, K., M. Furuse, K. Fujimoto & S. Tsukita. 1999. Claudin multigene family encoding four-transmembrane domain protein components of tight junction strands. *Proc. Natl. Acad. Sci. USA* **96:** 511–516.
4. Mineta, K., Y. Yamamoto, Y. Yamazaki, *et al.* 2011. Predicted expansion of the claudin multigene family. *FEBS Lett.* **585:** 606–612.
5. Ikenouchi, J., M. Furuse, K. Furuse, *et al.* 2005. Tricellulin constitutes a novel barrier at tricellular contacts of epithelial cells. *J. Cell Biol.* **171:** 939–945.
6. Steed, E., N.T. Rodrigues, M.S. Balda & K. Matter. 2009. Identification of MarvelD3 as a tight junction-associated transmembrane protein of the occludin family. *BMC Cell Biol.* **10:** 95.
7. Farquhar, M.G. & G.E. Palade. 1963. Junctional complexes in various epithelia. *J. Cell Biol.* **17:** 375–412.

8. Staehelin, L.A., T.M. Mukherjee & A.W. Williams. 1969. Freeze-etch appearance of tight junctions in the epithelium of small and large intestine of mice. *Protoplasma* **67:** 165–184.

9. Claude, P. 1978. Morphological factors influencing transepithelial permeability: a model for the resistance of the zonula occludens. *J. Membr. Biol.* **39:** 219–232.

10. Stevenson, B.R., J.M. Anderson, D.A. Goodenough & M.S. Mooseker. 1988. Tight junction structure and ZO-1 content are identical in two strains of Madin-Darby canine kidney cells which differ in transepithelial resistance. *J. Cell Biol.* **107:** 2401–2408.

11. Walker, D.C., A. MacKenzie, W.C. Hulbert & J.C. Hogg. 1985. A re-assessment of the tricellular region of epithelial cell tight junctions in trachea of guinea pig. *Acta Anat. (Basel)* **122:** 35–38.

12. Ebnet, K. 2008. Organization of multiprotein complexes at cell-cell junctions. *Histochem. Cell Biol.* **130:** 1–20.

13. Mandell, K.J. & C.A. Parkos. 2005. The JAM family of proteins. *Adv. Drug Deliv. Rev.* **57:** 857–867.

14. Lemmers, C., D. Michel, L. Lane-Guermonprez, *et al.* 2004. CRB3 binds directly to Par6 and regulates the morphogenesis of the tight junctions in mammalian epithelial cells. *Mol. Biol. Cell* **15:** 1324–1333.

15. Coyne, C.B. & J.M. Bergelson. 2005. CAR: a virus receptor within the tight junction. *Adv. Drug Deliv. Rev.* **57:** 869–882.

16. Russ, P.K., C.J. Pino, C.S. Williams, *et al.* 2011. Bves modulates tight junction associated signaling. *PLoS One* **6:** e14563.

17. Aijaz, S., M.S. Balda & K. Matter. 2006. Tight junctions: molecular architecture and function. *Int. Rev. Cytol.* **248:** 261–298.

18. Guillemot, L., S. Paschoud, P. Pulimeno, *et al.* 2008. The cytoplasmic plaque of tight junctions: a scaffolding and signalling center. *Biochim. Biophys. Acta* **1778:** 601–613.

19. Hartsock, A. & W.J. Nelson. 2008. Adherens and tight junctions: structure, function and connections to the actin cytoskeleton. *Biochim. Biophys. Acta* **1778:** 660–669.

20. Miyoshi, J., Y. Takai. 2005. Molecular perspective on tight-junction assembly and epithelial polarity. *Adv. Drug Deliv. Rev.* **57:** 815–855.

21. Raleigh, D.R., A.M. Marchiando, Y. Zhang, *et al.* 2010. Tight junction-associated MARVEL proteins marveld3, tricellulin, and occludin have distinct but overlapping functions. *Mol. Biol. Cell* **21:** 1200–1213.

22. Krause, G., L. Winkler, S.L. Mueller, *et al.* 2008. Structure and function of claudins. *Biochim. Biophys. Acta* **1778:** 631–645.

23. Mankertz, J., J.S. Waller, B. Hillenbrand, *et al.* 2002. Gene expression of the tight junction protein occludin includes differential splicing and alternative promoter usage. *Biochem. Biophys. Res. Comm.* **298:** 657–666.

24. Loh, Y.H., A. Christoffels, S. Brenner, *et al.* 2004. Extensive expansion of the claudin gene family in the teleost fish, Fugu rubripes. *Genome Res.* **14:** 1248–1257.

25. Stiffler, M.A., J.R. Chen, V.P. Grantcharova, *et al.* 2007. PDZ domain binding selectivity is optimized across the mouse proteome. *Science* **317:** 364–369.

26. Lal-Nag, M. & P.J. Morin. 2009. The claudins. *Genome Biol.* **10:** 235.1–235.7.

27. Günzel, D. & M. Fromm. 2012. Claudins and other tight junction proteins. *Compreh. Physiol.* **2:** 1–32.

28. Blasig, I.E., C. Bellmann, J. Cording, *et al.* 2011. Occludin protein family—oxidative stress and reducing conditions. *Antioxid. Redox Signal.* **15:** 1195–1219.

29. Krug, S.M., D. Günzel, M.P. Conrad, *et al.* 2012. Charge-selective claudin channels. *Ann. N. Y. Acad. Sci.* **1257:** 20–28.

30. Schultz, S.G. 1972. Electrical potential differences and electromotive forces in epithelial tissues. *J. Gen. Physiol.* **59:** 794–798.

31. Amasheh, S., M. Fromm & D. Günzel. 2011. Claudins of intestine and nephron—a correlation of molecular tight junction structure and barrier function. *Acta Physiol.* **201:** 133–140.

32. Inai, T., J. Kobayashi & Y. Shibata. 1999. Claudin-1 contributes to the epithelial barrier function in MDCK cells. *Eur. J. Cell Biol.* **78:** 849–855.

33. Milatz, S., S.M. Krug, R. Rosenthal, *et al.* 2010. Claudin-3 acts as a sealing component of the tight junction for ions of either charged and uncharged solutes. *Biochim. Biophys. Acta Biomembr.* **1798:** 2048–2057.

34. Furuse, M., H. Sasaki & S. Tsukita. 1999. Manner of interaction of heterogeneous claudin species within and between tight junction strands. *J. Cell Biol.* **147:** 891–903.

35. Mitchell, L.A., C.E. Overgaard, C. Ward, *et al.* 2011. Differential effects of claudin-3 and claudin-4 on alveolar epithelial barrier function. *Am. J. Physiol. Lung Cell Mol. Physiol.* **301:** L40–L49.

36. Coyne, C.B., T.M. Gambling, R.C. Boucher, *et al.* 2003. Role of claudin interactions in airway tight junctional permeability. *Am. J. Physiol. Lung Cell Mol. Physiol.* **285:** L1166–L1178.

37. Fontijn, R.D., J. Rohlena, J. van Marle, *et al.* 2006. Limited contribution of claudin-5-dependent tight junction strands to endothelial barrier function. *Eur. J. Cell Biol.* **85:** 1131–1144.

38. Amasheh, S., T. Schmidt, M. Mahn, *et al.* 2005. Contribution of caludin-5 to barrier properties in tight junctions of epithelial cells. *Cell Tiss. Res.* **321:** 89–96.

39. Gow, A., C.M. Southwood, J.S. Li, *et al.* 1999. CNS myelin and sertoli cell tight junction strands are absent in Osp/claudin-11 null mice. *Cell* **99:** 649–659.

40. Gow, A., C. Davies, C.M. Southwood, *et al.* 2004. Deafness in claudin 11-null mice reveals the critical contribution of basal cell tight junctions to stria vascularis function. *J. Neurosci.* **24:** 7051–7062.

41. Wessells, H., C.J. Sullivan, Y. Tsubota, *et al.* 2009. Transcriptional profiling of human cavernosal endothelial cells reveals distinctive cell adhesion phenotype and role for claudin 11 in vascular barrier function. *Physiol. Genomics* **39:** 100–108.

42. Ben-Yosef, T., I.A. Belyantseva, T.L. Saunders, *et al.* 2003. Claudin 14 knockout mice, a model for autosomal recessive deafness DFNB29, are deaf due to cochlear hair cell degeneration. *Hum. Mol. Genet.* **12:** 2049–2061.

43. Miyamoto, T., D. Morita, D. Takemoto, *et al.* 2005. Tight junctions in Schwann cells of peripheral myelinated axons:

a lesson from claudin-19-deficient mice. *J. Cell Biol.* **169:** 527–538.

44. Hou, J., A. Renigunta, A.S. Gomes, *et al.* 2009. Claudin-16 and claudin-19 interaction is required for their assembly into tight junctions and for renal reabsorption of magnesium. *Proc. Natl. Acad. Sci. U. S. A.* **106:** 15350–15355.

45. Angelow, S., R. El-Husseini, S.A. Kanzawa & A.S. Yu. 2007. Renal localization and function of the tight junction protein, claudin-19. *Am. J. Physiol. Renal Physiol.* **293:** F166–F177.

46. Krug, S.M., S. Amasheh, J.F. Richter, *et al.* 2009. Tricellulin forms a barrier to macromolecules in tricellular tight junctions without affecting ion permeability. *Mol. Biol. Cell* **20:** 3713–3724.

47. Riazuddin, S., Z.M. Ahmed, A.S. Fanning, *et al.* 2006. Tricellulin is a tight-junction protein necessary for hearing. *Am. J. Hum. Genet.* **79:** 1040–1051.

48. Furuse, M., K. Furuse, H. Sasaki & S. Tsukita. 2001. Conversion of zonulae occludentes from tight to leaky strand type by introducing claudin-2 into Madin-Darby canine kidney I cells. *J. Cell Biol.* **153:** 236–272.

49. Amasheh, S., N. Meiri, A.H. Gitter, *et al.* 2002. Claudin-2 expression induces cation-selective channels in tight junctions of epithelial cells. *J. Cell Sci.* **115:** 4969–4976.

50. Rosenthal, R., S. Milatz, S.M. Krug, *et al.* 2010. The tight junction protein claudin-2 forms a paracellular water channel. *J. Cell Sci.* **123:** 1913–1921.

51. Günzel, D., M. Stuiver, P.J. Kausalya, *et al.* 2009. Claudin-10 exists in six alternatively spliced isoforms which exhibit distinct localization and function. *J. Cell Sci.* **122:** 1507–1517.

52. Van Itallie, C.M., S. Rogan, A. Yu, *et al.* 2006. Two splice variants of claudin-10 in the kidney create paracellular pores with different ion selectivities. *Am. J. Physiol. Renal Physiol.* **291:** F1288–F1299.

53. Fujita, H., H. Chiba, H. Yokozaki, *et al.* 2006. Differential expression and subcellular localization of claudin-7, -8, -12, -13, and -15 along the mouse intestine. *J. Histochem. Cytochem.* **54:** 933–944.

54. Van Itallie, C.M., A.S. Fanning & J.M. Anderson. 2003. Reversal of charge selectivity in cation or anion-selective epithelial lines by expression of different claudins. *Am. J. Physiol. Renal Physiol.* **285:** F1078–F1084.

55. Tamura, A., H. Hayashi, M. Imasato, *et al.* 2011. Loss of claudin-15, but not claudin-2, causes Na$^+$ deficiency and glucose malabsorption in mouse small intestine. *Gastroenterology* **140:** 913–923.

56. Krug, S.M., D. Günzel, M.P. Conrad, *et al.* 2012. Claudin-17 forms tight junction channels with distinct anion selectivity. *Cell. Mol. Life Sci.* doi: 10.1007/s00018-012-0949-x

57. Shen, L., C.R. Weber, D.R. Raleigh, *et al.* 2011. Tight junction pore and leak pathways: a dynamic duo. *Annu. Rev. Physiol.* **73:** 283–309.

58. Hou, J., A. Renigunta, J. Yang & S. Waldegger. 2010. Claudin-4 forms paracellular chloride channel in the kidney and requires claudin-8 for tight junction localization. *Proc. Natl. Acad. Sci. U. S. A.* **107:** 18010–18015.

59. Le Moellic, C., S. Boulkroun, D. González-Nunez, *et al.* 2005. Aldosterone and tight junctions: modulation of claudin-4 phosphorylation in renal collecting duct cells. *Am. J. Physiol. Cell Physiol.* **289:** C1513–C1521.

60. Kirschner, N. & J.M. Brandner. 2012. Barriers and more – functions of tight junction proteins in the skin. *Ann. N. Y. Acad. Sci.* **1257:** 158–166.

61. Hou, J., A.S. Gomes, D.L. Paul & D.A. Goodenough. 2006. Study of claudin function by RNA interference. *J. Biol. Chem.* **281:** 36117–36123.

62. Alexandre, M.D., Q. Lu & Y.H. Chen. 2005. Overexpression of claudin-7 decreases the paracellular Cl- conductance and increases the paracellular Na$^+$ conductance in LLC-PK1 cells. *J. Cell Sci.* **118:** 2683–2693.

63. Yu, A.S.L., A.H. Enck, W.I. Lencer & E.E. Schneeberger. 2003. Claudin-8 expression in Madin-Darby canine kidney cells augments the paracellular barrier to cation permeation. *J. Biol. Chem.* **278:** 17350–17359.

64. Yu, A.S. 2009. Molecular basis for cation selectivity in claudin-2-based pores. *Ann. N. Y. Acad. Sci.* **1165:** 53–57.

65. Saitou, M., M. Furuse, H. Sasaki, *et al.* 2000. Complex phenotype of mice lacking occludin, a component of tight junction strands. *Mol. Biol. Cell* **11:** 4131–4142.

66. Schulzke, J.D., A.H. Gitter, J. Mankertz, *et al.* 2005. Epithelial transport and barrier function in occludin-deficient mice. *Biochim. Biophys. Acta* **1669:** 34–42.

67. Van Itallie, C.M., A.S. Fanning, J. Holmes & J.M. Anderson. 2010. Occludin is required for cytokine-induced regulation of tight junction barriers. *J. Cell Sci.* **123:** 2844–2852.

68. Turksen, K., T.C. Troy. 2002. Permeability barrier dysfunction in transgenic mice overexpressing claudin 6. *Development* **129:** 1775–1784.

69. Sas, D., M. Hu, O.W. Moe & M. Baum. 2008. Effect of claudins 6 and 9 on paracellular permeability in MDCK II cells. *Am. J. Physiol. Regul. Integ. Comp. Physiol.* **295:** 1713–1719.

70. Abuazza, G., A. Becker, S.S. Williams, *et al.* 2006. Claudins 6, 9, and 13 are developmentally expressed renal tight junction proteins. *Am. J. Physiol. Renal Physiol.* **291:** F1132–F1141.

71. Kitajiri, S.I., M. Furuse, K. Morita, *et al.* 2004. Expression patterns of claudins, tight junction adhesion molecules, in the inner ear. *Hear. Res.* **187:** 25–34.

72. Nakano, Y., S.H. Kim, H.M. Kim, *et al.* 2009. A claudin-9-based ion permeability barrier is essential for hearing. *PLoS Genet.* **5:** e1000610.

73. Brandner, J.M., S. McIntyre, S. Kief, *et al.* 2003. Expression and localization of tight junction-associated proteins in human hair follicles. *Arch. Dermatol. Res.* **295:** 211–221.

74. Ohtsuki, S., S. Sato, H. Yamaguchi, *et al.* 2007. Exogenous expression of claudin-5 induces barrier properties in cultured rat brain capillary endothelial cells. *J. Cell Physiol.* **210:** 81–86.

75. Niimi, T., K. Nagashima, J.M. Ward, *et al.* 2001. Claudin-18, a novel downstream target gene for the T/EBP/NKX2.1 homeodomain transcription factor, encodes lung- and stomach-specific isoforms through alternative splicing. *Mol. Cell Biol.* **21:** 7380–7390.

76. Jovov, B., C.M. Van Itallie, N.J. Shaheen, *et al.* 2007. Claudin-18: a dominant tight junction protein in Barrett's

esophagus and likely contributor to its acid resistance. *Am. J. Physiol. Gastrointest. Liver Physiol.* **293:** G1106–G1113.

77. Lim, T.S., S.R. Vedula, W. Hunziker & C.T. Lim. 2008. Kinetics of adhesion mediated by extracellular loops of claudin-2 as revealed by single-molecule force spectroscopy. *J. Mol. Biol.* **381:** 681–691.

78. Piontek, J., L. Winkler, H. Wolburg, *et al.* 2008. Formation of tight junction: determinants of homophilic interaction between classic claudins. *FASEB J.* **22:** 146–158.

79. Piontek, J., S. Fritzsche, J. Cording, *et al.* 2011. Elucidating the principles of the molecular organization of heteropolymeric tight junction strands. *Cell. Mol. Life Sci.* **68:** 3903–3918.

80. Daugherty, B.L., C. Ward, T. Smith, *et al.* 2007. Regulation of heterotypic claudin compatibility. *J. Biol. Chem.* **282:** 30005–30013.

81. Van Itallie, C.M., L.L. Mitic & J.M. Anderson. 2011. Claudin-2 forms homodimers and is a component of a high molecular weight protein complex. *J. Biol. Chem.* **286:** 3442–3450.

82. Blasig, I.E., L. Winkler, B. Lassowski, *et al.* 2006. On the self-association potential of transmembrane tight junction proteins. *Cell. Mol. Life Sci.* **63:** 505–514.

83. Westphal, J.K., M.J. Dörfel, S.M. Krug, *et al.* 2010. Tricellulin forms homomeric and heteromeric tight junctional complexes. *Cell. Mol. Life. Sci.* **67:** 2057–2068.

84. Inai, T., T. Kamimura, E. Hirose, *et al.* 2010. The protoplasmic or exoplasmic face association of tight junction particles cannot predict paracellular permeability or heterotypic claudin compatibility. *Eur. J. Cell Biol.* **89:** 547–556.

85. Hou, J., A. Renigunta, M. Konrad, *et al.* 2008. Claudin-16 and claudin-19 interact and form a cation-selective tight junction complex. *J. Clin. Invest.* **118:** 619–628.

86. Matsuda, M., A. Kubo, M. Furuse & S. Tsukita. 2004. A peculiar internalization of claudins, tight junction-specific adhesion molecules, during the intercellular movement of epithelial cells. *J. Cell Sci.* **117:** 1247–1257.

87. Shen, L., C.R. Weber & J.R. Turner. 2008. The tight junction protein complex undergoes rapid and continuous molecular remodeling at steady state. *J. Cell Biol.* **181:** 683–695.

88. Sakakibara, A., M. Furuse, M. Saitou, *et al.* 1997. Possible involvement of phosphorylation of occludin in tight junction formation. *J. Cell Biol.* **137:** 1393–1401.

89. Raleigh, D.R., D.M. Boe, D. Yu, *et al.* 2011. Occludin S408 phosphorylation regulates tight junction protein interactions and barrier function. *J. Cell Biol.* **193:** 565–582.

90. Suzuki, T., B.C. Elias, A. Seth, *et al.* 2009. PKCη regulates occludin phosphorylation and epithelial tight junction integrity. *Proc. Natl. Acad. Sci. U. S. A.* **106:** 61–66.

91. Sheth, P., N. Delos Santos, A. Seth, *et al.* 2007. Lipopolysaccharide disrupts tight junctions in cholangiocyte monolayers by a c-Src-, TLR4-, and LBP-dependent mechanism. *Am. J. Physiol. Gastrointest. Liver Physiol.* **293:** 308–318.

92. Nunbhakdi-Craig, V., T. Machleidt, E. Ogris, *et al.* 2002. Protein phosphatase 2A associates with and regulates atypical PKC and the epithelial tight jucntion complex. *J. Cell Biol.* **158:** 967–978.

93. Sallee, J.L. & K. Burridge. 2009. Density-enhanced phosphatase 1 regulates phosphorylation of tight junction proteins and enhances barrier function of epithelial cells. *J. Biol. Chem.* **284:** 14997–15006.

94. Dörfel, M.J., J.K. Westphal & O. Huber. 2009. Differential phosphorylation of occludin and tricellulin by CK2 and CK1. *Ann. N. Y. Acad. Sci.* **1165:** 69–73.

95. Fujibe, M., H. Chiba, T. Kojima, *et al.* 2004. Thr203 of claudin-1, a putative phosphorylation site for MAP kinase, is required to promote the barrier function of tight junctions. *Exp. Cell Res.* **295:** 36–47.

96. Yamauchi, K., T. Rai, K. Kobayashi, *et al.* 2004. Disease-causing mutant WNK4 increases paracellular chloride permeability and phosphorylates claudins. *Proc. Natl. Acad. Sci. U. S. A.* **101:** 4690–4694.

97. D'Souza, T., R. Agarwal & P.J. Morin. 2005. Phosphorylation of claudin-3 at threonine 192 by cAMP-dependent protein kinase regulates tight junction barrier function in ovarian cancer cells. *J. Biol. Chem.* **280:** 26233–26240.

98. Ikari, A., C. Okude, H. Sawada, *et al.* 2008. Activation of a polyvalent cation-sensing receptor decreases magnesium transport via claudin-16. *Biochim. Biophys. Acta* **1778:** 283–290.

99. Van Itallie, C.M., T.M. Gambling, J.L. Carson & J.M. Anderson. 2005. Palmitoylation of claudins is required for efficient tight-junction localization. *J. Cell Sci.* **118:** 1427–1436.

100. Benedicto, I., F. Molina-Jiménez, B. Bartosch, *et al.* 2009. The tight junction-associated protein occludin is required for a postbinding step in hepatitis C virus entry and infection. *J. Virol.* **83:** 8012–8020.

101. Meertens, L., C. Bertaux, L. Cukierman, *et al.* 2008. The tight junction proteins claudin-1, -6, and -9 are entry cofactors for hepatitis C virus. *J. Virol.* **82:** 3555–3560.

102. Zheng, J., Y. Xie, R. Campbell, *et al.* 2005. Involvement of claudin-7 in HIV infection of CD4(-) cells. *Retrovirology* **2:** 79.

103. Gao, F., X. Duan, X. Lu, *et al.* 2010. Novel binding between pre-membrane protein and claudin-1 is required for efficient dengue virus entry. *Biochem. Biophys. Res. Commun.* **391:** 952–957.

104. Barton, E.S., J.C. Forrest, J.L. Connolly, *et al.* 2001. Junction adhesion molecule is a receptor for reovirus. *Cell* **104:** 441–451.

105. Cohen, C.J., J. Gaetz, T. Ohman & J.M. Bergelson. 2001. Multiple regions within the coxsackievirus and adenovirus receptor cytoplasmic domain are required for basolateral sorting. *J. Biol. Chem.* **276:** 25392–25398.

106. Katahira, J., N. Inoue, Y. Horiguchi, *et al.* 1997. Molecular cloning and functional characterization of the receptor for Clostridium perfringens enterotoxin. *J. Cell Biol.* **136:** 1239–1247.

107. Winkler, L., C. Gehring, A. Wenzel, *et al.* 2009. Molecular determinants of the interaction between Clostridium perfringens enterotoxin fragments and claudin-3. *J. Biol. Chem.* **284:** 18863–18872.

108. Bagnoli, F., L. Buti, L. Tompkins, *et al.* 2005. *Helicobacter pylori* CagA induces a transition from polarized to invasive phenotypes in MDCK cells. *Proc. Natl. Acad. Sci. U. S. A.* **102:** 16339–16344.

109. Sandle, G.I. 2005. Pathogenesis of diarrhea in ulcerative colitis: new views on an old problem. *J. Clin. Gastroenterol.* **39:** S49–S52.

110. Zeissig, S., N. Bürgel, D. Günzel, *et al.* 2007. Changes in expression and distribution of claudin 2, 5 and 8 lead to discontinuous tight junctions and barrier dysfunction in active Crohn's disease. *Gut* **56:** 61–72.

111. Oshima, T., H. Miwa & T. Takashi Joh. 2008. Changes in the expression of claudins in active ulcerative colitis. *J. Gastroenterol. Hepatol.* **23**(Suppl. 2): S146–S150.

112. Funakoshi, K., K. Sugimura, T. Sasakawa, *et al.* 1995. Study of cytokines in ulcerative colitis. *J. Gastroenterol.* **30**(Suppl. 8): 61–63.

113. Mennigen, R., K. Nolte, E. Rijcken, *et al.* 2009. Probiotic mixture VSL#3 protects the epithelial barrier by maintaining tight junction protein expression and preventing apoptosis in a murine model of colitis. *Am. J. Physiol. Gastrointest. Liver Physiol.* **296:** G1140–G1149.

114. Bauer, C., P. Duewell, C. Mayer, *et al.* 2010. Colitis induced in mice with dextran sulfate sodium (DSS) is mediated by the NLRP3 inflammasome. *Gut* **59:** 1192–1199.

115. Amasheh, S., S. Dullat, M. Fromm, *et al.* 2009. Inflamed pouch mucosa possesses altered tight junctions indicating recurrence of inflammatory bowel disease. *Int. J. Colorectal Dis.* **24:** 1149–1156.

116. Häuser, W., C. Schmidt & A. Stallmach. 2011. Depression and mucosal proinflammatory cytokines are associated in patients with ulcerative colitis and pouchitis—a pilot study. *J. Crohns Colitis.* **5:** 350–353.

117. Patel, R.T., I. Bain, D. Youngs & M.R. Keighley. 1995. Cytokine production in pouchitis is similar to that in ulcerative colitis. *Dis. Colon Rectum* **38:** 831–837.

118. Bürgel, N., C. Bojarski, J. Mankertz, *et al.* 2002. Mechanisms of diarrhea in collagenous colitis. *Gastroenterology* **123:** 433–443.

119. Tagkalidis, P.P., P.R. Gibson & P.S. Bhathal. 2007. Microscopic colitis demonstrates a T helper cell type 1 mucosal cytokine profile. *J. Clin. Pathol.* **60:** 382–387.

120. Ståhle-Bäckdahl, M., J. Maim, B. Veress, *et al.* 2000. Increased presence of eosinophilic granulocytes expressing transforming growth factor-beta1 in collagenous colitis. *Scand. J. Gastroenterol.* **35:** 742–746.

121. De Benedetto, A., N.M. Rafaels, L.Y. McGirt, *et al.* 2011. Tight junction defects in patients with atopic dermatitis. *J. Allergy Clin. Immunol.* **127:** 773–786.

122. Markov, A.G., M.A. Voronkova, G.N. Volgin, *et al.* 2011. Tight junction proteins contribute to barrier properties in human pleura. *Resp. Physiol. Neurobiol.* **175:** 331–335.

123. De Smedt, A., E. Vanderlinden, C. Demanet, *et al.* 2004. Characterisation of pleural inflammation occurring after primary spontaneous pneumothorax. *Eur. Respir. J.* **23:** 896–900.

124. Szakál, D.N., H. Gyŏrffy, A. Arató, *et al.* 2010. Mucosal expression of claudins 2, 3 and 4 in proximal and distal part of duodenum in children with coeliac disease. *Virchows Arch.* **456:** 245–250.

125. Manavalan, J.S., L. Hernandez, J.G. Shah, *et al.* 2010. Serum cytokine elevations in celiac disease: association with disease presentation. *Hum. Immunol.* **71:** 50–57.

126. Sander, G.R., A.G. Cummins & B.C. Powell. 2005. Rapid disruption of intestinal barrier function by gliadin involves altered expression of apical junctional proteins. *FEBS Lett.* **579:** 4851–4855.

127. Roxas, J.L., A. Koutsouris, A. Bellmeyer, *et al.* 2010. Enterohemorrhagic E. coli alters murine intestinal epithelial tight junction protein expression and barrier function in Shiga toxin independent manner. *Lab. Invest.* **90:** 1152–1168.

128. Epple, H.J., T. Schneider, H. Troeger, *et al.* 2009. Impairment of the intestinal barrier is evident in untreated but absent in suppressively treated HIV-infected patients. *Gut* **58:** 220–227.

129. Amasheh, M., A. Fromm, S.M. Krug, *et al.* 2010. TNFα-induced and berberine-antagonized tight junction barrier impairment via tyrosine kinase, pAkt, and NFkB signaling. *J. Cell Sci.* **123:** 4145–4155.

130. Amasheh, M., I. Grotjohann, S. Amasheh, *et al.* 2009. Regulation of mucosal structure and barrier function in rat colon exposed to tumor necrosis factor alpha and interferon gamma in vitro: a novel model for studying the pathomechanisms of inflammatory bowel disease cytokines. *Scand. J. Gastroent.* **44:** 1226–1235.

131. Heller, F., A. Fromm, A.H. Gitter, *et al.* 2008. Epithelial apoptosis is a prominent feature of the epithelial barrier disturbance in intestinal inflammation: effect of proinflammatory interleukin-13 on epithelial cell function. *Mucosal Immunol.* **1**(Suppl. 1): S58–S61.

132. Kojima, T., K. Takano, T. Yamamoto, *et al.* 2008. Transforming growth factor-beta induces epithelial to mesenchymal transition by down-regulation of claudin-1 expression and the fence function in adult rat hepatocytes. *Liver Int.* **28:** 534–545.

133. Hering, N.A., S. Andres, A. Fromm, *et al.* 2011. Transforming growth factor β, a whey protein component, strengthens the intestinal barrier by upregulating claudin-4 in HT-29/B6 cells. *J. Nutr.* **141:** 783–789.

134. Schmitz, H., C. Barmeyer, M. Fromm, *et al.* 1999. Altered tight junction structure contributes to the impaired epithelial barrier function in ulcerative colitis. *Gastroenterology* **116:** 301–309.

135. Heller, F., P. Florian, C. Bojarski, *et al.* 2005. Interleukin-13 is the key effector Th2 cytokine in ulcerative colitis that affects epithelial tight junctions, apoptosis, and cell restitution. *Gastroenterology* **129:** 550–564.

136. Gitter, A.H., K. Bendfeldt, J.D. Schulzke & M. Fromm. 2000. Leaks in the epithelial barrier caused by spontaneous and TNF-alpha-induced single-cell apoptosis. *Faseb J.* **14:** 1749–1753.

137. Schumann, M., D. Günzel, N. Buergel, *et al.* 2011. Cell polarity-determining proteins Par-3 and PP-1 are involved in epithelial tight junction defects in coeliac disease. *Gut* **61:** 220–228.

138. Krug, S.M., C. Bojarski, A. Fromm, *et al.* 2010. Tricellulin in Crohn's disease and ulcerative colitis. *FASEB J.* **24:** 998.1.

139. Mankertz, J., M. Amasheh, S.M. Krug, *et al.* 2009. TNFalpha up-regulates claudin-2 expression in epithelial HT-29/B6 cells via phosphatidylinositol-3-kinase signaling. *Cell Tiss. Res.* **336:** 67–77.

140. Prasad, A.S., B. Bao, F.W. Beck, *et al.* 2004. Antioxidant effect of zinc in humans. *Free Radic. Biol. Med.* **37:** 1182–1190.

141. Blair, S.A., S.V. Kane, D.R. Clayburgh & J.R. Turner. 2006. Epithelial myosin light chain kinase expression and activity are upregulated in inflammatory bowel disease. *Lab. Invest.* **86:** 191–201.

142. Al-Sadi, R., D. Ye, H.M. Said & T.Y. Ma. 2011. Cellular and molecular mechanism of interleukin-1beta modulation of Caco-2 intestinal epithelial tight junction barrier. *J. Cell. Mol. Med.* **15:** 970–982.

143. Shen, L., E.D. Black, E.D. Witkowski, *et al.* 2006. Myosin light chain phosphorylation regulates barrier function by remodeling tight junction structure. *J. Cell Sci.* **119:** 2095–2106.

144. Yu, D., A.M. Marchiando, C.R. Weber, *et al.* 2010. MLCK-dependent exchange and actin binding region-dependent anchoring of ZO-1 regulate tight junction barrier function. *Proc. Natl. Acad. Sci. U. S. A.* **107:** 8237–8241.

145. Marchiando, A.M., L. Shen, W.V. Graham, *et al.* 2010. Caveolin-1-dependent occludin endocytosis is required for TNF-induced tight junction regulation in vivo. *J. Cell Biol.* **189:** 111–126.

146. Bruewer, M., M. Utech, A.I. Ivanov, *et al.* 2005. Interferon-gamma induces internalization of epithelial tight junction proteins via a macropinocytosis-like process. *FASEB J.* **19:** 923–933.

147. Clark, E., C. Hoare, J. Tanianis-Hughes, *et al.* 2005. Interferon gamma induces translocation of commensal Escherichia coli across gut epithelial cells via a lipid raft-mediated process. *Gastroenterology* **128:** 1258–1267.

148. Wapenaar, M.C., A.J. Monsuur, A.A. van Bodegraven, *et al.* 2008. Associations with tight junction genes PARD3 and MAGI2 in Dutch patients point to a common barrier defect for coeliac disease and ulcerative colitis. *Gut* **57:** 463–467.

149. Barrett, J.C., J.C. Lee, C.W. Lees, *et al.* 2009. Genome-wide association study of ulcerative colitis identifies three new susceptibility loci, including the HNF4A region. *Nat. Genet.* **41:** 1330–1334.

150. Battle, M.A., G. Konopka, F. Parviz, *et al.* 2006. Hepatocyte nuclear factor 4alpha orchestrates expression of cell adhesion proteins during the epithelial transformation of the developing liver. *Proc. Natl. Acad. Sci. U. S. A.* **103:** 8419–8424.

151. Darsigny, M., J.P. Babeu, A.A. Dupuis, *et al.* 2009. Loss of hepatocyte-nuclear-factor-4alpha affects colonic ion transport and causes chronic inflammation resembling inflammatory bowel disease in mice. *PLoS One* **4:** e7609.

152. Ahn, S.H., Y.M. Shah, J. Inoue, *et al.* 2008. Hepatocyte nuclear factor 4alpha in the intestinal epithelial cells protects against inflammatory bowel disease. *Inflamm. Bowel Dis.* **14:** 908–920.

153. Sakaguchi, T., X. Gu, H.M. Golden, *et al.* 2002. Cloning of the human claudin-2 5'-flanking region revealed a TATA-less promoter with conserved binding sites in mouse and human for caudal-related homeodomain proteins and hepatocyte nuclear factor-1alpha. *J. Biol. Chem.* **277:** 21361–21370.

154. Lala, S., Y. Ogura, C. Osborne, *et al.* 2003. Crohn's disease and the NOD2 gene: a role for paneth cells. *Gastroenterology* **125:** 47–57.

155. Bühner, S., C. Buning, J. Genschel, *et al.* 2006. Genetic basis for increased intestinal permeability in families with Crohn's disease: role of CARD15 3020insC mutation? *Gut* **55:** 342–347.

156. Wehkamp, J., J. Harder, M. Weichenthal, *et al.* 2004. NOD2 (CARD15) mutations in Crohn's disease are associated with diminished mucosal alpha-defensin expression. *Gut* **53:** 1658–1664.

157. Sellon, R.K., S. Tonkonogy, M. Schultz, *et al.* 1998. Resident enteric bacteria are necessary for development of spontaneous colitis and immune system activation in interleukin-10-deficient mice. *Infect. Immun.* **66:** 5224–5231.

158. Hudcovic, T., R. Stepankova, J. Cebra & H. Tlaskalova-Hogenova. 2001. The role of microflora in the development of intestinal inflammation: acute and chronic colitis induced by dextran sulfate in germ-free and conventionally reared immunocompetent and immunodeficient mice. *Folia Microbiol. (Praha)* **46:** 565–572.

159. Chapman, M.A., M.F. Grahn, M.A. Boyle, *et al.* 1994. Butyrate oxidation is impaired in the colonic mucosa of sufferers of quiescent ulcerative colitis. *Gut* **35:** 73–76.

160. Tedelind, S., F. Westberg, M. Kjerrulf & A. Vidal. 2007. Anti-inflammatory properties of the short-chain fatty acids acetate and propionate: a study with relevance to inflammatory bowel disease. *World J. Gastroenterol.* **13:** 2826–2832.

161. Lewis, K., F. Lutgendorff, V. Phan, *et al.* 2010. Enhanced translocation of bacteria across metabolically stressed epithelia is reduced by butyrate. *Inflamm. Bowel Dis.* **16:** 1138–1148.

162. Rosenthal, R., M.S. Heydt, M. Amasheh, *et al.* 2012. Analysis of absorption enhancers in epithelial cell models. *Ann. N. Y. Acad. Sci.*

163. Sturniolo, G.C., V. Di Leo, A. Ferronato, *et al.* 2001. Zinc supplementation tightens "leaky gut" in Crohn's disease. *Inflamm. Bowel Dis.* **7:** 94–98.

164. Zhang, B. & Y. Guo. 2009. Supplemental zinc reduced intestinal permeability by enhancing occludin and zonula occludens protein-1 (ZO-1) expression in weaning piglets. *Br. J. Nutr.* **102:** 687–693.

165. Amasheh, M., S. Schlichter, S. Amasheh, *et al.* 2008. Quercetin enhances epithelial barrier function and increases claudin-4 expression in Caco-2 cells. *J. Nutr.* **138:** 1067–1073.

166. Zhou, H. & S. Mineshita. 2000. The effect of berberine chloride on experimental colitis in rats in vivo and in vitro. *J. Pharmacol. Exp. Ther.* **294:** 822–829.

167. Roselli, M., A. Finamore, M.S. Britti & E. Mengheri. 2006. Probiotic bacteria Bifidobacterium animalis MB5 and Lactobacillus rhamnosus GG protect intestinal Caco-2 cells from the inflammation-associated response induced by enterotoxigenic Escherichia coli K88. *Br. J. Nutr.* **95:** 1177–1184.

168. Rembacken, B.J., A.M. Snelling, P.M. Hawkey, *et al.* 1999. Non-pathogenic Escherichia coli versus mesalazine for the treatment of ulcerative colitis: a randomised trial. *Lancet* **354:** 635–639.

169. Ukena, S.N., A. Singh, U. Dringenberg, *et al.* 2007. Probiotic Escherichia coli Nissle 1917 inhibits leaky gut by enhancing mucosal integrity. *PLoS One* **12:** e1308.

170. Colegio, O.R., C.M. Van Itallie, H.J. McCrea, *et al.* 2002. Claudins create charge-selective channels in the paracellular pathway between epithelial cells. *Am. J. Physiol. Cell Physiol.* **283:** C142–C147.

171. Yu, A.S.L., M.H. Cheng, S. Angelow, *et al.* 2009. Molecular basis for cation selectivity in claudin-2-based paracellular pores: identification of an electrostatic interaction site. *J. Gen. Physiol.* **133:** 111–127.

172. Elkouby-Naor, L. & T. Ben-Yosef. 2010. Functions of claudin tight junction proteins and their complex interactions in various physiological systems. *Int. Rev. Cell Mol. Biol.* **279:** 1–32.

173. Masuda, S., Y. Oda, H. Sasaki, *et al.* 2011. LSR defines cell corners for tricellular tight junction formation in epithelial cells. *J. Cell Sci.* **124:** 548–855.

174. Mandel, L.J., R. Bacallao & G. Zampighi. 1993. Uncoupling of the molecular 'fence' and paracellular 'gate' functions in epithelial tight junctions. *Nature* **361:** 552–555.

175. Takakuwa, R., Y. Kokai, T. Kojima, *et al.* 2000. Uncoupling of gate and fence functions of MDCK cells by the actin-depolymerizing reagent mycalolide B. *Exp. Cell Res.* **257:** 238–244.

176. González-Mariscal, L., E. Garay & M. Quirós. 2010. Regulation of claudins by posttranslational modifications and cell-signaling cascades. *Curr. Top. Membr.* **65:** 113–150.

177. Van Itallie, C.M. & J.M. Anderson. 2012. Caveolin binds independently to claudin-2 and occludin. *Ann. N.Y. Acad. Sci.* **1257:** 103–107.

178. Amasheh, S., S. Milatz, S.M. Krug, *et al.* 2009. Na$^+$ absorption defends from paracellular back-leakage by claudin-8 upregulation. *Biochem. Biophys. Res. Commun.* **378:** 45–50.

179. Zhang, J., J. Piontek, H. Wolburg, *et al.* 2010. Establishment of a neuroepithelial barrier by Claudin5a is essential for zebrafish brain ventricular lumen expansion. *Proc. Natl. Acad. Sci. U. S. A.* **107:** 1425–1430.

180. Escudero-Esparza, A., W.G. Jiang & T.A. Martin. 2011. Claudin-5 participates in the regulation of endothelial cell motility. *Mol. Cell. Biochem.* **362:** 71–85.

181. Watson, A.J.M. & K.R. Hughes. 2012. Intestinal epithelial cell shedding and intestinal barrier function. *Ann. N.Y. Acad. Sci.*

Ann. N.Y. Acad. Sci. ISSN 0077-8923

ANNALS OF THE NEW YORK ACADEMY OF SCIENCES
Issue: *Barriers and Channels Formed by Tight Junction Proteins*

Charge-selective claudin channels

Susanne M. Krug,[1] Dorothee Günzel,[1] Marcel P. Conrad,[1] In-Fah M. Lee,[1] Salah Amasheh,[1] Michael Fromm,[1] and Alan S. L. Yu[2]

[1]Institute of Clinical Physiology, Charité – Universitätsmedizin Berlin, Freie Universität and Humboldt Universität, Berlin, Germany. [2]Division of Nephrology and Hypertension and the Kidney Institute, University of Kansas Medical Center, Kansas City, Kansas

Address for correspondence: Susanne M. Krug, Institute of Clinical Physiology, Charité Berlin, Campus Benjamin Franklin, Hindenburgdamm 30, 12203 Berlin, Germany. susanne.m.krug@charite.de

Claudins are the main determinants of barrier properties of the tight junction. Many claudins have been shown to act by tightening the paracellular pathway, but several function as paracellular channels. While some depend on the endogenous claudin background of the analyzed cell line, for other claudins, a distinct charge-selectivity has been shown. This paper portrays cation-selective (claudin-2, claudin-10b, claudin-15) and anion-selective (claudin-10a, claudin-17) claudins and claudins with debatable channel properties (claudin-4, claudin-7, claudin-16). It also describes molecular properties determining the observed charge-selectivity and pore properties in general. In leaky tissues, they widely determine overall transport characteristics by providing paracellular ion-selective pathways. In small intestine, claudin-2 and claudin-15 replace each other in the developing gut. In kidney proximal tubules, claudin-2, claudin-10, and claudin-17 allow for paracellular reabsorption of sodium, chloride, and water.

Keywords: claudin-2; claudin-10a; claudin-10b; claudin-15; claudin-17; pore properties

Tight junctions (TJs) contribute to epi- and endothelial barrier function, restricting the paracellular diffusion of solutes with size and charge selectivity. The specific properties of this barrier are mainly determined by the combination of the protein family of claudins. In mammals, 27 different claudins have been reported, some of which also have splice variants.[1]

Many claudins have been functionally analyzed already, and several have been shown to increase the tightness of the barrier—for example, claudin-1,[2] claudin-3[3] or claudin-5.[4]

Unlike these purely sealing claudins, other claudins forming paracellular channels have been identified. While some are very dependent on the endogenous claudin background of the cell line used for analysis, other claudins have been shown to possess clear selectivity preferences: claudin-2, which is by far the most extensively studied, forms channels permeable for small inorganic or organic cations,[5,6]

and also for water.[7] Claudin-15 is not only a paracellular channel for cations but also indirectly involved in intestinal glucose absorption.[8] Claudin-10b is one of several splice variants of claudin-10 and is also a cation-selective claudin.[9,10]

Presently, two clearly anion-selective claudins have been identified: the anion-selective paracellular channel claudin-10a,[9,10] which shows anion-selective effects of different extents within different cell systems used, and the recently identified distinct anion channel-former claudin-17.[11] For some claudins, like claudin-7, claudin-16, and claudin-4, function as ion-selective paracellular channels is less clear. They seem not only to be dependent on the system used for analysis, but may also interact or lead to displacement of other TJ proteins and, hence, results are contradicting even within one cell type.

In the following sections, the above-mentioned claudins are portrayed in detail and determinants for charge-selectivity are discussed.

doi: 10.1111/j.1749-6632.2012.06555.x

Claudins with debatable channel properties

Claudin-7

Claudin-7 has been shown to act as cation channel in LLC-PK$_1$ cells,[12] but decreased Na$^+$ permeability in MDCK II cells.[13] Regarding general function, a distinct role in electrolyte regulation has been demonstrated for claudin-7, as claudin-7–deficient mice exhibit lethal renal salt wasting.[14] Claudin-7 is expressed in the kidney and in various other tissues, including intestine and ovary.[13,15,16] In frozen tissue sections, claudin-7 is expressed at the TJ and basolateral membrane in different segments of the distal nephron, including the distal convoluted tubule and collecting duct,[13,17] and possibly, the thick ascending limb of Henle's loop.[13] However, in a study of manually dissected rabbit kidneys, claudin-7 showed only cytosolic staining in Henle's loop, which was thought to be nonspecific.[15] The finding of basolateral localization raised speculation that one function of claudin-7 could be participating in cell–cell and cell–matrix adhesion at the basolateral membrane.[15,17]

Claudin-16

Claudin-16 was initially suggested as a channel selective for divalent cations,[18] but overexpression in different cell lines did not support this hypothesis or only yielded marginal effects.[19–21] Increases in permeability to a monovalent cation were found upon overexpression in LLC-PK$_1$[22] but not MDCK C7 cells.[20] Effects or even localization within the TJ were dependent on interaction with claudin-19,[23] the Ca^{2+}-sensing receptor (CaSR),[21,24] or phosphorylation[25,26] of claudin-16 in some studies. Contradicting results, also in knockout or knockdown mouse models make it difficult to give claudin-16 a clear role as cation channel.[27,28]

Claudin-4

For claudin-4, it was recently shown that under certain conditions it may act as an anion channel when expressed together with claudin-8.[29] However, originally, it had been reported to be a sealing claudin.[30,31] Overexpression studies in the cation-selective cell line MDCK II showed an increased transepithelial resistance and a decreased Na$^+$ permeability, whereas no effect was observed after overexpression in the anion-selective cell line LLC-PK$_1$.[32,33]

Claudin-4 shows high homology to claudin-3, which has been shown to be a claudin sealing the TJ barrier.[3] Sixty-seven percent of amino acids are identical between both claudins, further the first extracellular loop (ECL1) is up to 94% identical, suggesting very similar action as a paracellular sealing protein.[3]

Nevertheless, in M-1 and mIMCD3 cells, knockdown of claudin-4 led to an increase in transepithelial resistance, which was based on decreased ion permeability.[29] Here, a change of the ratio P_{Na}/P_{Cl} was reported, which would indicate a function as an anion-preferring channel.

Importantly, interaction of claudin-4 with claudin-8 has been reported to be necessary for correct localization of claudin-4 within the TJ:[29] when claudin-8 was knocked down in M-1 or mIMCD3 cells, claudin-4 was not found in the TJ. Thus, it might be speculated whether, similar to claudin-10a, the function of claudin-4 is dependent on the endogenous claudin expression background within the respective cell lines. However, claudin-4 can be found in TJs of several tissues or cell lines that do not express claudin-8, for example, the skin[34] or CMT-93 cells.[25] Thus, claudin-8 may not to be the only important interaction partner for claudin-4.

Claudin-4 is expressed in many tissues, predominantly in lung, intestine, and the renal collecting duct.[35,36] It is controversial whether claudin-4 (together with claudin-8) is expressed within an extremely tight epithelium, such as the urinary bladder,[34] where it can be excluded that claudin-4 acts as paracellular channel of any kind.

In conclusion, there is only circumstantial evidence for claudin-4 forming an anion channel, while the majority of experimental data are in accordance with a predominant barrier function.

Cation-selective claudins

Claudin-2

Claudin-2 was the first claudin described as a channel-forming TJ protein and has been the most extensively studied of the channel-forming claudins. Its resistance-reducing property was first described in 2001 by the Tsukita group.[5] It was then discovered that claudin-2 forms channels for small cations by overexpression in the high-resistance cell line MDCK C7.[6] Additional comparison of these MDCK C7 cells and another MDCK subtype, the low-resistance MDCK C11 cells, showed that both cell lines differ markedly in their endogenous expression of claudin-2. Similarly, expression differences have been observed in comparison of the cell lines MDCK I and MDCK II and CMT-93 I and CMT-93 II.[37] Presence of claudin-2 was connected to lower resistance and increased paracellular permeability. This increase in paracellular permeability applies to inorganic or organic cations but not to anions or uncharged molecules, such as mannitol, lactulose, or 4 kDa dextran.[6,37,38]

More recently, it was found that claudin-2 is not only responsible for cation permeability, but also for paracellular water transport.[7] Here, MDCK C7 cells with and without exogenous claudin-2 as well as MDCK C11 cells with endogenous claudin-2 were analyzed in the presence of osmotic or Na^+ gradients. Under all conditions, water transport was increased in the claudin-2 expressing cells.

Conversely, an osmotic gradient was able to stimulate paracellular Na^+ transport in the absence of a transepithelial electrochemical gradient for Na^+. Paracellular water transport has been only connected to claudin-2 thus far. In contrast, overexpression of claudin-10b, which is also a pore-forming claudin, did not enhance water transport.

Claudin-2 is present in leaky epithelia such as kidney proximal convoluted tubule (PCT)[35,39] and small intestine.[36,40] Additionally, claudin-2 is up-regulated under several pathological conditions as acute or chronic inflammation[41–43] and several cancers.[35,44,45]

Claudin-15

Claudin-15 is another clear-cut cation-selective channel-forming claudin. First, overexpression studies in the anion-selective cell line LLC-PK$_1$, showed increase of cation permeability. In MDCK II cells, which are already cation-selective, no effect had been observed.[33,46]

Claudin-15 is expressed in the intestine,[47–49] and claudin-15–deficient mice show a very interesting intestinal phenotype:[31] although the mice grow normally, they develop a mega-intestine. The duodenum and jejunum are nearly twice the normal length, and their diameter is doubled. Additionally, these mice have enlarged villi. The development of a mega-intestine was interpreted as a compensatory regulation.[31]

Comparison of claudin-15 and claudin-2 deficiency in 2-week and 8- to 16-week-old mice[8] revealed that in infant mice claudin-15 is restricted to

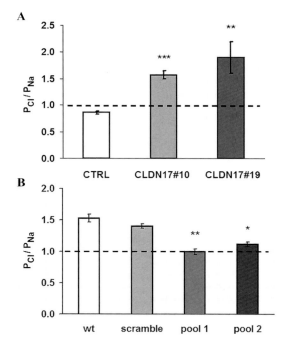

Figure 1. Charge selectivity changes by claudin-17 in overexpression and knockdown experiments. Permeability coefficients for Cl^- and Na^+ were determined from dilution potentials under HCO_3^--free conditions. A ratio $P_{Cl}/P_{Na} > 1$ indicates anion selectivity. (A) Overexpression of claudin-17 in MDCK C7 cells, a cell line genuinely devoid of claudin-17. The increased ratio revealed charge-preference for anions in claudin-17–expressing clones #10 and #19 (**$P < 0.01$, ***$P < 0.001$, $n = 6$). (B) Claudin-17 knockdown (KD) experiments in LLC-PK1 cells. Wild-type (WT) and scramble contained genuine claudin-17. The P_{Cl}/P_{Na} ratio was changed to lowered preference for anions after claudin-17 KD (pool 1 and 2) (*$P < 0.05$, $n = 4–6$).[11]

the crypts of the small intestine, while claudin-2 was equally expressed in villi and crypts. In contrast, in adult mice, claudin-15 was found both in villi and in crypts, while claudin-2 was expressed mainly in crypts at that age. In adult mice, therefore, claudin-15 is in close proximity to the sodium–glucose symporter SGLT1 and may contribute to recirculation of Na^+. This was evidenced by the observation that claudin-15–deficient mice exhibited reduced glucose uptake and an impaired oral glucose tolerance test.[8]

Claudin-10b

Claudin-10 exists in several splice variants.[9,10] After alternative splicing, the two main variants claudin-10a and claudin-10b differ in their first exon, which

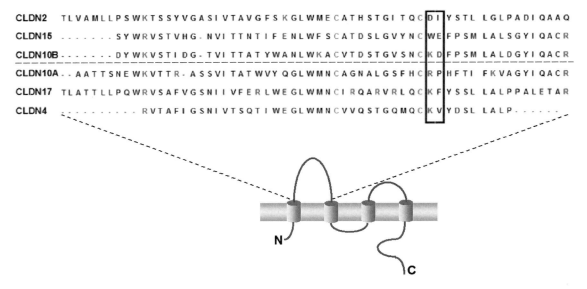

Figure 2. ECL1 amino acid sequence alignment of channel-forming claudins. ECL1 sequences of human cation-selective claudin channels were compared to ECL1 of anion-selective claudins using PRALINE.[54] Red, positively charged amino acids; blue, negatively charged amino acids; green, conserved cysteins of the claudin motif. The black square indicates the amino acid position comparable to D65 in claudin-2, which is crucial for ion-selectivity and also in other channel-forming claudins. The cartoon in the lower part represents a typical claudin, featuring four transmembrane domains, two extracellular loops (ECL1, ECL2), one intracellular loop, and the intracellularly located N and C terminals.

leads to differences in the first transmembrane region and ECL1. In general, the ECL1 is determines the specific interactions of claudins that lead to tightening or selective pore-formation.[19,50] The differences in ECL1 in claudin-10a and claudin-10b also have an impact on their function. One may also turn the argument around: Since claudin-10a and claudin-10b have opposite charge selectivities but differ solely in their ECL1 sequence, this speaks to the importance of ECL1 for determining ion specificity.

Claudin-10b has been described as a cation-selective claudin channel when overexpressed within MDCK C7 or MDCK II cells.[9,10] Claudin-10b is not permeable to water as claudin-2.[7] In addition, both pores seem to be different in their kind of interaction with the permeating cation. As claudin-10b shows strong preference for Na^+ over Li^+ and K^+, whereas claudin-2 shows nearly no discrimination between these ions,[7,38] it may be concluded that a stronger interaction between the claudin-10b pore and the hydration shell of the ion occurs and it is therefore tempting to speculate that claudin-10b may form pores smaller in diameter than claudin-2.

Claudin-10b is expressed in many tissues,[9,10] including brain, lung, and kidney. Within the kidney, it can be found within the medullary thick ascending limb of Henle's loop (mTAL), the outer (OMCD) and inner medullary collecting duct (IMCD), but not in PCT or the cortical collecting duct (CCD).[9]

Anion-selective claudins

Claudin-10a

While claudin-10b was found to be a cation-selective, channel-forming protein, claudin-10a leans to the opposite charge selectivity.

When claudin-10a was transfected into the low-resistance and cation-selective cell line MDCK II, it increased permeability for Cl^- and NO_3^-.[9,10] When expressed in MDCK C7 cells,[9] only permeability for NO_3^- was increased, whereas permeability for Cl^- remained unchanged and permeability for pyruvate was decreased.

From these findings, it was concluded that, on the one hand, interaction between the formed pore and the permeating ion was increased, and, on the other hand, the extent of anion permeability caused by claudin-10a is dependent on the endogenous claudin expression background of the respective cell line.

Claudin-10a is present within epithelia of the uterus and the nephron.[9] In the kidney, it can be found in the PCT,[9] including the S2 segment of the PCT,[51] the medullary thick ascending limb of Henle's loop and CCD, and possibly in the outer but not IMCD.[9] The extent to which claudin-10a functions as an anion channel in the kidney might be comparable to the situation in MDCK II cells, because those possess barrier properties of the PCT.

Claudin-17

For claudin-17, neither the organ-specific expression nor its functional properties were known until recently, but now it has been reported to form a paracellular anion channel.[11] Overexpression in high-resistance MDCK C7 cells did not change the TJ ultrastructure as seen in freeze fracture electron microscopy but caused a large decrease in paracellular resistance, which was measured by two-path impedance spectroscopy.[52] Dilution and bi-ionic potential measurements revealed this to be predominantly based on a threefold increased permeability for anions like chloride or bicarbonate. By this, the ratio P_{Cl}/P_{Na} was dramatically changed from cation-selective to clearly anion-selective (Fig. 1A). This result was confirmed by transient knockdown of claudin-17 in LLC-PK$_1$ cells, which possess high endogenous expression of that claudin (Fig. 1B). The apparent permeability to larger molecules than fluorescein (332 Da) and to water was not altered. Thus, claudin-17 possesses a clear-cut anion selec-

tivity, switching a cell line from cation to anion selectivity.[11]

By mutagenesis, it was shown that claudin-17 anion selectivity critically depends on a positively charged amino acid at position 65. From a plot of the relation between unhydrated diameters of different anions and their respective permeability, a pore diameter of 9–10 Å was estimated. This fits with the observation that claudin-17 overexpression does not alter the permeability for fluorescein.

Claudin-17 is expressed predominantly within the kidney and to much less extent within the brain. Within the murine and human nephron, expression was intense in proximal tubules and gradually decreased toward distal segments. No expression was detectable in the collecting duct. The proximal nephron exhibits substantial, though molecularly so far undefined paracellular chloride reabsorption. As claudin-17 is predominantly expressed in that segment, it is suggested that claudin-17 provides the molecular basis of paracellular chloride (and bicarbonate) transport in the proximal nephron.[11]

Comparison of the ECL1s

The ECL1s of claudins are thought to be responsible for the tightening or pore character of claudins.[10,29,51] It is also supposed to determine the charge-selectivity gained by interaction of ECL1s of same or different claudins. Thus, it can be speculated whether pore-forming claudins exhibit conserved amino acids within the ECL1 that determine the respective charge preference.

Mutagenesis studies of the ECL1 have been performed for several claudins already. For claudin-2,

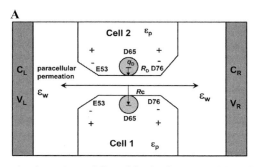
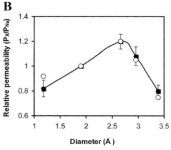

Figure 3. Brownian dynamics modeling of the claudin-2 pore. (A) Model of the claudin-2 pore. E53 and D76 are positioned near the pore entrances. Residue D65 is placed in the middle and is represented by a charged sphere of size R_D, charge q_D, the distance between the pore centerline, and the sphere center, R_C. (B) Alkali metal cation selectivity (from left to right: Li$^+$, Na$^+$, K$^+$, Rb$^+$, and Cs$^+$, given as permeability relative to that of Na$^+$) determined by the bi-ionic potentials obtained in numerical simulations (filled squares) or experimental values (open circles).[38]

the prototypic cation-selective paracellular channel,[5,6] a negatively charged amino acid at position 65 (aspartate) has been found to be essential for charge selectivity.[38,53] Charge-neutralizing mutation of this position led to loss of cation-selectivity of the claudin-2 pore. Furthermore, mutation of E64 in claudin-15, which also exhibits cation-selectivity,[46,33] changed the charge-selectivity from cation to anion preference.[46]

Alignment of the ECL1 of anion-selective claudins with the cation-selective claudins reveals that the respective amino acid comparable to position 65 in claudin-2 is positively charged in all three anion-selective claudins, while a negative charge can be found in the cation-selective claudins (Fig. 2). This suggests that charged amino acids at this position or in direct proximity are crucial for charge discrimination by the ECL1.

Mutagenesis of K65 in claudin-4,[29] or R62 in claudin-10a,[10] led to loss of anion-selectivity underlining this hypothesis. Although no mutations of the respective positions have been performed in claudin-17 or claudin-10b, it may be speculated that mutation to oppositely charged amino acids may lead to loss or even reversal of the original charge preference, because both claudins show sequences similar to the already analyzed claudins. Interestingly, in claudin-10b, two charged residues are located next to each other, lysin and the negative charged residue aspartate. Whether the influence of both or only aspartate is essential for the cation-channel properties of claudin-10b can only be speculated at this time.

Additionally, other positions may be of importance for charge selectivity of claudin channels, for example, mutation of R33 of claudin-10a[10] or

Table 1. Properties of channel-forming claudins

	Localization	Ref.	Channel selectivity	Estimated pore diameter	Permeability properties	Cell type (approach used)	Ref.
Claudin-2	Leaky epithelia like kidney PCT and small intestine	35, 36, 39, 40	cations			MDCK I (OE)	5
			cations			MDCK C7 (OE)	6
			cations	6.5 Å	K > Rb > Na > Li >> Cs	MDCK I (TetOff)	38
			water		Na = K = Li > Rb > Cs	MDCK C7 (OE)	7
Claudin-10b	Brain, lung, kidney tubule: mTAL, OMCD, IMCD	9, 10	cations		Na > Li > K > Rb > Cs	MDCK C7 (OE)	9
			cations			MDCK II (OE)	10
			cations		4 kDa FITC-dextran decreased	MDCK I (OE)	56
					not for water	MDCK C7 (OE)	7
Claudin-15	Intestine	47–49	cations			LLC-PK$_1$ (OE)	46
			cations		4 kDa FITC-dextran increased	MDCK I (OE)	56
			cations			mice (KO)	8
Claudin-10a	Uterus and kidney tubule: PCT, including S2 segment, mTAL, CCD, possibly OMCD	9, 52	anions	> 4.2 Å < 6.5 Å	NO$_3^-$ increased, Cl$^-$ unchanged, pyruvate decreased	MDCK C7 (OE)	9
			anions		NO$_3^-$, Cl$^-$ increased	MDCK II (OE)	9, 10
Claudin-17	Kidney: most abundant in PCT, decreasing toward distal segments; marginal in brain	11	anions	9–10 Å	Cl > Br > I > F, pyruvate increased	MDCK C7 (OE) and LLC-PK$_1$ (KD)	11
					not for water	MDCK C7 (OE)	11

OE, overexpression; TetOff, tetracycline-controlled transcriptional inhibition; KO, knockout; KD, knockdown.

mutation of D55 in claudin-15,[46] which both lead to loss of charge preference.

The claudin-2 pore as a general model

Claudin-2 is the most extensively studied paracellular channel to date. The alignment of the ECL1, together with several mutagenesis experiments, already suggests the importance of the charge around position 65 (in claudin-2) in general; there may also be other claudin-2 findings regarding the structural basis that could be used to gain insight into other paracellular channels.

Simulations of Brownian dynamics were performed to model the behavior of the claudin-2 pore.[38] Under the assumption of a hexameric character of the claudin-2 channel, a simple model was calculated in which the pore was assumed to be a 6.5 Å diameter cylinder with conical vestibules, and the negatively charged side chain of D65 positioned close to the center and facing into the lumen (Fig. 3A). The parameters of the model were fit to the observed values for Na^+ selectivity. Simulations were then run to examine model behaviors such as alkali metal selectivity. Relative selectivities for Li^+, Na^+, K^+, Rb^+, and Cs^+ obtained from numerical simulations and from experimental values were in excellent agreement with each other, indicating the validity of the model (Fig. 3B).

Similar models could be created for other claudin channels, but are not yet available. For example, the first rough estimation of the narrowness of claudin-10b was made comparing the permeabilities and Eisenman sequences for alkali metal ions of claudin-2 with those of claudin-10b.[7] Based on these differences, channels formed by claudin-10b were suggested to be narrower than those formed by claudin-2.

As calculated models can only roughly reflect the real pore, cysteine scanning mutagenesis of several amino acids within the ECL1 of claudin-2 has been employed to locate amino acid positions and identify the function within the pore:[53] the positions of interest were mutated to cysteine and afterwards screened for their accessibility to thiol-reactive reagents. For the D65C position, discussed earlier as a determinant of charge-selectivity, intermolecular dimerization was observed, which led to loss of charge and size selectivity. From this, it was concluded that claudin-2 pores are multimeric and the aspartate of position 65 lies close to a protein–protein interface. For the other residues analyzed in that study, functional activity remained.

Methanethiosulfonate reagents' coupling to the introduced cysteines was analyzed using different charges and sizes of those reagents, leading to the conclusion that I66 was buried deep within a narrow area of the channel with its side group facing into the lumen because conductance was decreased with increasing size of the coupled reagent. Y35C and H57C were unaffected by the coupling reagents, suggesting that these amino acids are located at the protein surface outside of the pore.[53]

In addition to the findings of localization within the paracellular pore or their role in charge selectivity, amino acids may also function as stabilizing residues of the ECL1. Here, conserved cysteins and tryptophan that are conserved among all claudins may be of importance.

Concluding remarks

Paracellular channels formed by claudins possess common features, for example, a charge-selectivity that is mainly based on charged amino acids at comparable position within the ECL1. Nevertheless, same charge-specificity cannot be generalized, as structural differences influence the observed selectivity. Channel-forming claudins can be differentiated in clear-cut anion-selective, cation-selective, and also in less-clear charge-selective channels, which depend on the expression background causing heterogeneous interaction with other proteins.[55]

Cation- and anion-selective claudins are present in various epi- and endothelia (Table 1). If they predominate, the paracellular pathway becomes more ion-conductive than the transcellular pathway. In such "leaky" epi- and endothelia, overall transport characteristics are widely determined by providing paracellular pathways for cations, anions, and, in the case of claudin-2, also for water. Raising two examples from the small intestine, the two cation-selective proteins claudin-2 and claudin-15 replace each other in the developing gut; in kidney proximal tubules, the concerted action of the cation and water channel former claudin-2, the anion channel former claudin-17, and also the two forms of claudin-10, allow for effective paracellular reabsorption of Na^+, Cl^-, and water.

Acknowledgments

This work was supported by grants of the Deutsche Forschungsgemeinschaft Research Unit FOR 721 and the Sonnenfeld-Stiftung Berlin.

Conflicts of interest

The authors declare no conflicts of interest.

References

1. Mineta, K., Y. Yamamoto, Y. Yamazaki, *et al.* 2011. Predicted expansion of the claudin multigene family. *FEBS Lett.* **585:** 606–612.

2. Furuse, M., M. Hata, K. Furuse, *et al.* 2002. Claudin-based tight junctions are crucial for the mammalian epidermal barrier: a lesson from claudin-1-deficient mice. *J. Cell Biol.* **156:** 1099–1111.

3. Milatz, S., S.M. Krug, R. Rosenthal, *et al.* 2010. Claudin-3 acts as a sealing component of the tight junction for ions of either charge and uncharged solutes. *Biochim. Biophys. Acta Biomembr.* **1798:** 2048–2057.

4. Amasheh, S., T. Schmidt, M. Mahn, *et al.* 2005. Contribution of claudin-5 to barrier properties in tight junctions of epithelial cells. *Cell Tiss. Res.* **321:** 89–96.

5. Furuse, M., K. Furuse, H. Sasaki & S. Tsukita. 2001. Conversion of zonulae occludentes from tight to leaky strand type by introducing claudin-2 into Madin-Darby canine kidney I cells. *J. Cell Biol.* **153:** 236–272.

6. Amasheh S., N. Meiri, A.H. Gitter, *et al.* 2002. Claudin-2 expression induces cation-selective channels in tight junctions of epithelial cells. *J. Cell Sci.* **115:** 4969–4976.

7. Rosenthal R., S. Milatz, S.M. Krug, *et al.* 2010. Claudin-2, a component of the tight junction, forms a paracellular water channel. *J. Cell Sci.* **123:** 1913–1921.

8. Tamura A., H. Hayashi, M. Imasato, *et al.* 2010. Loss of claudin-15, but not claudin-2, causes Na^+ deficiency and glucose malabsorption in mouse small intestine. *Gastroenterology* **140:** 913–923.

9. Günzel D., M. Stuiver, P.J. Kausalya, *et al.* 2009. Claudin-10 exists in six alternatively spliced isoforms which exhibit distinct localization and function. *J. Cell Sci.* **122:** 1507–1517.

10. Van Itallie C.M., S. Rogan, A. Yu, *et al.* 2006. Two splice variants of claudin-10 in the kidney create paracellular pores with different ion selectivities. *Am. J. Physiol. Renal Physiol.* **291:** F1288–F1299.

11. Krug S.M., D. Günzel, M.P. Conrad, *et al.* 2012. Claudin-17 forms tight junction channels with distinct anion selectivity. *Cell. Mol. Life Sci.* **69:** doi:10.1007/s00018-012-0949-x.

12. Hou J., A.S. Gomes, D.L. Paul & D.A. Goodenough. 2006. Study of claudin function by RNA interference. *J. Biol. Chem.* **281:** 36117–36123.

13. Alexandre M.D., Q. Lu & Y.H. Chen. 2005. Overexpression of claudin-7 decreases the paracellular Cl^- conductance and increases the paracellular Na^+ conductance in LLC-PK1 cells. *J. Cell Sci.* **118:** 2683–2693.

14. Tatum R., Y. Zhang, K. Salleng, *et al.* 2010. Renal salt wasting and chronic dehydration in claudin-7-deficient mice. *Am. J. Physiol. Renal Physiol.* **298:** F24–F34.

15. Gonzalez-Mariscal L., C. Namorado Mdel, D. Martin, *et al.* 2006. The tight junction proteins claudin-7 and -8 display a different subcellular localization at Henle's loops and collecting ducts of rabbit kidney. *Nephrol. Dial. Transplant.* **9:** 2391–2398.

16. Mendoza-Rodríguez C.A., L. González-Mariscal & M. Cerbón. 2005. Changes in the distribution of ZO-1, occludin, and claudins in the rat uterine epithelium during the estrous cycle. *Cell Tiss. Res.* **319:** 315–330.

17. Li W.Y., C.L. Huey & A.S. Yu. 2004. Expression of claudin-7 and -8 along the mouse nephron. *Am. J. Physiol. Renal Physiol.* **286:** F1063–F1071.

18. Simon D.B., Y. Lu, K.A. Choate, *et al.* 1999. Paracellin-1, a renal tight junction protein required for paracellular Mg^{2+} resorption. *Science* **285:** 103–106.

19. Hou J., D.L. Paul & D.A. Goodenough. 2005. Paracellin-1 and the modulation of ion selectivity of tight junctions. *J. Cell Sci.* **118:** 5109–5118.

20. Kausalya P.J., S. Amasheh, D. Günzel, *et al.* 2006. Disease-associated mutations affect intracellular traffic and paracellular Mg^{2+} transport function of claudin-16. *J. Clin. Invest.* **116:** 878–891.

21. Günzel D., S. Amasheh, S. Pfaffenbach, *et al.* 2009. Claudin-16 affects transcellular Cl^- secretion in MDCK cells. *J. Physiol. (Lond.)* **587:** 3777–3793.

22. Hou J., A. Renigunta, M. Konrad, *et al.* 2008. Claudin-16 and claudin-19 interact and form a cation-selective tight junction complex. *J. Clin. Invest.* **118:** 619–628.

23. Hou J., A. Renigunta, A.S. Gomes, *et al.* 2009. Claudin-16 and claudin-19 interaction is required for their assembly into tight junctions and for renal reabsorption of magnesium. *Proc. Natl. Acad. Sci. USA* **106:** 15350–15355.

24. Ikari A., C. Okude, H. Sawada, *et al.* 2008. Activation of a polyvalent cation-sensing receptor decreases magnesium transport via claudin-16. *Biochim. Biophys. Acta* **1778:** 283–290.

25. Ikari A., S. Matsumoto, H. Harada, *et al.* 2006. Phosphorylation of paracellin-1 at Ser217 by protein kinase A is essential for localization in tight junctions. *J. Cell Sci.* **119:** 1781–1789.

26. Ikari A., K. Kinjo, K. Atomi, *et al.* 2010. Extracellular Mg^{2+} regulates the tight junctional localization of claudin-16 mediated by ERK-dependent phosphorylation. *Biochim. Biophys. Acta* **1798:** 415–421.

27. Hou J., Q. Shan, T. Wang, *et al.* 2007. Transgenic RNAi depletion of claudin-16 and the renal handling of magnesium. *J. Biol. Chem.* **282:** 17114–17122.

28. Will C., T. Breiderhoff, J. Thumfart, *et al.* 2010. Targeted deletion of murine Claudin-16 identifies extra- and intrarenal compensatory mechanisms of Ca^{2+} and Mg^{2+} wasting. *Am. J. Physiol. Renal Physiol.* **298:** F1152–F1161.

29. Hou J., A. Renigunta, J. Yang & S. Waldegger. 2010. Claudin-4 forms paracellular chloride channel in the kidney and requires claudin-8 for tight junction localization. *Proc. Natl. Acad. Sci. USA* **107:** 18010–18015.

30. Dhawan P., R. Ahmad, R. Chaturvedi, *et al.* 2011. Claudin-2 expression increases tumorigenicity of colon cancer cells: role of epidermal growth factor receptor activation. *Oncogene* **30:** 3234–3247.

31. Tamura A., Y. Kitano, M. Hata, *et al.* 2008. Megaintestine in claudin-15-deficient mice. *Gastroenterology* **134:** 523–534.

32. Van Itallie C., C. Rahner & J.M. Anderson. 2001. Regulated expression of claudin-4 decreases paracellular conductance through a selective decrease in sodium permeability. *J. Clin. Invest.* **107:** 1319–1327.

33. Van Itallie C.M., A.S. Fanning & J.M. Anderson. 2003. Reversal of charge selectivity in cation or anion selective epithelial lines by expression of different claudins. *Am. J. Physiol. Renal Physiol.* **285:** F1078–F1084.

34. Acharya P., J. Beckel, W.G. Ruiz, *et al.* 2004. Distribution of the tight junction proteins ZO-1, occludin, and claudin-4, -8, and -12 in bladder epithelium. *Am. J. Physiol. Renal Physiol.* **287:** F305–F318.

35. Kiuchi-Saishin Y., S. Gotoh, M. Furuse, *et al.* 2002. Differential expression patterns of claudins, tight junction membrane proteins, in mouse nephron segments. *J. Am. Soc. Nephrol* **13:** 875–886.

36. Rahner C., L.L. Mitic & J.M. Anderson. 2001. Heterogeneity in expression and subcellular localization of claudins 2, 3, 4, and 5 in the rat liver, pancreas, and gut. *Gastroenterology* **120:** 411–422.

37. Inai T., A. Sengoku, E. Hirose, *et al.* 2008. Comparative characterization of mouse rectum CMT93-I and -II cells by expression of claudin isoforms and tight junction morphology and function. *Histochem. Cell Biol.* **129:** 223–232.

38. Yu A.S.L., M.H. Cheng, S. Angelow, *et al.* 2009. Molecular basis for cation selectivity in claudin-2-based paracellular pores: identification of an electrostatic interaction site. *J. Gen. Physiol.* **133:** 111–127.

39. Enck A.H., U.V. Berger & A.S. Yu. 2001. Claudin-2 is selectively expressed in proximal nephron in mouse kidney. *Am. J. Physiol. Renal Physiol.* **281:** F966–F974.

40. Van Itallie C.M., J. Holmes, A. Bridges, *et al.* 2008. The density of small tight junction pores varies among cell types and is increased by expression of claudin-2. *J. Cell Sci.* **121:** 298–305.

41. Heller F., P. Florian, C. Bojarski, *et al.* 2005. Interleukin-13 is the key effector Th2 cytokine in ulcerative colitis that affects epithelial tight junctions, apoptosis, and cell restitution. *Gastroenterology* **129:** 550–564.

42. Zeissig S., N. Bürgel, D. Günzel, *et al.* 2007. Changes in expression and distribution of claudin 2, 5 and 8 lead to discontinuous tight junctions and barrier dysfunction in active Crohn's disease. *Gut* **56:** 61–72.

43. Weber C.R., S.C. Nalle, M. Tretiakova, *et al.* 2008. Claudin-1 and claudin-2 expression is elevated in inflammatory bowel disease and may contribute to early neoplastic transformation. *Lab. Invest.* **88:** 1110–1120.

44. Tabariès S., Z. Dong, M.G. Annis, *et al.* 2011. Claudin-2 is selectively enriched in and promotes the formation of breast cancer liver metastases through engagement of integrin complexes. *Oncogene* **30:** 1318–1328.

45. Kinugasa T., Q. Huo, D. Higashi, *et al.* 2007. Selective up-regulation of claudin-1 and claudin-2 in colorectal cancer. *Anticancer Res.* **27:** 3729–3734.

46. Colegio O.R., C.M. Van Itallie, H.J. McCrea, *et al.* 2002. Claudins create chargeselective channels in the paracellular pathway between epithelial cells. *Am. J. Physiol. Cell Physiol.* **283:** C142–C147.

47. Fujita H., H. Chiba, H. Yokozaki, *et al.* 2006. Differential expression and subcellular localization of claudin-7, -8, -12, -13, and -15 along the mouse intestine. *J. Histochem. Cytochem.* **54:** 933–944.

48. Inai T., A. Sengoku, X. Guan, *et al.* 2005. Heterogeneity in expression and subcellular localization of tight junction proteins, claudin-10 and -15, examined by RT-PCR and immunofluorescence microscopy. *Arch. Histol. Cytol.* **68:** 349–360.

49. Laukoetter M.G., P. Nava, W.Y. Lee, *et al.* 2007. JAM-A regulates permeability and inflammation in the intestine in vivo. *J. Exp. Med.* **204:** 3067–3076.

50. Wen, H., D.D. Watry, M.C.G. Marcondes & H.S. Fox. 2004. Selective decrease in paracellular conductance of tight junctions: role of the first extracellular domain of claudin-5. *Mol. Biol. Cell* **24:** 8408–8417.

51. Muto, S., M. Hata, J. Taniguchi, *et al.* 2010. Claudin-2-deficient mice are defective in the leaky and cation-selective paracellular permeability properties of renal proximal tubules. *Proc. Natl. Acad. Sci. USA* **107:** 8011–8016.

52. Krug, S.M., M. Fromm & D. Günzel. 2009. Two-path impedance spectroscopy for measuring paracellular and transcellular epithelial resistance. *Biophys. J.* **97:** 2202–2211.

53. Angelow, S. & A.S. Yu. 2009. Structure-function studies of claudin extracellular domains by cysteine scanning mutagenesis. *J. Biol. Chem.* **284:** 29205–29017.

54. Simossis, V.A., J. Kleinjung & J. Heringa. 2005. Homology-extended sequence alignment. *Nucleic Acid Res.* **33:** 816–824.

55. Günzel, D. & M. Fromm. 2012. Claudins and other tight junction proteins. *Compreh. Physiol.* **2:** doi:10.1002/cphy.c110045.

56. Inai, T., T. Kamimura, E. Hirose, *et al.* 2010. The protoplasmic or exoplasmic face association of tight junction particles cannot predict paracellular permeability or heterotypic claudin compatibility. *Eur. J. Cell Biol.* **89:** 547–556.

Ann. N.Y. Acad. Sci. ISSN 0077-8923

Claudin-derived peptides are internalized via specific endocytosis pathways

Denise Zwanziger,[1,*] Christian Staat,[1,*] Anuska V. Andjelkovic,[2] and Ingolf E. Blasig[1]

[1]Leibniz Institut für Molekulare Pharmakologie, Berlin-Buch, Germany. [2]Department of Pathology, University of Michigan, Medical School, Ann Arbor, Michigan

Address for correspondence: Ingolf E. Blasig, Leibniz Institut für Molekulare Pharmakologie, Robert-Rössle-Str. 10, 13125, Berlin-Buch, Germany. iblasig@fmp-berlin.de

Claudin proteins are involved in the paracellular tightening of epithelia and endothelia. Their internalization, which can be modulated by extracellular stimuli, for example, proinflammatory cytokines, is a prerequisite for the regulation of the paracellular barrier to allow, for instance, cell migration or drug delivery. The internalization of peptide sequences of claudins is completely unknown. Here, we studied the internalization of two peptides, TAMRA-claudin-1 and TAMRA-claudin-5, derivatives of the extracellular loop of claudin-1 and -5, respectively, in either epithelial or endothelial cells. The cellular uptake of the claudin-1 peptide follows the clathrin-mediated endocytosis as indicated by inhibitors and respective tracers for colocalization. In addition, macropinocytosis and caveolae-mediated endocytosis of the peptide was observed. In contrast, the claudin-5 peptide is mainly internalized via the caveolae-mediated endocytosis evidenced by the colocalization with respective tracers and vesicle markers, whereas the nonselective macropinocytosis seems to be involved in a less effective manner. In conclusion, the assumption is supported that claudin peptides can be internalized by specific and nonspecific pathways.

Keywords: internalization; endocytosis; tight junction; claudin peptides; clathrin mediated; caveolae mediated

Introduction

Transmembrane tight junction (TJ) proteins seal the paracellular cleft between epithelial and endothelial cells.[1–3] Moreover, TJ proteins regulate the paracellular flux of fluids and solutes.[1–3] However, little is known about their intracellular function. Members of these proteins are claudin proteins, occludin-like proteins, and junctional adhesion molecules (JAM). Consequently, proteins are located at cell–cell contact membranes of opposing cells and form a paracellular barrier. In contrast to the distinct organization, there is a continuous flux of the proteins by endocytosis followed by either recycling and incorporation back to the membrane or degradation and *de novo* synthesis.[4] Endocytosis of the TJ proteins can influence the paracellular barrier integrity either due to one subtype, which is internalized, or due to different subtypes, such as claudins and occludin.[4,5]

The intracellular uptake of TJ proteins can take place by three different endocytosis pathways, caveolae-mediated, clathrin-mediated, or more unspecifically, by macropinocytosis.[6–8] The endocytosis pathways differ between TJ proteins and can be influenced by an extracellular stimulus, for example proinflammatory cytokines.[9] The tumor necrosis factor-α or interferon-γ induce internalization of claudin-1, zonula occludens protein 1 (ZO-1) or JAM-A by clathrin-mediated endocytosis and by macropinocytosis.[10,11] By using a nonphysiological stimulus, like Ca^{2+} depletion, the clathrin-mediated endocytosis of several TJ proteins, for instance occludin and JAM-1, is initiated.[12] In contrast, caveolae-mediated endocytosis and macropinocytosis have been observed for occludin and claudin-5 by the *Escherichia coli* toxin cytotoxic necrotizing factor-1 or the chemokine ligand 2.[4,13] However, nothing is known about the

---

*These authors contributed equally.

doi: 10.1111/j.1749-6632.2012.06567.x
Ann. N.Y. Acad. Sci. 1257 (2012) 29–37 © 2012 New York Academy of Sciences.

internalization pathway of peptides designed from claudin segments.

The 27 members of the claudin family are the major component of TJ strands.[14] Their expression pattern varies between different tissues and these proteins are regulating the tissue-, charge-, and size-selectivity.[15,16] In the blood–brain barrier, claudin-5 is the significant claudin subtype,[17] whereas claudin-1 is essential for the tightness of the epidermal barrier.[18] Claudins contain four transmembrane helices, an intracellular N- and C-terminus as well as one intracellular loop and two extracellular loops (ECL). The first ECL is with ~50 amino acids much longer than the second one with ~10–20 amino acids. They can interact with each other between the cell membranes of opposing cells (*trans*-interaction).[19] Claudins are able to interact in a homophilic (same claudin subtype) or heterophilic (different claudin subtypes) manner. However, the complete mechanism of interaction is still unclear but both ECL seem to be involved.[19–21]

In this study, we investigated two fluorescently labeled peptides derived from the ECL of claudin-1, the 5,6-carboxytetramethylrhodamine (TAMRA)-claudin-1 peptide, and derived from the ECL of claudin-5, the TAMRA-claudin-5 peptide, with respect to their internalization behavior. As claudin-1 is a key claudin for the epithelium and claudin-5 for the endothelium epithelial and endothelial cell lines were used. Both peptides are internalized within few minutes and appear vesicle-like distributed. The claudin-1–derived peptide enters the cell via the clathrin-mediated endocytosis and macropinocytosis as well as to lower amounts by the caveolae-mediated endocytosis. In contrast, a caveolae-mediated endocytosis but no clathrin-mediated endocytosis, and low macropinocytosis could be observed for the uptake of the claudin-5-derived peptide. Furthermore, our results predominantly agree with data regarding the internalization of the full-length proteins of claudin-1 and claudin-5.

Methods

Peptide synthesis

Peptides were generated automatically (ACTIVO-P11, Activotec.com, Cambridge, UK) by solid-phase peptide synthesis using Fmoc/^{t}Bu chemistry. Synthesis followed in a batch-wise mode at a distinct temperature with coupling times of 10 min at 70 °C and the Fmoc-removal for seven minutes at 70 °C on Peg-resins (SRAM, 0.2 mmol/g, Rapp-Polymere [Sigma-Aldrich, Germany]). TAMRA was introduced N-terminally after Fmoc-removal to visualize the cellular pathway. Firstly, TAMRA (5 equiv. per Peg-resin, 0.1 mmol) was activated by 2-(6-chloro-1H-benzotriazole-1-yl)-1,1,3,3-tetramethylaminium hexafluoro-phosphate (0.1 mmol) and *N,N*-diisopropylethylamine (0.2 mmol) and then coupled for 40 min in dimethyl-sulfoxide with the respective resin. The cleavage from the resin occurred with trifluoroacetic acid (TFA)/H_2O (9/1) for three hours at room temperature. Then, peptides were precipitated with cold diethyl ether and purified by preparative reversed phase high-performance liquid chromatography (C-18 column) (Dionex, Sunnyvale, CA) with acetonitrile gradients in aqueous 0.1% TFA. Purified peptides were quantified and characterized by liquid chromatography (ACQUITY UPLC system, C18 column) electrospray time-of-flight mass spectrometry (LCT Premier, Waters, Germany), which showed a purity >95% (220 nm) and gave the expected masses. For *in vitro* experiments, TAMRA-labeled peptides were dissolved in dimethyl sulfoxide (final concentration of 0.1%), and afterward added to the corresponding medium containing pluronic® F-127 (0.012%) (Sigma-Aldrich, Germany).

Cell culture

Human embryonic kidney (HEK-293) cells[22] stably transfected with claudin-1-yellow fluorescence protein (YFP) and immortalized mouse brain endothelial cells (bEnd.3)[23] were grown at 37 °C and 10% CO_2. The transfection was thought to visualize colocalization between the claudin and the peptide tested. HEK-293 claudin-1-YFP cells were cultured in Dulbecco's modified Eagle's medium (DMEM) containing 10% fetal calf serum, 100 units/mL penicillin, 100 µg/mL streptomycin, and 4 mM L-analyl-L-glutamine. bEnd.3 cells and bEnd.3 caveolin-1-green fluorescent protein (GFP) cells were cultured in DMEM containing 10% fetal calf serum, 1× nonessential amino acids, and 2 mM L-glutamine.

Inhibitor study

HEK-293 claudin-1-YFP cells were transferred into 35 mm tissue culture dishes containing poly-L-lysine-coated cover-slips and grown to subconfluence. Cells were pretreated with the respective inhibitor: chlorpromazine,[24] as inhibitor of the clathrin-mediated endocytosis with 15 μM for 30 min, wortmannin[25] with 250 nM for 60 min or LY294002 with 100 μM for 60 min as macropinocytosis inhibitors, and filipin[26] as inhibitor of the caveolae-mediated endocytosis with 1 μg/mL for 30 minutes. Afterward, cells were treated with the TAMRA-claudin-1 peptide (40 μM) and the respective inhibitor for 35 minutes. Nuclei were stained with 5 μM Hoechst 33342 (Life Technologies, Germany). Then, cells were washed three times with Hank's buffered salt solution and analyzed by quantification of the intracellular amount of peptide with the laser scanning microscope 510 Meta confocal microscope (Zeiss, Germany).

Tracer study

Colocalization studies of TAMRA-claudin-1: HEK-293-claudin-1-YFP cells were seeded as described above. The clathrin internalization tracer Alexa680-transferrin (25 μg/mL) (Invitrogen, Carlsbad, CA) and 40 μM peptide were dissolved into medium and added to the apical side of HEK-293 claudin-1-YFP cells. After 25 min at 37 °C, samples were analyzed by confocal microscopy. Colocalization studies of TAMRA-claudin-5: The macropinocytosis marker, lysine-fixable Alexa Fluor 488-dextran[12] (1 mg/mL), the caveolae internalization marker, bodipy-FL[4] (5 μM), and the clathrin internalization marker, FITC-transferrin[27] (5 μg/mL) (Invitrogen), were dissolved in ice-cold medium (DMEM w/o phenol red) and added to the apical side of bEnd.3 cell monolayers together with the TAMRA-claudin-5 peptide (1–200 μM). Monolayers were first kept at 4 °C for 30 min to allow tracer accumulation on the cell surface and then incubated for 0–30 min at 37 °C. All samples were viewed on a confocal laser scanning microscope.

Immunofluorescence

The following antibodies were used: anti-Rab34 (Abcam, Boston, MA), -caveolin-1, and -α-adaptin antibodies (BD Biosciences, Franklin Lakes, NJ) and secondary antibodies antimouse-FITC and -rabbit-FITC (Vector laboratories, Burlingame, CA).[28] Samples were fixed in 4% paraformaldehyde and then preincubated in the blocking solution containing 5% normal goat serum, 5% bovine serum albumine, and 0.05% Tween 20 in phosphate-buffered saline. Samples were then incubated with primary

antibodies overnight at 4 °C. Reactions were visualized by fluorescence-conjugated antimouse and/or antirabbit antibodies. All samples were viewed on a confocal laser scanning microscope.

Evaluation of the endocytosis rate

To quantify the endocytosis rate of the peptide with or without inhibitors, cells from four independent experiments were analyzed. For visualization of TAMRA-labeled peptides, excitation was performed at 543 nm. All wavelengths greater than 560 nm were collected as emission signal. For each treatment at least 15 pictures with an average number of 24.1 ± 0.4 cells were taken and the fluorescence intensity of the internalized TAMRA-labeled peptide was determined by Zeiss LSM Image Browser Software (Zeiss, Germany). For quantitation of the endocytosis, the emission signals of TAMRA were used. The analysis of each picture provided by the LSM software resulted in a histogram of the emission intensities (0–255 relative fluorescence units, rfu), in which the background (<150 rfu) and saturated signals (>254 rfu) were excluded. All signal events of a histogram were added up to the number of signal events of a picture. To allow comparison between different pictures the events were normalized to the number of the cells analyzed.

Evaluation of colocalization

To quantify the colocalization of claudin-5 peptide and tracers or vesicle markers, each z-optical section was analyzed using the colocalization finder plug-in of ImageJ (National Institutes of Health, Bethesda, MD). Cells from three independent experiments and three areas per experiment were analyzed. The background contribution to colocalization was corrected using the formula: corrected colocalization = measured colocalization − background colocalization/1 − background colocalization/100. The colocalization probability was estimated by Pearson's correlation coefficient (R_r). The Pearson coefficient ranges between −1 (perfect negative correlation between two images) to +1 (perfect positive correlation between two images). A coefficient of 0 means no correlation between two images.

Statistical analysis

Results are shown as mean ± SEM as well as mean ± SD obtained by the Mann–Whitney test (two-tailed) and Bonferroni test (two-way ANOVA). Differences were considered as significant if $P < 0.05$.

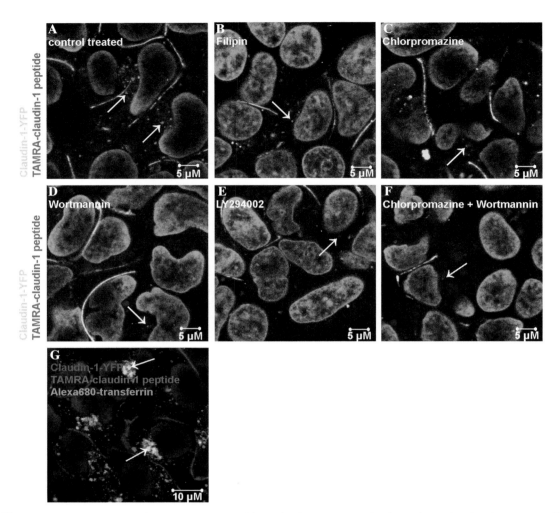

Figure 1. The internalization of the claudin-1 peptide follows the clathrin-mediated and caveolae-mediated endocytosis as well as macropinocytosis. (A) Epithelial human embryonic kidney (HEK)-293 claudin-1 yellow fluorescent protein (YFP) cells were treated with the 5,6-carboxytetramethylrhodamine (TAMRA)-labeled claudin-1 peptide (40 μM) for 35 minutes. The picture shows a clear intracellular distribution of the claudin-1 peptide (red) (arrows). (B–F) HEK-293 claudin-1-YFP cells were pretreated with the corresponding endocytosis inhibitor or inhibitor cocktail. Then, cells were incubated with the claudin-1 peptide (40 μM) and the inhibitor or inhibitor cocktail for 35 minutes. Arrows are indicating intracellular localization of the claudin-1 peptide. (B) Filipin, an inhibitor of the caveolae-mediated endocytosis shows a reduction of the cellular uptake of the claudin-1 peptide. (C) Chlorpromazine, an inhibitor of clathrin-mediated endocytosis leads to a higher decrease of intracellular claudin-1 peptide. (D) An inhibition of internalized claudin-1 peptide could be obtained for wortmannin, an inhibitor of macropinocytosis. (E) LY294002, another inhibitor of macropinocytosis, decreases the uptake of the claudin-1 peptide in a lesser amount as compared to wortmannin. (F) The cocktail of the inhibitors chlorpromazine and wortmannin leads to the highest reduction in the cellular uptake of the claudin-1 peptide. (G) Epithelial HEK-293 claudin-1-YFP cells were cotreated with the claudin-1 peptide (red) and Alexa680-tranferrin (green) to determine colocalization. There is a clear colocalization (yellow, arrows) with Alexa680-transferrin, a tracer for the clathrin-mediated endocytosis and the claudin-1 peptide. Nuclei were stained by Hoechst33342 (cyan).

Results

We synthesized two peptides, TAMRA-claudin-1 and TAMRA-claudin-5, derived from the ECL of the corresponding claudin proteins. The fluorescence dye TAMRA has been introduced N-terminally. The internalization of the claudin-1–derived peptide has been investigated on epithelial HEK-293 cells stably transfected with claudin-1-YFP, whereas the

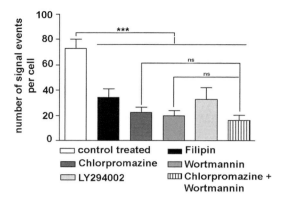

Figure 2. The endocytosis of the claudin-1 peptide decreases by specific pathway inhibitors. For quantification, the number of signal events of the claudin-1 peptide per cell was determined 35 min after addition of the peptide. All wavelengths greater then 560 nm were collected as emission signal. For the evaluation, all signal event intensities between 150 and 254 relative fluorescence units were calculated as number of signal events per cell by using the Zeiss LSM Browser Software (Zeiss, Germany). Data are represented as mean ± SEM; $n = 15$, $^*P <$ 0.05 (Mann–Whitney test, one-tailed), ns = not significant.

claudin-5 peptide has been studied on the brain endothelial cell line bEnd.3, endogenously expressing claudin-5.[23] The HEK cells were used as they exhibit membrane localization of claudin-1 enriched in cell–cell contacts and as the claudin-1 peptide shows a relationship to its full-length protein. The endothelial cells were selected as they express claudin-5 in the cell contacts and as the claudin-5 peptide is related to its mother protein. To study the internalization, we used pharmacological pathway inhibitors to block or decrease the internalization of peptides or applied specific tracers and vesicle markers to determine colocalization of them and the claudin peptide of interest.

For the TAMRA-claudin-1 peptide, Figure 1 (A–G) and Figure 2 show the internalization of the claudin-1 peptide with and without pathway inhibitors into transfected HEK-293 cells, the colocalization of the claudin-1 peptide and of a specific tracer as well as the calculated number of signal events per cell. A clear peptide uptake could be observed after incubation of the cells with 40 μM TAMRA-claudin-1 peptide for 35 minutes (Fig. 1A, 70–80 events/cell). The uptake of the peptide by the HEK claudin-1 cells was time-dependent. After two hours, ∼200 and after five hours, ∼250 events/cell

were registered. Saturation was estimated between 300 and 350 events/cell after 6–24 hours.

To find out the uptake mechanism, different pathway inhibitors were applied and the intracellular amount of claudin-1 peptide was quantified by the determination of signal events per cell (for details see the Methods section) (Fig. 2). The lowest difference in the intracellular amount of the claudin-1 peptide compared to nontreated cells was observed for filipin, an inhibitor of the caveolae-mediated endocytosis (Fig. 1B) an uptake pathway expressed in the HEK cells used.[33] Chlorpromazine, an inhibitor of the clathrin-mediated pathway, significantly decreased the amount of internalized claudin-1 peptide, much higher than filipin (Fig. 1C). Nearly the same amount of internalized claudin-1 peptide was obtained by the inhibition of macropinocytosis using wortmannin (Fig. 1D) as nonreversible inhibitor. Otherwise, the reversible macropinocytosis inhibitor LY294002 showed a less efficacy than wortmannin (Fig. 1E). However, there was still internalized claudin-1 peptide by using pathway inhibitors. The additive effect of chlorpromazine and wortmannin revealed in the highest reduction of intracellularly located claudin-1 peptide (Fig. 1F). To confirm that the clathrin-mediated endocytosis is involved in the peptide internalization, cells were coincubated with the claudin-1 peptide and transferrin a tracer for this endocytosis pathway. A colocalization of the peptide and the tracer was observed (Fig. 1G) with a correlation coefficient between 0.4 and 0.5. However, not all red fluorescent peptide internalized showed overlay with the internalized green fluorescent transferrin due to the other uptake mechanisms found, such as macropinocytosis and the caveolae pathway.

For the TAMRA-claudin-5 peptide, Figure 3 (A–D) and Figure 4(A–D) show the internalization and colocalization of the claudin-5 peptide in bEnd.3 cells and tracers as well as vesicle markers and the calculated Pearson's correlation coefficients. An adequate peptide uptake could be observed for incubation with 40 μM of TAMRA-claudin-5 for 10–30 minutes. To determine the endocytosis pathway several tracers have been used and the colocalization of the corresponding tracer and the claudin-5 peptide has been investigated. No colocalization was determined for transferrin, a tracer for clathrin-mediated endocytosis (Fig. 3A). On the other hand, the claudin-5 peptide

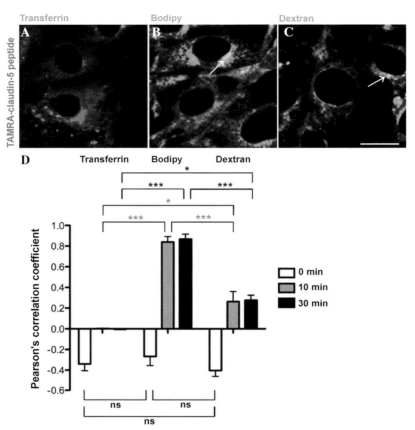

Figure 3. The claudin-5 peptide is colocalized with tracers for the caveolae-mediated endocytosis and macropinocytosis. (A–D) Immortalized mouse brain endothelial cells (bEnd.3) were cotreated with the 5,6-carboxytetramethylrhodamine (TAMRA)-labeled claudin-5 peptide (red) and the corresponding tracer (green) to determine colocalization. (A) bEnd.3 cells were treated with the claudin-5 peptide (40 μM) for 10 and 30 minutes. The picture shows no colocalization of the claudin-5 peptide and transferrin, a tracer for clathrin-mediated endocytosis. (B) Bodipy, a tracer for caveolae-mediated endocytosis is clearly colocalized with the claudin-5 peptide (arrow). (C) A less efficient colocalization is given for dextran, a tracer of macropinocytosis (arrow). (D) Evaluation of the Pearson's correlation coefficients of the tracers and the claudin-5 peptide. Data are represented as mean ± SD, $n = 9$; *, **, ***, $P < 0.05, 0.01$, and 0.001, respectively; ns, not significant, $P > 0.05$ (Bonferroni test, two-way ANOVA). Scale bar, 20 μm.

clearly colocalized with the tracer bodipy, which is internalized via caveolae-mediated endocytosis (Fig. 3B). In a smaller amount, a colocalization of the macropinocytosis tracer dextran and the claudin-5 peptide was observed (Fig. 3C). The less the concentration of peptide used the less was the amount of macropinocytosis as endocytosis pathway (data not shown). To have a closer look to the intracellular compartment of internalized claudin-5 peptide, different specific vesicle markers have been used. The α-adaptin marker for the clathrin-mediated endocytosis revealed no colocalization with the claudin-5 peptide (Fig. 4B). In contrast, there was a significant correlation between the caveolin-1-GFP and Rab34 marker with the claudin-5 peptide, which confirmed the endocytosis pathways determined by the tracer studies (Fig. 4A and C). In separate experiments, HEK-293 cells transfected with claudin-5-YFP showed an internalization of the claudin-5 peptide, whereas, the uptake by the nontransfected cells was negligible. This is in agreement with the imagination that the peptide internalization is related to that of the claudin protein.

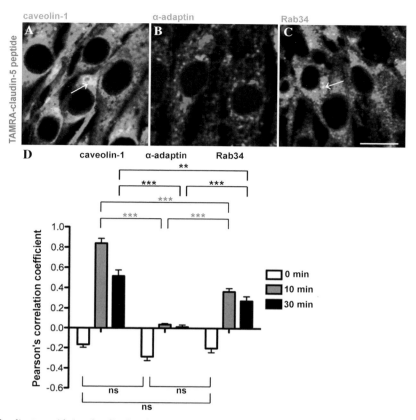

Figure 4. The claudin-5 peptide is colocalized with vesicle markers for the caveolae-mediated endocytosis and macropinocytosis. (A–D) Immortalized mouse brain endothelial cells (bEnd.3) were cotreated with the 5,6-carboxytetramethylrhodamine (TAMRA)-labeled claudin-5 peptide (red) and the corresponding vesicle marker (green) to determine colocalization. (A) The vesicle marker caveolin-1 is colocalized with the claudin-5 peptide (arrow). (B) No colocalization is shown for α-adaptin, a tracer for the clathrin-mediated endocytosis and the peptide. (C) A less efficient colocalization is given for the macropinocytosis vesicle marker Rab34 and the claudin-5 peptide (arrow). (D) Evaluation of the Pearson's correlation coefficients of the vesicle markers and the claudin-5 peptide. Data are represented as mean $\pm$ SD, $n = 9$; **, ***$P < 0.01$ and 0.001, respectively; ns, not significant, $P > 0.05$ (Bonferroni test, two-way ANOVA). Scale bar, 20 μm.

Discussion

Here we demonstrate that a claudin-1–derived peptide and a claudin-5–derived peptide are internalized within the first 35 min into epithelial and endothelial cells expressing the corresponding claudin-subtype; that the claudin-1–derived peptide is internalized by the clathrin-mediated endocytosis and macropinocytosis as well as to a lower amount by the caveolae-mediated pathway; and that the claudin-5–derived peptide is taken up by caveolae-mediated endocytosis and in a lesser amount by macropinocytosis, but not by the clathrin-mediated pathway. These results are predominantly in agreement with earlier studies that showed the same pathways of the respective full-length claudin proteins after extracellular stimuli.[4–11] Only the partly caveolae-mediated pathway for claudin-1 has not been described in earlier studies. Otherwise, here we present the internalization of claudin peptides derived from the respective ECL instead of the full-length claudin proteins and could determine an internalization without any extracellular stimulus. The finding that the peptides use the same pathways as their full-length proteins supports the assumption that the peptides are internalized together with their proteins. This conclusion is in agreement with the reports that the ECL, where the peptides are derived from, mediates the cell-to-cell association.[21,29] Nevertheless, it cannot be excluded that the uptake of the peptide segments may also occur independently from the uptake of the whole protein.

The internalization and redistribution of TJ proteins is an important function for the migration of epithelial and endothelial cells as well as vessels and tubuli formed by epithelia and endothelia. Furthermore, for selective paracellular barrier opening, which can enhance and/or allow drug delivery, it is necessary to understand the exact mechanism of TJ protein endocytosis. Moreover, to focus on the internalization behavior of claudin proteins, claudin-1, which can be overexpressed in colon cancer cells[30] and claudin-5, which is a major TJ protein of the blood–brain barrier are promising claudin targets.[17,28] Thus, claudin-1 peptides conjugated with cytostatic agents could be a suitable approach to be tested as antitumor agent. Claudin-5 peptides could have the potential to interfere with the blood–brain barrier to facilitate drug delivery.

Recently, it has been described that the internalization of TJ proteins is not only depending on an extracellular stimulus given, but also on the cell line and the TJ protein.[4] In this study, peptides of two different claudin subtypes are internalized in different pathways which confirm the relevance of claudin-subtype specificity in terms of the internalization. In addition, it can be assumed that after inhibition of a specific endocytosis pathway, substances (for example, peptides or proteins) can be internalized via another mechanism. This could explain our results of the claudin-1 peptide by using different inhibitors which shows still internalized claudin-1 peptide even if an endocytosis pathway is blocked. Otherwise, a cross-talk between different endocytosis pathways can also not be excluded.[4] Here, both claudin peptides are internalized by, at least, two endocytosis pathways. In recent findings of the claudin-5 internalization via caveolae-mediated endocytosis, a noncaveolae-typical early endosome antigen 1 vesicle localization has been observed.[4] The macropinocytosis is both a nonselective and an actin-dependent internalization mechanism.[7] The uptake of large macromolecules is often described via this endocytosis pathway in which it can take place by an extracellular stimulation or via a constitutive process.[31] This could explain the observation of this mechanism for both peptides, the claudin-1 and the claudin-5 peptide, which has also been described for several TJ proteins in the earlier literature.[4,10,12] The subcellular localization can take place in early endosomes, recycling endosomes, or lysosomes.[32] It has been found that, in most cases, TJ proteins are first located into early endosomes, followed by a recycling and a transport back to the cell membrane.[4] It seems to be that the redistribution of proteins plays a more important role than the degradation or *de novo* synthesis of them.[4]

In summary, we show that a claudin-1 and a claudin-5 peptide, derived from the ECL of their corresponding claudin subtypes, are rapidly internalized by distinct endocytosis pathways. The claudin-1 peptide is internalized by the clathrin-mediated endocytosis and macropinocytosis as well as to a lower amount by the caveolae-mediated endocytosis. The uptake of the claudin-5 peptide is caveolae-mediated, whereas macropinocytosis seems to be a less effective uptake mechanism.

Acknowledgments

We thank Dr. Michael Beyermann for peptide synthesis. This work was supported by the EU project JUSTBRAIN, Else Kröner-Fresenius-Stiftung 2010˙A 52, and DFG BL 408/7-4.

Conflicts of interest

The authors declare no conflicts of interest.

References

1. Morita, K. *et al.* 1999. Claudin multigene family encoding four-transmembrane domain protein components of tight junction strands. *Proc. Natl. Acad. Sci. USA* **96:** 511–516.

2. Furuse, M. *et al.* 1993. Occludin: a novel integral membrane protein localizing at tight junctions. *J. Cell Biol.* **123:** 1777–1788.

3. Wu, J. *et al.* 2006. Identification of new claudin family members by a novel PSI-BLAST based approach with enhanced specificity. *Proteins* **65:** 808–815.

4. Stamatovic, S.M. *et al.* 2009. Caveolae-mediated internalization of occludin and claudin-5 during CCL2-induced tight junction remodeling in brain endothelial cells. *J. Biol. Chem.* **284:** 19053–19066.

5. Shen, L. & J.R. Turner. 2005. Actin depolymerization disrupts tight junctions via caveolae-mediated endocytosis. *Mol. Biol. Cell* **16:** 3919–3936.

6. Marsh, M. & H.T. McMahon. 1999. The structural era of endocytosis. *Science* **285:** 215–220.

7. Amyere, M. *et al.* 2002. Origin, originality, functions, subversions and molecular signalling of macropinocytosis. *Int. J. Med. Microbiol.* **291:** 487–494.

8. Sandvig, K. *et al.* 2011. Clathrin-independent endocytosis: mechanisms and function. *Curr. Opin. Cell Biol.* **23:** 413–420.

9. Capaldo, C.T. & A. Nusrat. 2009. Cytokine regulation of tight junctions. *Biochim. Biophys. Acta.* **1788:** 864–871.

10. Bruewer, M. *et al.* 2005. Interferon-gamma induces internalization of epithelial tight junction proteins via a macropinocytosis-like process. *Faseb J.* **19:** 923–933.

11. Ivanov, A.I., A. Nusrat & C.A. Parkos. 2004. The epithelium in inflammatory bowel disease: potential role of endocytosis of junctional proteins in barrier disruption. *Novartis Found Symp.* **263:** 115–124; Discussion 124–132, 211–218.

12. Ivanov, A.I., A. Nusrat & C.A. Parkos. 2004. Endocytosis of epithelial apical junctional proteins by a clathrin-mediated pathway into a unique storage compartment. *Mol. Biol Cell.* **15:** 176–188.

13. Hopkins, A.M. *et al.* 2003. Constitutive activation of Rho proteins by CNF-1 influences tight junction structure and epithelial barrier function. *J. Cell Sci.* **116:** 725–742.

14. Mineta, K. *et al.* 2011. Predicted expansion of the claudin multigene family. *FEBS Lett.* **585:** 606–612.

15. Furuse, M. *et al.* 1998. A single gene product, claudin-1 or -2, reconstitutes tight junction strands and recruits occludin in fibroblasts. *J. Cell Biol.* **143:** 391–401.

16. Colegio, O.R. *et al.* 2003. Claudin extracellular domains determine paracellular charge selectivity and resistance but not tight junction fibril architecture. *Am. J. Physiol. Cell Physiol.* **284:** C1346–C1354.

17. Nitta, T. *et al.* 2003. Size-selective loosening of the blood-brain barrier in claudin-5-deficient mice. *J. Cell Biol.* **161:** 653–660.

18. Furuse, M. *et al.* 2002. Claudin-based tight junctions are crucial for the mammalian epidermal barrier: a lesson from claudin-1-deficient mice. *J. Cell Biol.* **156:** 1099–1111.

19. Krause, G. *et al.* 2008. Structure and function of claudins. *Biochim. Biophys. Acta.* **1778:** 631–645.

20. Piontek, J. *et al.* 2011. Elucidating the principles of the molecular organization of heteropolymeric tight junction strands. *Cell Mol. Life Sci.* **68:** 3903–3918.

21. Piontek, J. *et al.* 2008. Formation of tight junction: determinants of homophilic interaction between classic claudins. *FASEB J.* **22:** 146–158.

22. Graham, F.L. *et al.* 1977. Characteristics of a human cell line transformed by DNA from human adenovirus type 5. *J. Gen. Virol.* **36:** 59–74.

23. Omidi, Y. *et al.* 2003. Evaluation of the immortalised mouse brain capillary endothelial cell line, b.End3, as an in vitro blood-brain barrier model for drug uptake and transport studies. *Brain Res* **990:** 95–112.

24. Wiranowska, M., L.O. Colina & J.O. Johnson. Clathrin-mediated entry and cellular localization of chlorotoxin in human glioma. *Cancer Cell Int.* **11:** 27.

25. Magzoub, M. *et al.* 2006. *N*-terminal peptides from unprocessed prion proteins enter cells by macropinocytosis. *Biochem. Biophys. Res. Commun.* **348:** 379–385.

26. Sabah, J.R. *et al.* 2007. Transcytotic passage of albumin through lens epithelial cells. *Invest. Ophthalmol. Vis. Sci.* **48:** 1237–1244.

27. Gottlieb, T.A. *et al.* 1993. Actin microfilaments play a critical role in endocytosis at the apical but not the basolateral surface of polarized epithelial cells. *J. Cell Biol.* **120:** 695–710.

28. Stamatovic, S.M., R.F. Keep & A.V. Andjelkovic. Tracing the endocytosis of claudin-5 in brain endothelial cells. *Methods Mol. Biol.* **762:** 303–320.

29. Daugherty, B.L. *et al.* 2007. Regulation of heterotypic claudin compatibility. *J. Biol. Chem.* **282:** 30005–30013.

30. Singh, A.B. *et al.* 2011. Claudin-1 up-regulates the repressor ZEB-1 to inhibit E-cadherin expression in colon cancer cells. *Gastroenterology* **141:** 2140–2153.

31. Kerr, M.C. & R.D. Teasdale. 2009. Defining macropinocytosis. *Traffic* **10:** 364–371.

32. Kirkham, M. & R.G. Parton. 2005. Clathrin-independent endocytosis: new insights into caveolae and non-caveolar lipid raft carriers. *Biochim. Biophys. Acta.* **1745:** 273–286.

33. Cha, S.-K. *et al.* 2011. Calcium-sensing receptor decreases cell surface expression of the inwardly rectifying K$^+$ channel Kir4.1. *J. Biol. Chem.* **286:** 1828–1835.

Ann. N.Y. Acad. Sci. ISSN 0077-8923

ANNALS OF THE NEW YORK ACADEMY OF SCIENCES

Issue: *Barriers and Channels Formed by Tight Junction Proteins*

A phosphorylation hotspot within the occludin C-terminal domain

Max J. Dörfel and Otmar Huber

Department of Biochemistry II, Jena University Hospital, Friedrich-Schiller-University Jena, Jena, Germany

Address for correspondence: Dr. Otmar Huber, Universitätsklinikum Jena, Institut für Biochemie II, Nonnenplan 2–4, 07743 Jena, Germany. otmar.huber@mti.uni-jena.de

Tight junctions (TJs) form paracellular barriers defining the permeability characteristics of epithelial and endothelial cell layers in our body. Tetraspanin integral membrane proteins, including occludin, tricellulin, MarvelD3, and a set of claudins, form a network of anastomosing strands bringing the membranes of neighboring cells into close contact. Occludin is assumed to play an important role in the regulation of TJ formation, structure, and function, and is tightly regulated by phosphorylation. We here summarize the role of occludin phosphorylation on assembly/disassembly and function of TJs and specifically focus on a cluster of 11 amino acids in the C-terminal cytoplasmic domain of occludin (Tyr398–Ser408), including highly conserved phosphorylation sites for c-Src, PKCs, and CK2. Phosphorylation by these kinases affects occludin localization, dynamics, and interaction with other TJ proteins. Interestingly, this phosphorylation hotspot is localized in an unstructured region close to the ZO-1 binding site, and a cysteine residue which is involved in intermolecular disulfide-bond formation thus contributing to occludin dimerization. We discuss potential consequences and open questions in respect to the functional role of this phosphorylation hotspot.

Keywords: occludin; phosphorylation; tight junction; barrier function

Introduction

The paracellular permeability characteristics of epithelial and endothelial tissues are defined by the protein composition of their tight junctions (TJs). Occludin was the first identified integral transmembrane protein within TJs,[1] and now, together with tricellulin[2] and MarvelD3,[3,4] forms the tight junction-associated MARVEL (MAL and related proteins for vesicle trafficking and membrane link) protein family (TAMPs). All three proteins are inserted into the membrane by four transmembrane domains connected by two extracellular loops and one intracellular loop, which together form the MARVEL domain.[5] Occludin and MarvelD3 are equally distributed in bicellular TJs, whereas tricellulin is preferentially localized at tricellular junctions forming a barrier for macromolecules.[6] Recent studies revealed that the three TAMPs have overlapping but nonredundant functions that cannot be mutually compensated.[3] After knockdown of the different TAMP members, tight junctional structures were still detectable indicating that none of the TAMPs is essential for TJ formation.[3,4,7] The N- and C-terminal domains of full-length TAMPs are oriented into the cytoplasm and play an essential role in the regulation of transport, stability, and dynamics of these TJ proteins.

In addition, up to 27 claudins have been identified that make up the major TJ-forming components. Similar to the TAMPs, claudins are integrated into the membrane by four transmembrane domains but differ in their short intracellular N- and C-terminal cytoplasmic tails. There is evidence that different claudins can interact specifically and thereby form homo- and heteromeric complexes.[8] These interactions define the barrier characteristics of individual tissues. Both *cis*-interactions within the same membrane and trans-interactions between opposing cells have been reported. Detailed analyses

doi: 10.1111/j.1749-6632.2012.06536.x

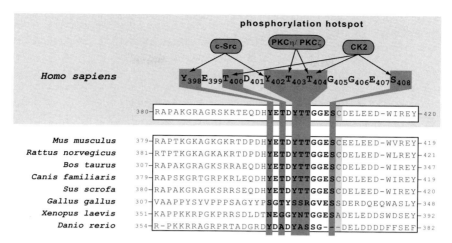

Figure 1. The phosphorylation hotspot in the C-terminal cytosolic domain of occludin. Alignment of amino acid sequences around the phosphorylation hotspot in occludin from different species. Amino acids targeted by the indicated kinases are marked in orange. A conserved cysteine residue marked in blue and located next to the C-terminus of the hotspot region was shown to be involved in occludin homodimerization by disulfide bond formation.[54]

identified tightening and pore-forming activities of individual claudins, as shown for claudin-1 or -4 and claudin-2 or -10, respectively.[9,10]

Further, more than 40 TJ-associated proteins have been identified, including additional transmembrane proteins, such as members of the immunglobulin superfamily of adhesion receptors including the junctional adhesion molecules (JAMs),[11] coxsackie adenovirus receptor (CAR),[12] and endothelial cell-selective adhesion molecule (ESAM).[13] These proteins cannot assemble TJ strands by themselves but have been shown to be involved in the assembly and modulation of TJ function. The cytosolic domains of all these transmembrane proteins form a platform for the assembly of a multitude of intracellular adaptor, cytoskeletal, and regulatory proteins important for TJ structure and function.[14]

Occludin as TJ regulator

Although occludin was the first reported integral membrane protein of TJ strands and is now known for nearly 20 years, its function is still not completely understood. Human occludin initially was characterized as a 522 amino acid protein expressed in epithelial and endothelial cells but is not detectable in fibroblasts. Western blot analyses detected a major band with an apparent molecular mass of 60 kDa and a set of additional bands

defined as phosphorylation or alternative splicing products.[15] Expression of occludin in Madin-Darby canine kidney (MDCK) cells induced an increase in transepithelial resistance (TER) and unexpectedly enhanced paracellular flux.[16,17] These initial findings suggested that occludin is essential for TJ formation and function. Therefore, it was surprising that mice after inactivation of occludin by gene knockout were viable, formed normal TJ strands, and showed only minor signs of pathological disorders.[18,19] Similarly, knockdown of occludin in MDCK cells did not prevent TJ formation, however, changes in the expression of specific claudins and reduced extrusion of apoptotic cells was observed.[7]

It is now well accepted that claudins compose the major TJ constituents, whereas occludin is postulated to play a regulatory role for TJ assembly/disassembly and function. Thereby occludin forms a kind of signaling platform for different stimuli and signaling pathways including growth factors, cytokines, or oxidative stress (for review, see Refs. 20–22). In line with the regulatory role of occludin, different proteins associated around the cytosolic domains of occludin, which are involved in the formation of the tight junctional plaque, have a dual function at the TJs as well as within the nucleus where they modulate gene transcription. These tight junctional proteins including

Table 1. Amino acid residues within the human occludin C-terminal phosphorylation hotspot, which were identified as specific kinase targets

Amino acid(s) modified by kinase	Effects induced by phosphorylation					Ref.
	TER	Permeability	Junctional occludin	TJ-assembly	Interaction with TJ proteins	
c-Src[a]	↓	↑ (inulin)	↓	↓	↓ (ZO-1, -2, -3)	32,34,55
Y398			↓			34
Y402			↓			
Y398 Y402	↓	↑ (inulin)	↓	↓	↓ (ZO-1)	34
nPKCη[b]	↑	↓ (inulin)	↑	↑		41,48
T403			↑			48
T404			↑			48
T403 T404			↑	↑		48
CK2	↓	↑ (Na⁺)	↓		↓ (ZO-1,Cld-1, -2) ↑ (Occ)	43,49–51
T400						
T404	→		mobility[c] →			50
T404 S408			mobility[c] ↑		↓ (ZO-1,Cld-1, -2) ↑ (Occ)	50
S408	↓		↓ mobility[c] ↑		↓ (ZO-1,Cld-1, -2) ↑ (Occ)	50

Note: The effects on tight junction structure and function induced by overexpression/stimulation of kinase are summarized for each kinase in the lines with gray background. In addition, effects induced by phosphorylation of specific amino acid residues are shown. For more detail, see references.
[a]Results obtained by mutation of corresponding amino acids in the mouse are comparable.
[b]Same amino acids also appear to be phosphorylated by aPKCζ.[42]
[c]Mobility of occludin measured by FRAP.

ZO-1, ZO-2, symplekin, ZONAB, and huASH1 were defined as the nucleus and adhesion complexes (NACos) proteins (for review, see Ref. 23). Cingulin, another tight junctional plaque protein, can associate with molecules regulating RhoA and Rac1 signaling.[24,25] The regulatory role of occludin was emphasized when it was recently shown that occludin is required for cytokine-induced modulations of barrier properties.[26,27] There is clear evidence that phosphorylation/dephosphorylation of occludin is essential in the regulation of TJ structure and function.

Phosphorylation of the occludin C-terminal domain

Early studies by Sakakibara et al.[28] already indicated that phosphorylation of occludin is of central importance for TJ formation. Meanwhile, different kinases and phosphatase have been identified that directly target occludin and thereby modulate barrier structure and function (for review, see Ref. 29). The Src-family kinases c-Yes[30,31] and c-Src[32] have been shown to bind to the occludin C-terminal domain. Tyrosine residues corresponding to Tyr398, Tyr402, and Tyr474 in human occludin have been identified as c-Src phosphorylation sites,[33,34] whereas no specific sites targeted by c-Yes have been reported. Currently, the precise role of occludin tyrosine phosphorylation is not fully understood. Nevertheless, much data suggest that occludin tyrosine phosphorylation contributes to TJ destabilization and disruption probably by abrogation of the interaction between the occludin C-terminus and ZO-1.[32] Moreover, Tyr473 in mouse occludin recruits the p85a subunit of PI3K to the leading edge, activates it, and augments lamellipodia formation during cell migration.[33] In addition, an interaction of the cytosolic tyrosine kinase, the focal adhesion kinase (FAK), was observed. However, since a FAK phosphorylation site in occludin has not been identified, it is currently not clear whether the barrier disruption observed in response to knockdown of FAK[35] is a direct effect or a consequence of cytoskeletal rearrangements.

Two further phosphorylation sites in the occludin C-terminal domain have been identified for Rho kinase (RhoK) at amino acids T382 and S507 in mouse occludin.[36] In human occludin, a corresponding amino acid to T382 is missing. Inhibition of RhoK activity was shown to reduce occludin phosphorylation and monocyte transmigration across the blood–brain barrier[37] and to attenuate LPA-induced increases in endothelial cell permeability.[38]

Extensive studies have been performed to analyze the role of protein kinases C (PKCs) on TJ function. Apparently, depending on the inhibitor/activator and the cellular system used, results obtained differed and sometimes were difficult to interpret. This can be explained by the fact that different PKC isoforms are expressed and become active on tight junctional proteins including occludin.[39] There is evidence now that classical (cPKCs) and novel PKCs (nPKCs) have antagonistic effects on TJ assembly and target different amino acids in the occludin C-terminal domain.[40,41] In this respect, Ser338 in mouse occludin was identified as a cPKC phosphorylation site.[40] Threonines 403, 404, 424, and 438 in human occludin were reported as potential PKCζ sites.[42] Mutation of threonine residues 424 and 438 to alanine resulted in delayed assembly of occludin into TJs in Ca^{2+}-switch experiments.

Association of occludin with CK1 (formerly casein kinase 1) as another Ser/Thr kinase was shown in coimmunoprecipitation and pull-down assays and in vitro phosphorylation experiments suggested that occludin is a direct target of CK1.[43,44] In addition to the occludin C-terminal domain, we also have evidence that the N-terminal domain includes putative CK1 phosphorylation sites.[43] Whether all CK1 isoforms bind to occludin is not clear at the moment. Up to now, binding has only been shown for CK1α and ε.[44] Preliminary evidence suggests that multiple sites can be phosphorylated by CK1. It is not clear, however, whether different CK1 isoforms phosphorylate different sites in occludin.

Phosphorylation of Ser490 in response to vascular endothelial growth factor (VEGF) or platelet derived growth factor (PDGF) treatment attracted some research attention. Although the kinase involved is not identified yet, interesting functional consequences with respect to Ser490 phosphorylation have been observed. Expression of an occludin-S490D mutated protein resulted in diminished ZO-1 binding.[45] Furthermore, phosphorylation of Ser490 enhances binding of the ubiquitin-ligase Itch and subsequent occludin ubiquitination.[46] Interestingly, it was recently reported that Ser490-phosphorylation of occludin increases during mitosis and regulates mitotic entry. Ser490-phosphorylated occludin colocalizes with γ-tubulin in centrosomes in mitotic cells and affects centrosome separation. Moreover, expression of a phosphomimetic occludin-S490D construct enhanced cell proliferation, whereas occludin-S490A had the opposite effect.[47]

A phosphorylation cluster within the occludin C-terminal domain

In addition to the phosphorylation sites summarized above, a sequence of 11 amino acids from Tyr398 to Ser408 in human occludin was identified, which includes six residues that are targeted by different kinases (Fig. 1). The previously mentioned c-Src sites Tyr398 and Tyr402[34] lie within this cluster, as well as Thr403 and Thr404 functioning as sites for the novel PKCη.[48] Overexpression of PKCη or expression of a phosphomimetic variant of occludin (Occ-T403D/T404D) enhanced tight junctional localization of occludin, increased TER, and reduced paracellular permeability (Table 1). In contrast, expression of occludin-Y398D/Y402D resulted in its reduced junctional localization and TER, impaired ZO-1 binding, and delayed assembly of TJs in Ca^{2+}-switch experiments.

Amino acid residues Thr400, Thr404, and Ser408 have been identified as casein kinase 2 (CK2) phosphorylation sites.[43,49–51] Inhibition or knockdown of CK2 resulted in increased TER, reduced paracellular Na^+-flux and enhanced tight junctional localization of occludin.[50] An important finding in this context was, that inhibition of CK2 reduces the mobile fraction of occludin. Moreover,

CK2-dependent phosphorylation appears to affect occludin dimerization and heterodimerization with claudin-1 and -2. Taken together CK2 appears to induce mobile occludin homodimers, whereas inhibition of CK2-dependent phosphorylation promotes formation of heteromeric complexes with ZO-1, claudin-1, and -2.[50] Our own studies using an occludin-T400D/T404D/S408D construct surprisingly resulted in an increase in the paracellular resistance (unpublished data). Currently, we have no explanation for these discrepancies, which may be caused by the use of different cell lines. All observations available in the context of the Tyr398-Ser408 sequence motif and its kinase-dependent posttranslational modification suggest that this region may be a phosphorylation hotspot for occludin.

Future perspectives

A number of open questions remain with respect to the role of this phosphorylation hotspot for occludin. The Tyr398-Ser408 motif is located close to the coiled-coil sequence regions in the occludin C-terminal domain that mediate occludin dimerization and the interaction with ZO-1.[52] It is assumed that changes in surface charge distribution and concomitant structural alterations induced by the phosphorylation of amino acids within this hotspot region modulate occludin homomeric and heteromeric interactions. Structural analysis of the distal C-terminal domain of occludin (amino acids 383–522) revealed that the N-terminal 34 amino acids of this construct including the phosphorylation hotspot region of occludin are disordered.[53] Since the recombinant protein used for structural analyses was expressed in *Escherichia coli,* it is assumed that none of the amino acids within the hotspot region was phosphorylated. It is currently not know if phosphorylation of amino acids within the hotspot region induces a switch to an ordered structure. Moreover, it would be interesting to know if and how phosphorylation of the hotspot region affects the ordered structure of occludin residues 416–522, which form three α-helices that build two antiparallel coiled-coils with an N-terminal loop.[53]

It is completely unclear at the moment whether phosphorylation of Tyr residues within this sequence cluster excludes Ser/Thr phosphorylation and vice versa, or if both types of phosphorylation can occur simultaneously, may be by sequen-

tial actions of kinases. In consequence, the different phospho-signatures of occludin may define its dynamic and binding behavior.

Another highly interesting aspect in this context is the role of Cys408 in mouse occludin (Cys409 in human occludin), which was shown to be involved in occludin dimerization by forming disulfide bridges.[54] Thus redox-dependent changes in occludin dimerizations are assumed to contribute to changes in TJ assembly and maintenance under physiological and pathological conditions. It will be interesting to see whether and how phosphorylation of this occludin kinase hotspot affects the redox-sensitivity of occludin dimerization.

Finally, it cannot be excluded that further kinases can target amino acid residues within the Tyr398-Ser408 cluster. At least at the sequence level, a potential consensus motif for CK1 can be found within this cluster.

Acknowledgment

This work was supported by the Deutsche Froschungsgemeinschaft (FOR 721 TP3).

Conflicts of interest

The authors declare no conflicts of interest.

References

1. Furuse, M. *et al.* 1993. Occludin: a novel integral membrane protein localizing at tight junctions. *J. Cell Biol.* **123:** 1777–1788.
2. Ikenouchi, J. *et al.* 2005. Tricellulin constitutes a novel barrier at tricellular contacts of epithelial cells. *J. Cell Biol.* **171:** 939–945.
3. Raleigh, D.R. *et al.* 2010. Tight junction-associated MARVEL proteins marveld3, tricellulin, and occludin have distinct but overlapping functions. *Mol. Biol. Cell* **21:** 1200–1213.
4. Steed, E. *et al.* 2009. Identification of MARVELD3 as a tight junction-associated transmembrane protein of the occludin family. *BMC Cell Biol.* **10:** 95.
5. Sanchez-Pulido, L. *et al.* 2002. MARVEL: a conserved domain involve in membrane apposition events. *Trends Biochem. Sci.* **27:** 599–601.
6. Krug, S.M. *et al.* 2009. Tricellulin forms a barrier to macromolecules in tricellular tight junctions without affecting ion permeability. *Mol. Biol. Cell* **20:** 3713–3724.
7. Yu, A.S.L. *et al.* 2005. Knockdown of occludin expression leads to diverse phenotypic alterations in epithelial cells. *Am. J. Physiol. Cell Physiol.* **288:** C1231-C1241.
8. Hou, J. *et al.* 2009. Claudin-16 and claudin-19 interaction is required for their assembly into tight junctions and for renal reabsorption of magnesium. *Proc. Natl. Acad. Sci. USA* **106:** 15350–15355.

9. Krause, G. *et al.* 2008. Structure and function of claudins. *Biochim. Biophys. Acta* **1778:** 631–645.

10. van Itallie, C.M. & J.M. Anderson. 2005. Claudins and epithelial paracellular transport. *Annu. Rev. Physiol.* **68:** 403–429.

11. Bazzoni, G. 2011. Pathobiology of junctional adhesion molecules. *Antioxid. Redox Signal.* **15:** 1221–1234.

12. Cohen, C.J. *et al.* 2001. The coxsackie virus and adenovirus receptor is a transmembrane component of the tight junction. *Proc. Natl. Acad. Sci. USA* **98:** 15191–15196.

13. Nasdala, I. *et al.* 2002. A transmembrane tight junction protein selectively expressed on endothelial cells and platelets. *J. Biol. Chem.* **277:** 16294–16303.

14. González-Mariscal, L. *et al.* 2008. Crosstalk of tight junction components with signaling pathways. *Biochim. Biophys. Acta* **1778:** 729–756.

15. Mankertz, J. *et al.* 2002. Gene expression of the tight junction protein occludin includes differential splicing and alternative promoter usage. *Biochem. Biophys. Res. Commun.* **298:** 657–666.

16. Balda, M.S. *et al.* 1996. Functional dissociation of paracellular permeability and transepithelial electrical resistance and disruption of the apical-basolateral intramembrane diffusion barrier by expression of a mutant tight junction protein. *J. Cell Biol.* **134:** 1031–1049.

17. McCarthy, K.M. *et al.* 1996. Occludin is a functional component of the tight junction. *J. Cell Sci.* **109:** 2287–2298.

18. Saitou, M. *et al.* 2000. Complex phenotype of mice lacking occludin, a component of tight junction strands. *Mol. Biol. Cell.* **11:** 4131–4142.

19. Schulzke, J.D. *et al.* 2005. Epithelial transport and barrier function in occludin-deficient mice. *Biochim. Biophys. Acta* **1669:** 34–42.

20. Blasig, I.E. *et al.* 2011. Occludin protein family – oxidative stress and reducing conditions. *Antioxid. Redox Signal.* **15:** 1195–1219.

21. Capaldo, C.T. & A. Nusrat. 2009. Cytokine regulation of tight junctions. *Biochim. Biophys. Acta* **1788:** 864–871.

22. Matter, K. *et al.* 2005. Mammalian tight junctions in the regulation of epithelial differentiation and proliferation. *Curr. Opin. Cell Biol.* **17:** 453–458.

23. Balda, M.S. & K. Matter. 2008. Tight junctions and the regulation of gene expression. *Biochim. Biophys. Acta* **1788:** 761–767.

24. Aijaz, S. *et al.* 2005. Binding of GEF-H1 to the tight junction-associated adapter cingulin results in inhibition of Rho signaling and G1/S phase transition. *Dev. Cell* **8:** 777–786.

25. Terry, S.J. *et al.* 2011. Saptially restricted activation of RhoA signalling at epithelial junctions by p114RhoGEF drives junction formation and morphogenesis. *Nat. Cell Biol.* **13:** 159–166.

26. Marchiando, A.M. *et al.* 2010. Caveolin-1-dependent occludin endocytosis is required for TNF-induced tight junction regulation in vivo. *J. Cell Biol.* **189:** 111–126.

27. van Itallie, C.M. *et al.* 2010. Occludin is required for cytokine-induced regulation of tight junction barriers. *J. Cell Sci.* **123:** 2844–2852.

28. Sakakibara, A. *et al.* 1997. Possible involvement of phosphorylation of occludin in tight junction formation. *J. Cell Biol.* **137:** 1393–1401.

29. Dörfel, M.J. & O. Huber. 2012. Modulation of tight junction structure and function by kinases and phosphatases targeting occludin. *J. Biomed. Biotechnol.* **2012:** 807356.

30. Nusrat, A. *et al.* 2000. The coiled-coil domain of occludin can act to organize structural and functional elements of the epithelial tight junction. *J. Biol. Chem.* **275:** 29816–29822.

31. Xiao, X. *et al.* 2011. c-Yes regulates cell adhesion at the blood-testis barrier and the apical extoplasmic specialization in the seminiferous epithelium of rat testes. *Int. J. Biochem. Cell Biol.* **43:** 651–665.

32. Kale, G. *et al.* 2003. Tyrosine phosphorylation of occludin attenuates its interactions with ZO-1, ZO-2, and ZO-3. *Biochem. Biophys. Res. Commun.* **302:** 324–329.

33. Du, D. *et al.* 2010. The tight junction protein, occludin, regulates the directional migration of epithelial cells. *Dev. Cell* **18:** 52–63.

34. Elias, B.C. *et al.* 2009. Phosphorylation of Tyr-398 and Tyr-402 in occludin prevents its interaction with ZO-1 and destabilizes its assembly at the tight junction. *J. Biol. Chem.* **284:** 1559–1569.

35. Siu, E.R. *et al.* 2009. Focal adhesion kinase is a blood-testis barrier regulator. *Proc. Natl. Acad. Sci. USA* **106:** 9298–9303.

36. Yamamoto, M. *et al.* 2008. Phosphorylation of claudin-5 and occludin by rho kinase in brain endothelial cells. *Am. J. Pathol.* **172:** 521–533.

37. Persidsky, Y. *et al.* 2006. Rho-regulation of tight junctions during monocyte migration across the blood-brain barrier in HIV-1 encephalitis (HIVE). *Blood* **107:** 4770–4780.

38. Hirase, T. *et al.* 2001. Regulation of tight junction permeability and occludin phosphorylation by RhoA-p160ROCK-dependent and -independent mechanisms. *J. Biol. Chem.* **276:** 10423–10431.

39. Rao, R. 2009. Occludin phosphorylation in regulation of epithelial tight junctions. *Ann. N.Y. Acad. Sci.* **1165:** 62–68.

40. Andreeva, A.Y. *et al.* 2001. Protein kinase C regulates the phosphorylation and cellular localization of occludin. *J. Biol. Chem.* **276:** 38480–38486.

41. Andreeva, A.Y. *et al.* 2006. Assembly of tight junction is regulated by the antagonism of conventional and novel protein kinase C isoforms. *Int. J. Biochem. Cell Biol.* **38:** 222–233.

42. Jain, S. *et al.* 2011. Protein kinase Czeta phosphorylates occludin and promotes assembly of epithelial tight junctions. *Biochem. J.* **437:** 289–299.

43. Dörfel, M.J. *et al.* 2009. Differential phosphorylation of occludin and tricellulin by CK2 and CK1. *Ann. N.Y. Acad. Sci.* **1165:** 69–73.

44. McKenzie, J.A.G. *et al.* 2006. Casein kinase Iε associates with and phosphorylates the tight junction protein occludin. *FEBS Lett.* **580:** 2388–2394.

45. Sundstrom, J.M. *et al.* 2009. Identification and analysis of occludin phosphosites: a combined mass spectrometry and bioinformatics approach. *J. Proteome Res.* **8:** 808–817.

46. Murakami, T. *et al.* 2009. Occludin phosphorylation and ubiquitination regulate tight junction trafficking and vascular endothelial growth factor-induced permeability. *J. Biol. Chem.* **284:** 21036–21046.

47. Runkle, E.A. *et al.* 2011. Occludin localizes to centrosomes and modifies mitotic entry. *J. Biol. Chem.* **286:** 30847–30858.

48. Suzuki, T. *et al.* 2009. PKCη regulates occludin phosphorlyation and epithelial tight junction integrity. *Proc. Natl. Acad. Sci. USA* **106:** 61–66.

49. Cordenonsi, M. *et al.* 1999. Xenopus laevis occludin: identification of in vitro phosphorylation sites by protein kinase CK2 and association with cingulin. *Eur. J. Biochem.* **264:** 374–384.

50. Raleigh, D.R. *et al.* 2011. Occludin S408 phosphorylation regulates tight junction protein interactions and barrier function. *J. Cell Biol.* **193:** 565–582.

51. Smales, C. *et al.* 2003. Occludin phosphorylation: identification of an occludin kinase in brain and cell extracts as CK2. *FEBS Lett.* **545:** 161–166.

52. Müller, S.L. *et al.* 2005. The tight junction protein occludin and the adherens junction protein α-catenin share a common interaction mechanism with ZO-1. *J. Biol. Chem.* **280:** 3747–3756.

53. Li, Y. *et al.* 2005. Structure of the conserved cytoplasmic C-terminal domain of occludin: identification of the ZO-1 binding surface. *J. Mol. Biol.* **352:** 151–164.

54. Walter, J.K. *et al.* 2009. Redox-sensitivity of the dimerization of occludin. *Cell Mol. Life Sci.* **66:** 3655–3662.

55. Basuroy, S. *et al.* 2003. Expression of kinase-inactive c-Src delays oxidative stress-induced disassembly and accelerates calcium-mediated reassembly of tight junctions in the Caco-2 cell monolayer. *J. Biol. Chem.* **278:** 11916–11924.

Ann. N.Y. Acad. Sci. ISSN 0077-8923

Determinants contributing to claudin ion channel formation

Anna Veshnyakova,[1] Susanne M. Krug,[2] Sebastian L. Mueller,[1] Jörg Piontek,[1] Jonas Protze,[1] Michael Fromm,[2] and Gerd Krause[1]

[1]Leibniz-Institut für Molekulare Pharmakologie (FMP), Berlin, Germany. [2]Institut für Klinische Physiologie, Charité - Universitätsmedizin Berlin, Berlin, Germany

Address for correspondence: Gerd Krause, Leibniz-Institut für Molekulare Pharmakologie, Robert-Roessle-Str. 10, 13125 Berlin, Germany. gkrause@fmp-berlin.de

Pore-forming properties of claudins (Cld) are likely defined by residues of their first extracellular loop (ECL1). Detailed mechanisms are unclear. MDCK cells overexpressing FLAG-Cld-1 wild-type and mutants were characterized by transepithelial resistance (TER) and ion permeability measurements. Replacing ECL1 residues of sealing Cld-1 by corresponding Cld-2 residues we aimed to identify new determinants responsible for sealing and/or pore formation. We found that E48K and S53E substitutions in human Cld-1 strongly reduced TER and increased permeability for Na^+ and Cl^-. In contrast, K65D, D68S, and other single substitutions showed no significant change of TER and permeability for Na^+ and Cl^-. Double substitution S53E/K65D did not change TER and ion permeability, whereas S53E/D68S decreased TER, albeit weaker than S53E. Ratio of permeabilities for Na^+ and Cl^- revealed no clear charge specificity of the pore induced by S53E or S53E/D68S in Cld-1, suggesting that primarily S53 and potentially D68 in Cld-1 are involved in sealing of the paracellular cleft and that charge-unselective pores may be induced by substituting S53E.

Keywords: claudins; first extracellular loop; paracellular permeability; pore formation

Introduction

Claudins (Cld) form the backbone of tight junction (TJ) strands and by this they define the tightness of epithelial and endothelial barriers. Several members of the claudin family have been shown to seal the TJ (Cld-1, -3, -5, -11, -14, -19), while others form paracellular channels permeable for ions.

To our present knowledge, ion-conducting claudins do not substantially discriminate between single ion species (e.g., Na^+ vs. K^+) but rather between positively and negatively charged ions and according to ion size.[1,2] A basic measure of the permeability for all ions present in a solution (times their concentrations) is the transepithelial conductance, usually expressed as its reciprocal, the transepithelial resistance (TER). However, TER does not distinguish between cations and anions, so that dilution potential or ion flux measurements have to be added to get information on charge selectiv-

ity. A permeability ratio of Na^+ over Cl^- (P_{Na}/P_{Cl}) above unity indicates cation selectivity and vice versa.

Properties of cation-selective claudins (Cld-2, -10b, -15) and anion-selective claudins (Cld-10a and -17) are described in detail in this volume by Krug *et al.*[3] The best investigated channel-forming claudin so far is Cld-2. It has been demonstrated to lower the resistance of the TJ,[4] which is based on a dramatic increase in permeability to small inorganic and organic cations but not to anions or large molecules.[5] Concomitantly, it is also permeable to water.[6] Claudins form polymeric TJ strands across the paracellular cleft by homo- and heterophilic *cis*- as well as *trans*-interactions.[7,8] These interactions are essential for tightening[9] and assumed to participate in formation of a selective pore as well.

It has been described that the second extracellular loop (ECL2) of claudins is involved in homo- and heterophilic *trans*-interactions between the claudins

doi: 10.1111/j.1749-6632.2012.06566.x

narrowing the paracellular cleft, enabling the barrier properties of TJ.[10]

It was shown that the first extracellular loop (ECL1) of claudins is critical for paracellular sealing and, on the other hand, for selective paracellular ion permeability.[11,12,13] ECL1 consists of ~50 amino acids and contains highly conserved amino acids (W......GLW...C...C), involved in the barrier function.[14] The two cysteines are thought to be involved in protein conformation through formation of disulfide bridges. The variable residues may determine selective paracellular ion permeability of different claudins. Some of the negatively[5,11,15] and positively[16,17] charged amino acids of claudin's ECL1 have been already shown to contribute to pore formation. It has been suggested that charged residues interact with the passing ion.[18]

In the low-resistance cell line MDCK-II, replacement of basic residues by acidic ones in ECL1 of Cld-4 leads to the increase of cation permeability, while exchange of acidic to basic residues in ECL1 of Cld-15 reversed paracellular charge selectivity from preference for cations to anions.[19] Expression of Cld-2 in the high-resistance cell line MDCK-I increases the paracellular cation permeability. Mutation D65N introduced in ECL1 of exogenously expressed Cld-2 reverse the pore-forming effect, thus confirming the hypotheses that ECL1 of claudins contains key residues for pore formation.[3,20]

There are several successful approaches for characterizing the channel function of claudins: in many studies high-resistance cell layers (often MDCK-I or MDCK-C7) were transfected with the respective pore-forming claudin[5,18,21] or a less permeable mutant of that claudin, and then TER or ion permeability was measured between transfected and untransfected cells. In this study, we applied a reversed strategy as we started with a tightening protein, Cld-1, and compared it with mutations featuring anticipated higher permeability. For this, MDCK-I or Tet-On MDCK-C7 cells were transfected with wild-type (WT) Cld-1, or mutated forms of it that contained corresponding residues of the channel-forming protein Cld-2. Then we compared the functional properties of Cld-1 WT and the Cld-1 mutants.

The goal of this study was to identify determinants of ECL1 that are involved in paracellular pore formation. We hypothesized that in a course of potential intermolecular ECL1 interactions, the pore formation is supported by the spatial encounter of repulsing amino acids, while sealing might be caused by attracting amino acids. We found evidence suggesting S53E to be sufficient to induce charge-unselective pores in Cld-1. This adds to our understanding of the determinants of claudin ion permeability, because no genuine claudin appears to exist that forms ion channels without any charge selectivity.

Materials and methods

Generation and characterization of MDCK-I Cld-1 expressing cells

For construction of plasmids encoding N-terminally FLAG-tagged Cld-1 or Cld-2 fusion proteins, cDNA of human Cld-1 and human Cld-2 was amplified by PCR using pCR2.1-TOPO-Cld1 and pCR2.1-TOPO-Cld2 as templates and cloned into pcDNA3.1. Alternatively, 3xFLAG Cld-1 cDNA was cloned into pTRE2hyg vector (Clontech). Plasmids encoding FLAG-Cld-1 or 3xFLAG-Cld-1 with single or double mutations were generated by site-directed mutagenesis. The plasmids were used to transfect MDCK-I or MDCK-C7 Tet-On cells with Lipofectamine 2000 (Invitrogen). MDCK-I cells were maintained in minimum essential medium with Earle's salts (MEM), supplemented with 10% fetal calf serum, 100 units/mL penicillin, 100 μg/mL streptomycin, and 1% L-alanyl-L-glutamine; stable clones were selected in the presence of G418 disulfate (1 mg/mL) and picked using cloning cylinders. MDCK-C7 Tet-On cells were maintained in the same medium as MDCK-I cells with additional G418 disulfate (1 mg/mL) and stable clones were selected with hygromycin (0.8 mg/mL). Clones were seeded on cover slips in 24-well plates—three to four days later immunostained with rabbit anti-FLAG antibody (Invitrogen) and positive clones were identified by fluorescence microscopy (LSM 510 META-UV; Carl Zeiss Jena GmbH, Jena, Germany). Double labeling with mouse anti-Cld-4 (Invitrogen) or mouse anti-ZO1 (Invitrogen) was used to confirm correct localization of FLAG-tagged Cld-1 or its mutants to the TJs. For transient transfection, MDCK-I cells were seeded on cover slips and one day later transfected with either FLAG-Cld-1WT or transfected with double mutants thereof and stained three to four days later, as described above.

The effect of FLAG constructs on endogenously expressed Cld-1, -4, -7 in MDCK-I or on Cld-1, -3, -4 in MDCK-C7 was assessed by Western blot analysis. Cells were washed with PBS, harvested, lysed (20 mM Tris HCl in PBS, 5 mM $MgCl_2$, 1.25 mM EDTA, 0.75 mM EGTA), containing protein inhibitors (Roche). Membrane fractions were obtained by passing through a 26G3/8" needle, followed by a centrifugation at $200g$ for 5 min and subsequent centrifugation of the remaining supernatant at $43,000g$ for 30 minutes. Extracted protein was diluted in lysis buffer and quantified using BCA assay (Pierce, Rockford, IL) and a plate reader (Tecan, Grödig, Austria). Aliquots of protein were mixed with sodium dodecyl sulfate (SDS) containing buffer (Laemmli), denatured at 95 °C for 5 min, fractionated on SDS polyacrylamide gels, and analyzed by Western blot. Proteins were detected using specific antibodies against TJ proteins (Invitrogen) or the FLAG sequence (Sigma Aldrich, Taufkirchen, Germany) respectively, and visualized by luminescence imaging (Lumi-Imager, Roche, New York, NY).

Using chamber electrophysiological studies

Measurement of ion permeabilities was performed as described before.[22] Briefly, for TER and dilution potential measurements the chamber system was filled with 10 mL Ringer's solution (21 mM $NaHCO_3$, 119 mM NaCl, 5.4 mM KCl, 1 mM $MgSO_4$, 1.2 mM $CaCl_2$, 3 mM HEPES, 10 mM glucose) on each side; pH was maintained by constant equilibration with 95% O_2/5% CO_2. After acclimatization of cells, 5 mL of the basolateral or apical bathing solution was replaced by a modified Ringer's solution containing 238 mM mannitol instead of 119 mM NaCl. TER and voltage were recorded during the whole experiment, and permeability ratios for Na^+ and Cl^- (P_{Na}/P_{Cl}) were calculated using the Goldman–Hodgkin–Katz equation. Absolute permeabilities were calculated from relative permeabilities and TER.[22]

Statistics

Unless stated otherwise, results are shown as means±SEM. Statistical analyses were performed using GraphPad Prism version 5.0 (San Diego, CA). Data sets were analyzed using the unpaired two-tailed Student's *t*-test with Welch's correction. $P < 0.05$ was taken as significant.

Results and discussion

Analysis of single mutations within ECL1

Substitution S53E potentially enables pore formation. To find determinants responsible for paracellular pore formation, we created FLAG-tagged chimera mutants between sealing Cld-1 and pore-forming Cld-2. Residues have been chosen that noticeably differ in properties between the two claudins. Selected residues in ECL1 of Cld-1 were substituted by corresponding residues of Cld-2 (Fig. 1A). By site-directed mutagenesis Y33S, Y47S, E48K, S53E, Q57H, K65D, D68S, S74P, and R81Q mutants of FLAG-Cld-1 were created.

The resulting constructs were stably transfected into a high-resistance cell line, MDCK-I, which were later used to test TER and permeability for anions and cations. TER values of obtained clones were compared with values of MDCK-I cells transfected with FLAG-Cld-1 WT ($973 \pm 118 \ \Omega \ cm^2$) and FLAG-Cld-2 WT ($592 \pm 45 \ \Omega \ cm^2$) (Fig. 1B). The introduced mutations in Cld-1 resulted in different effects on TER and could, therefore, be divided into three groups: (i) mutants Y47S, D68S, and R81Q that did not change the TER values in comparison to Cld-1 WT; (ii) mutants Y33S, Q57H, and S74P that increased the TER values; while (iii) E48K and S53E significantly reduced the resistance. The substitution K65D decreased TER, but this change did not reach significance. Since certain substitutions in Cld-1 decreased the TER of MDCK-I cells, permeability measurements for Na^+ and Cl^- were conducted to investigate the ion specificity of pores, introduced by Cld-1 mutants. As expected,[5] cells transfected with Cld-2 developed a significantly higher permeability for Na^+ and a lower permeability for Cl^-, than those transfected with Cld-1 WT (Fig. 1C). Despite the fact that both E48K and S53E led to a significant decrease of TER values down to the level of Cld-2 WT (for E48K) or even lower (for S53E) (Fig. 1B), only the cells transfected with Cld-1 S53E mutant were able to build monolayers, which were significantly more permeable for Na^+ and Cl^- than Cld-1 WT transfected cells (Fig. 1C). In addition, E48K, K65D, and D68S substitutions showed at least a tendency toward increased permeability. However, none of

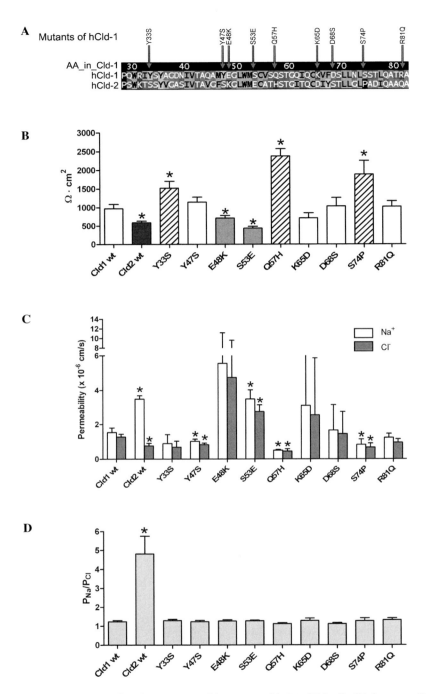

Figure 1. Functional characterization of single mutants in Cld-1, expressed in MDCK-I cells. (A) Sequence alignment of ECL1 of Cld-1 and -2 (ClustalW2; displayed using Geneious Pro v5.3.4). Positions numbered according to human Cld-1; green: bulky hydrophobic; dark green: small hydrophobic; dark blue: polar uncharged; blue: basic; magenta: acidic; cyan: Tyr; black: Pro; orange: Cys. Arrows: substitutions in Cld-1. (B) Effect of Cld-1 substitutions on transepithelial resistance (TER). Open bars: no effect in comparison to Cld-1 WT; striped bars: TER higher then Cld-1 WT; dark gray bar: Cld-2 WT; gray bars: decrease of TER down to Cld-2 WT level. (C) Permeability of Na^+ (open bars) and Cl^- (gray bars) for Cld-1 WT, Cld-2 WT, and Cld-1 single mutants. (D) P_{Na}/P_{Cl} for Cld-1 WT, Cld-2 WT, and Cld-1 single mutants. Two to six measurements were conducted for two to five clones of each construct. Mean ± SEM ($n \geq 4$); *$P < 0.05$ to Cld-1 WT.

the substitutions showed clear discrimination between Na$^+$ or Cl$^-$ ions (Fig. 1D), suggesting that pores, introduced by S53E, are not charge-selective, in contrast to Cld-2 pores, which are cation-selective.

In Cld-15, involvement of E46 (corresponding to E48 in Cld-1) in charge discrimination has been already investigated.[19] It has been demonstrated that replacement of E46 in ECL1 of Cld-15 by positively charged Lys had no effect on paracellular charge selectivity. Our study confirms these data for the corresponding residue in Cld-1, as similar substitution (E48K) also did not influence the ion discrimination (Fig. 1C and D).

Hence, the TER and permeability measurements indicate strong involvement of S53, a possible participation of E48, K65, and D68 of Cld-1 in tightening of the paracellular cleft, whereas corresponding residues of Cld-2 (K48, E53, D65, and S68) might contribute to pore formation.

Mutations at positions Y33, Q57, and S74 tighten the paracellular cleft. Replacement of Y33, Q57, and S74 in Cld-1 by corresponding residues of Cld-2 significantly increased TER values in comparison to those of Cld-1 WT, indicating a stronger tightening effect (Fig. 1B). Permeability measurements conducted for Q57H and S74P mutants of Cld-1 showed a sealing effect against both Na$^+$ and Cl$^-$ (Fig. 1C) without any charge selectivity (Fig. 1D). These data suggest that introduction of bulkier residues at positions 57 and 74 led to an increase of resistance and corresponding charge-unselective tightening of paracellular cleft hampering both Na$^+$ and Cl$^-$ permeability. The findings concerning Q57H were consistent with involvement of Q57 in ECL-ECL interactions. In another study it has been reported that the pathogenic mutation Q57E in Cld-19 leads to hypomagnesemia, renal failure, and severe ocular abnormalities, since Q57E prevents homophilic interaction of Cld-19.[23]

The Y33S substitution increased TER but did not change the ion permeability significantly, whereas mutant Y47S showed no significant effect on resistance but lowered the permeability for Na$^+$ and Cl$^-$ more strongly than Cld-1 WT (Fig. 1C). Since these findings were not consistent, involvement of Y33 and Y47 in the sealing of paracellular barrier is questionable.

Analysis of double mutants

To enhance the effect on TER and verify involvement of the chosen residues in pore formation, double chimera mutants (E48K/S53E, S53E/K65D, and S53E/D68S) were created. Selection of these double mutants was based on decreased TER values by single mutations or their tendency to change TER and/or permeability for Na$^+$ and Cl$^-$.

Mutant E48K/K65D is not localized within the TJ. First, the ability of the created Cld-1 mutants to reach the plasma membrane was analyzed by immunostaining and later visualized by a laser-scanning microscope. S53E/K65D and S53E/D68S mutants of Cld-1 were transported to the plasma membrane (Fig. 2A). In contrast, the E48K/K65D mutant did not reach the plasma membrane of MDCK-I cells but showed intracellular localization (Fig. 2A). Since the trafficking defect was only detected for the E48K/K65D mutant, we hypothesize that the simultaneous substitutions of these two reversed charges specifically induces misfolding of the ECL1. In terms of structural stabilization, this also indicates that the side chains of position 48 and 65 are likely not interacting. The inability to reach the plasma membrane might be the reason for the failure in several attempts to establish a stable cell line overexpressing the E48K/K65D mutant of Cld-1.

Mutations S53E/K65D and S53E/D68S have no additive effect on pore formation. After establishment of stable lines, Western blot analysis was performed for FLAG-Cld-1 WT, S53E/K65D, and S53E/D68S mutants. It was revealed that expression of mentioned constructs led to higher levels of Cld-1, but did not affect the expression of Cld-4 and -7, known to be expressed endogenously in MDCK-I cells (Fig. 2B).

Obtained clones were characterized by TER and ion permeability measurements. While TER of S53/K65D was not significantly changed; it was reduced for S53E/D86S (Fig. 2C). These results indicate that the strong TER decrease versus Cld-1 WT caused by the single S53E mutation in Cld-1 (Fig. 1B) was neutralized or weakened in double mutants of Cld-1: introduction of two negatively charged residues (S53E/K65D mutant) did not decrease TER, while simultaneous introduction of a negatively charged amino acid at one position and its removal at another position (S53E/D68S

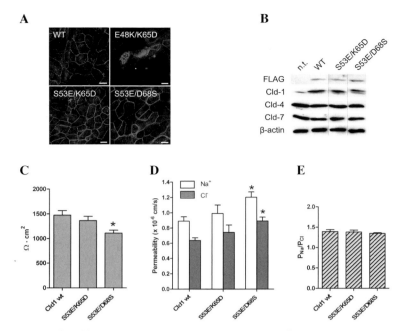

Figure 2. Characterization of double mutants in Cld-1, expressed in MDCK-I cells. (A) Subcellular localization of transiently transfected FLAG-tagged Cld-1 WT, E48K/K65D, S53E/K65D, and S53E/D68S mutants of Cld-1. The expression and distribution of FLAG-Cld-1 constructs was tested by immunofluorescence staining with antibodies to FLAG. Confocal images show that WT, S53E/K65D, and S53E/D68S mutants of Cld-1 localized predominantly in the plasma membrane, whereas E48K/K65D localized intracellularly. Scale bar 10 μm. (B) Effect of FLAG-Cld-1 constructs appearance on expression of other claudins. Proteins of membrane fractions were immunoblotted with antibodies against the indicated claudin isoforms or β-actin, as loading control. Gray lines indicate that intervening lanes have been spliced out. Transepithelial resistance (C), Na$^+$ and Cl$^-$ permeabilities (D) and permeability ratio P_{Na}/P_{Cl} (E) in clones with Cld-1 WT, S53E/K65D, or S53E/D68S expression. Eight to twelve measurements were conducted for three clones of Cld-1 WT, two clones of S53E/K65D mutant, and for three clones of S53E/D68S mutant. **Mean ± SEM** ($n \geq 20$); $^*P < 0.05$ to Cld-1 WT.

mutant) weakly decreased TER versus Cld-1 WT (Fig. 2C).

Permeability analyses revealed that in the case of the double mutant S53E/D68S a decrease of TER was accompanied by higher permeability for both, Na$^+$ and Cl$^-$ (Fig. 2D). Ion permeability as well as TER of the S53E/K65D mutant was similar to those of Cld-1 WT. However, the S53E/D68S mutant as well as S53E/K65D did not change the ratio P_{Na}/P_{Cl}, indicating no alteration of charge selectivity (Fig. 2E).

Previously, the role of the amino acid residue at position 65 has been described for Cld-2. Cysteine-scanning mutagenesis studies suggested that the charge-selective pore introduced by Cld-2 has a diameter in the range of 6.6–7 Å. D65 was shown to be responsible for ion selectivity of Cld-2 and was located in the most narrow part of this pore.[21] Yu *et al.*[18] created another mutant at position 65 of Cld-2 as well where D65 (negatively charged) was re-

placed with Asn (polar, uncharged). As a result, the point mutation D65N in Cld-2 strongly decreased Na$^+$ permeability.[18] In another study it was demonstrated that replacement of Lys (positively charged) by Asp (negatively charged) at this position in Cld-4 considerably diminished the ability of Cld-4 to discriminate against Na$^+$ when expressed in MDCK-II cells.[19] Using MDCK-I cells in this study, we could not obtain similar results with reverse K65D mutant of Cld-1. Introduction of a negative charge at position 65 alone (K65D) or together with S53E substitution did not change paracellular charge selectivity of Cld-1.

Inducible systems confirm position 53 as sensitive for sealing and pore formation

To verify the functional effect of the pore introduced by S53E substitution in Cld-1, and to avoid strong clonal variation, the constructs of interest (Cld-1 WT, S53E, and S53E/D68S mutants thereof) having

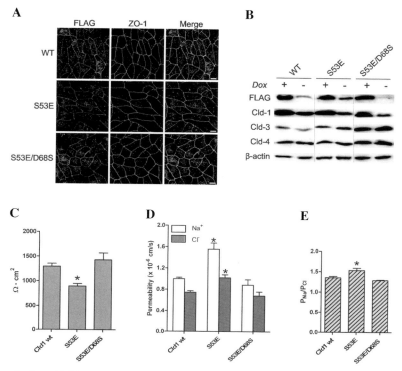

Figure 3. Characterization of WT, S53E, and S53E/D68S mutants of Cld-1, stably transfected in MDCK-C7 Tet-On cells. (A) The expression and distribution of induced FLAG-Cld-1 constructs was visualized by double-immunofluorescence staining with antibodies to FLAG and to the TJ marker, ZO-1. Confocal images induced by doxycycline cells show that WT and mutant Cld-1 colocalize predominantly with ZO-1 to the TJ but are also found intracellularly. Scale bar 10 μm. (B) Effect of FLAG-Cld-1 constructs induction on expression of endogenous Cld-1, -3, -4. Cells have been cultured in the presence (+) or absence (−) of doxycycline (Dox). Proteins of membrane fractions were immunoblotted with antibodies against the indicated claudin isoforms or β-actin as loading control. Gray lines indicate that intervening lanes have been spliced out. Transepithelial resistance (C), Na^+ and Cl^- permeabilities (D), and permeability ratio of Na^+ to Cl^- (P_{Na}/P_{Cl}) (E) in induced clones expressing Cld-1 WT, S53E or S53E/D68S. Three to four measurements were conducted for three clones of each analyzed construct. Mean ± SEM ($n = 9$); $^*P < 0.05$ to Cld-1 WT.

3xFLAG-tag were stably transfected in a Tet-On system of the high-resistance cell line MDCK-C7. After selection, localization of 3xFLAG-Cld-1 WT, S53E, or S53E/D68S mutants in the plasma membrane was verified (Fig. 3A). Induction of expression of these constructs by doxycycline did not clearly influence the expression of endogenous Cld-3 and -4 (Fig. 3B). Unfortunately, the Tet-On system used was slightly leaky; the proteins of interest were also expressed under noninduced conditions (Fig. 3B). The difference in expression levels of noninduced and induced conditions was not strong enough to enable accurate comparison. Consequently, functional analysis of the cells was performed in most reliable, induced conditions. In this induced Tet-On system the single substitution S53E led to a decrease of TER (Fig. 3C) and increased permeability for Na^+

and Cl^- (Fig. 3D). Thus, the results were reproduced in two independent sets of clones (Fig. 1B and C and Fig. 3C and D).

In the Tet-On system, the pore introduced by S53E substitution exhibited cation-selective properties, as P_{Na}/P_{Cl} of the mutant was above that of Cld-1 WT (Fig. 3E). However, this charge selectivity was much smaller than that of the Cld-2 pore (Fig. 1D). Discrepancy in the permeability ratios P_{Na}/P_{Cl} concerning the single S53E substitution in both cell models might be due to minor differences in expression level or regulation of endogenous claudins. Data obtained with the double-mutant S53E/D68S supported previous results: S53E/D68S weakened the effect which was carried by S53E alone. This was confirmed by measurements conducted on the S53E/D68S mutant; although this mutant decreased the resistance (Fig. 2C) the induced pore was not ion selective (Fig. 3C–E).

Taken together, our major finding is consistent with the idea that particular residues in ECL1 are key players of an interacting but yet unknown structural entity that tightens the paracellular cleft in claudins. We provide evidence that S53 in Cld-1 is involved in sealing the paracellular cleft. In reverse, the corresponding E53 in Cld-2 might be involved in pore formation, since introducing the negative charged Glu at position 53 is sufficient to disturb the tightening of Cld-1 and therefore induces a pore. This is unlikely to be caused by a general disruption of the fold of ECL1 since Cld-1 S53E is not retarded in intracellular compartments (Fig. 3A). Nevertheless, future analysis of e.g. Cld-1 S53A might help to clarify the mechanism of this pore induction. However, the created pore is not charge selective, likely due to mismatching residues. On the other hand, introduction of D65, previously shown to be essential for charge selectivity of Cld-2, is alone not sufficient to induce a pore into Cld-1. It is likely that some residues in pore-forming claudins are responsible for opening the paracellular cleft, such as at position 53, and presumably, also at additional ones, whereas other residues, such as D65 in Cld-2, define charge selectivity of the pore. However, the mutations S53E and K65D and combinations in diverse double mutations were not capable to induce such an ion-selective pore, presumably, because these residues are not complementary with other intra- or intermolecular interactions of ECL1, which are necessary for the ion selective paracellular channel.

Finally, we provide further molecular determinants as parts of the puzzle for solving different molecular mechanisms distinguishing sealing and pore formation of claudins.

Conflicts of interest

The authors declare no conflicts of interest.

References

1. Diamond, J.M. 1978. Channels in epithelial cell membranes and junctions. *Fed. Proc.* **37:** 2639–2643.
2. Watson, C.J., M. Rowland & G. Warhurst. 2001. Functional modeling of tight junctions in intestinal cell monolayers using polyethylene glycol oligomers. *Am. J. Physiol. Cell Physiol.* **281:** C388–C397.
3. Krug, S.M., D. Günzel, M.P. Conrad, *et al.* 2012. Charge-selective claudin channels. *Ann. N.Y. Acad. Sci.* **1257:** 20–28.
4. Furuse, M., K. Furuse, H. Sasaki & S. Tsukita. 2001. Conversion of zonulae occludentes from tight to leaky strand type by introducing claudin-2 into Madin-Darby canine kidney I cells. *J. Cell Biol.* **153:** 263–272.
5. Amasheh, S., N. Meiri, A.H. Gitter, *et al.* 2002. Claudin-2 expression induces cation-selective channels in tight junctions of epithelial cells. *J. Cell Sci.* **115:** 4969–4976.
6. Rosenthal, R., S. Milatz, S.M. Krug, *et al.* 2010. Claudin-2, a component of the tight junction, forms a paracellular water channel. *J. Cell Sci.* **123:** 1913–1921.
7. Piontek, J., S. Fritzsche, J. Cording, *et al.* 2011. Elucidating the principles of the molecular organization of heteropolymeric tight junction strands. *Cell Mol. Life Sci.* **68:** 3903–3918.
8. Furuse, M., H. Sasaki & S. Tsukita. 1999. Manner of interaction of heterogeneous claudin species within and between tight junction strands. *J. Cell Biol.* **147:** 891–903.
9. Piehl, C., J. Piontek, J. Cording, *et al.* 2010. Participation of the second extracellular loop of claudin-5 in paracellular tightening against ions, small and large molecules. *Cell Mol. Life Sci.* **67:** 2131–2140.
10. Piontek, J., L. Winkler, H. Wolburg, *et al.* 2008. Formation of tight junction: determinants of homophilic interaction between classic claudins. *FASEB J.* **22:** 146–158.
11. Colegio, O.R., C. Van Itallie, C. Rahner & J.M. Anderson. 2003. Claudin extracellular domains determine paracellular charge selectivity and resistance but not tight junction fibril architecture. *Am. J. Physiol. Cell Physiol.* **284:** C1346–C1354.
12. Van Itallie, C.M. & J.M. Anderson. 2004. The molecular physiology of tight junction pores. *Physiology* **19:** 331–338.
13. Van Itallie, C.M., O.R. Colegio & J.M. Anderson. 2004. The cytoplasmic tails of claudins can influence tight junction barrier properties through effects on protein stability. *J. Membr. Biol.* **199:** 29–38.
14. Wen, H., D.D. Watry, M.C. Marcondes & H.S. Fox. 2004. Selective decrease in paracellular conductance of tight junctions: role of the first extracellular domain of claudin-5. *Mol. Cell Biol.* **24:** 8408–8417.
15. Hou, J., D.L. Paul & D.A. Goodenough. 2005. Paracellin-1 and the modulation of ion selectivity of tight junctions. *J. Cell Sci.* **118:** 5109–5118.
16. Van Itallie, C.M., S. Rogan, A. Yu, *et al.* 2006. Two splice variants of claudin-10 in the kidney create paracellular pores with different ion selectivities. *Am. J. Physiol. Renal Physiol.* **291:** F1288–F1299.
17. Alexandre, M.D., B.G. Jeansonne, R.H. Renegar, *et al.* 2007. The first extracellular domain of claudin-7 affects paracellular Cl- permeability. *Biochem. Biophys. Res. Commun.* **357:** 87–91.
18. Yu, A.S., M.H. Cheng, S. Angelow, *et al.* 2009. Molecular basis for cation selectivity in claudin-2-based paracellular pores: identification of an electrostatic interaction site. *J. Gen. Physiol.* **133:** 111–127.
19. Colegio, O.R., C.M. Van Itallie, H.J. McCrea, *et al.* 2002. Claudins create charge-selective channels in the paracellular pathway between epithelial cells. *Am. J. Physiol. Cell Physiol.* **283:** C142–C147.
20. Krause, G., L. Winkler, S.L. Mueller, *et al.* 2008. Structure and function of claudins. *Biochimica et Biophysica Acta-Biomembranes.* **1778:** 631–645.

21. Angelow, S. & A.S. Yu. 2009. Structure-function studies of claudin extracellular domains by cysteine-scanning mutagenesis. *J. Biol. Chem.* **284:** 29205–29217.

22. Günzel, D., M. Stuiver, P.J. Kausalya, *et al.* 2009. Claudin-10 exists in six alternatively spliced isoforms that exhibit distinct localization and function. *J. Cell Sci.* **122:** 1507–1517.

23. Hou, J., A. Renigunta, M. Konrad, *et al.* 2008. Claudin-16 and claudin-19 interact and form a cation-selective tight junction complex. *J. Clin. Invest.* **118:** 619–628.

Ann. N.Y. Acad. Sci. ISSN 0077-8923

ANNALS OF THE NEW YORK ACADEMY OF SCIENCES

Issue: *Barriers and Channels Formed by Tight Junction Proteins*

Lipolysis-stimulated lipoprotein receptor: a novel membrane protein of tricellular tight junctions

Mikio Furuse, Yukako Oda, Tomohito Higashi, Noriko Iwamoto, and Sayuri Masuda

Division of Cell Biology, Department of Physiology and Cell Biology, Kobe University Graduate School of Medicine, Kobe, Japan

Address for correspondence: Mikio Furuse, Ph.D., Department of Physiology and Cell Biology, Division of Cell Biology, Kobe University Graduate School of Medicine, 7-5-1 Kusunoki-cho, Chuo-ku, Kobe 650-0017, Japan. furuse@med.kobe-u.ac.jp

Tricellular tight junctions (tTJs) are specialized structural variants of tight junctions that restrict the free diffusion of solutes at the extracellular space of tricellular contacts. Their presence at cell corners, situated in the angles between three adjacent epithelial cells, was identified early by electron microscopy, but despite their potential importance, tTJs have been generally ignored in epithelial cell biology. Tricellulin was the first molecular component of tTJs shown to be involved in their formation and in epithelial barrier function. However, the precise molecular organization and function of tTJs are still largely unknown. Recently, we identified the lipolysis-stimulated lipoprotein receptor (LSR) as a tTJ-associated membrane protein. LSR recruits tricellulin to tTJs, suggesting that the LSR-tricellulin system plays a key role in tTJ formation. In this paper, we summarize the identification and characterization of LSR as a molecular component of tTJs.

Keywords: tricellular contact; tight junction; LSR; tricellulin

Introduction

A belt of tight junctions (TJs) circumscribes the apical-most regions of the lateral plasma membranes of epithelial cells.[1,2] Consequently, all routes through the intercellular space are sealed by TJs, and the cellular sheet is able to provide a barrier to the paracellular diffusion of solutes. However, this simple description ignores the frequent occurrence of tricellular contacts, where the vertices of three epithelial cells meet within the cellular sheet. Claudin-based TJs are formed between two plasma membranes, but it is not immediately obvious how the extracellular space is effectively excluded at tricellular contacts, where three plasma membranes assemble. Observations of these regions in vertebrate epithelial cells by freeze-fracture electron microscopy in the early 1970s identified a specialized type of tight junction, namely tricellular TJs (tTJs).[3–6] Images of freeze-fracture replicas suggest that the belt of TJs is not continuous at tricellular contacts. In these regions, the most apical elements of the TJ strands on each side of the junction join and turn to extend in the basal direction

(Fig. 1). These strands, which are considered to be the central sealing elements, are connected by short TJ strands extending from the TJs on both sides to form networks.[1] This results in three sets of central sealing elements assembling to exclude the extracellular space at the tricellular contacts. This arrangement is thought to impede the paracellular diffusion of solutes. These observations suggest that specialized proteins may be involved in the formation of tTJs. However, until recently, such molecules were not identified. Consequently, for a long period tTJs attracted little attention in the field of epithelial cell biology.

An integral membrane protein, tricellulin was the first molecular constituent of tTJs to be identified in 2005.[7] Tricellulin has four transmembrane domains and shows sequence similarity to occludin, a TJ-associated protein. Tricellulin is expressed in a variety of epithelial cells and is localized at the central sealing elements of tTJs. In RNAi-mediated suppression and overexpression studies on cultured epithelial cells, it was observed that tricellulin was required for full development of transepithelial electrical resistance (TER) and for the generation of

doi: 10.1111/j.1749-6632.2012.06486.x

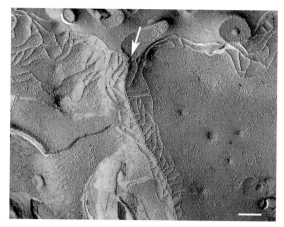

Figure 1. Freeze-fracture replica of tTJs in MDCK cells. TJ strands on both sides join to form the central sealing element (arrow) and extend to the basal direction. Scale bar: 200 nm.

the paracellular barrier to macromolecules.[7,8] In addition, exogenously expressed tricellulin increased the cross-linking of claudin-based TJ strands in claudin-expressing fibroblasts,[9] suggesting the possibility that tricellulin is also involved in the complex network of claudin-based TJ strands in tTJs. However, there are still many unsolved questions regarding tTJs. The detailed mechanism underlying the function of tricellulin has not been clarified at the molecular level. It is likely that the peculiar localization of tricellulin in tTJs at cell vertices is important for its function, but the basis of this localization is unknown. It is also likely that there are other molecular constituents specific to tTJs. It is necessary to identify and characterize the tTJ-associated molecules to fully understand their organization and function at the molecular level.

Identification of lipolysis-stimulated lipoprotein receptor as a novel molecular component of tTJs

TJs cannot be highly purified for biochemical analyses by subcellular fractionation. Therefore, to identify novel TJ-associated proteins, localization-based screening of monoclonal antibodies generated against a TJ-containing membrane fraction has often been used. Many of the key proteins of TJs, including ZO-1,[10] cingulin,[11] occludin,[12] and JAM,[13] were discovered in this way. However, a limitation of this approach is that it de-

pends on the antigenicity of putative junctional proteins, which cannot be controlled. To overcome this problem, we applied a localization-based expression cloning method, namely FL-REX (fluorescence localization-based retrovirus-mediated expression cloning), developed by Misawa and colleagues,[14] for the identification of novel molecular constituents of TJs, including tTJs. In this visual screening method, a green fluorescent protein (GFP)-fused cDNA library constructed in a retrovirus-based expression vector is first introduced into cells. The cells in which the exogenous GFP-fusion proteins are observed to be localized at certain cellular structures are then selected under a fluorescence microscope. Their cDNAs integrated into the genome are finally cloned by genomic PCR. Using an FL-REX screen using a GFP-fusion library constructed from T84 human colon carcinoma cell-derived poly(A)$^+$ RNA, we identified lipolysis-stimulated lipoprotein receptor (LSR) as a novel molecular constituent of tricellular contacts.[15] LSR is a single transmembrane protein of about 65 kD molecular mass. It contains an extracellular Ig-like domain, a transmembrane domain, and a long cytoplasmic domain. Immunofluorescence analyses of cultured epithelial cells and of frozen sections of various mouse tissues revealed that LSR was localized at tricellular contacts of most epithelial tissues (Fig. 2). Furthermore, immunoreplica labeling revealed that LSR was concentrated in the central sealing elements of tTJs observed by electron microscopy, that is, a similar location to tricellulin. These results indicate that LSR is a novel molecular component of tTJs.[15]

LSR is required for full barrier function of epithelial cells

To investigate the roles of LSR in tTJ formation and in epithelial barrier function, EpH4 mouse mammary epithelial cells with stable suppression of LSR expression by LSR-specific shRNA were established. Immunofluorescence staining of LSR-knockdown EpH4 cells with the antibody to occludin (an excellent marker of TJs) indicated that under subconfluent conditions tTJ formation was altered; occludin staining at tricellular contacts appeared to be discontinuous, and abnormal accumulation of occludin was often observed in these regions. Furthermore, LSR-knockdown in EpH4 cells grown on permeable filters under confluent conditions

Occludin   LSR

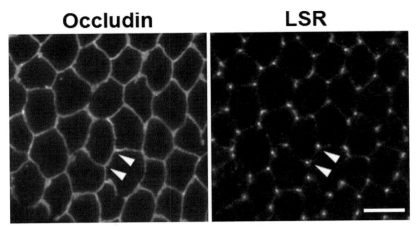

Figure 2. Immunofluorescence localization of LSR in epithelial cells in the mouse epidydimis. Occludin is also labeled to delineate TJs. LSR is highly concentrated in tricellular contacts. Scale bar: 10 μm.

showed reduced TER compared with parent EpH4 cells. Nevertheless, TER of LSR-knockdown EpH4 cellular sheets (~1000 ohm cm^2) was still much higher than that of confluent Madin-Darby canine kidney (MDCK) II cellular sheets (50–70 ohm cm^2).[16] The effects on tTJ formation and TER in LSR-knockdown EpH4 cells were canceled by re-expression of HA-tagged LSR.[15] These observations indicate that LSR is required for normal tTJ formation and to produce the high transepithelial electric resistance of the cellular sheet.

LSR recruits tricellulin to tTJs

The most significant feature of LSR is its interaction with tricellulin. In tricellulin-knockdown EpH4 cells, immunofluorescence staining showed that LSR was still located at tricellular contacts. In contrast, in LSR-knockdown EpH4 cells, tricellulin lost its tricellular localization and was distributed throughout the basolateral membrane. This phenotype was canceled by reexpression of RNAi-resistant LSR in LSR-knockdown cells (Fig. 3).[15] These observations suggest that LSR recruits tricellulin to tricellular contacts. When a deletion mutant of LSR lacking three-quarters of the C-terminal cytoplasmic domain was introduced into LSR-knockdown EpH4 cells, this construct was localized to tricellular contacts but tricellulin was absent, suggesting that the cytoplasmic region of LSR is necessary to recruit tricellulin.[15] LSR-mediated tricellulin recruitment was also reproduced in fibroblasts.[15] When LSR was stably expressed in L cells, exogenous LSR was as-

sembled into short lines of dots, part of which were cell–cell contacts. In contrast, tricellulin that was exogenously expressed in L cells was diffusely distributed throughout the plasma membrane. However, in LSR-expressing L cells, exogenous tricellulin was colocalized with LSR as dots, again suggesting that LSR recruits tricellulin. When N-terminal cytoplasmic domain-deleted or C-terminal cytoplasmic domain-deleted tricellulin was expressed in LSR-expressing L cells, the former, but not the latter, was colocalized with LSR. Furthermore, a non-TJ–associated membrane protein CD9 fused with the C-terminal cytoplasmic region of tricellulin was also colocalized with LSR, but CD9 was not.[15] Taken together, these results indicate that LSR recruits tricellulin to tTJs, and that the interaction between the cytoplasmic domain of LSR and the C-terminal cytoplasmic domain of tricellulin is required for this recruitment.

Perspective

Identification of LSR presents a possible model for tTJ formation. In this model, LSR assembles at the corners of epithelial cells to generate a landmark for tTJ formation, and tricellulin is recruited to tricellular contacts via its interaction with LSR. Because tricellulin seems to have an affinity for claudins within the plasma membrane,[9] it might recruit claudin-based TJ strands to tricellular contacts to form tTJs. Furthermore, the ability of tricellulin to increase cross-linking of claudin-based TJ strands in mouse L cells[9] suggests that it might help to connect the

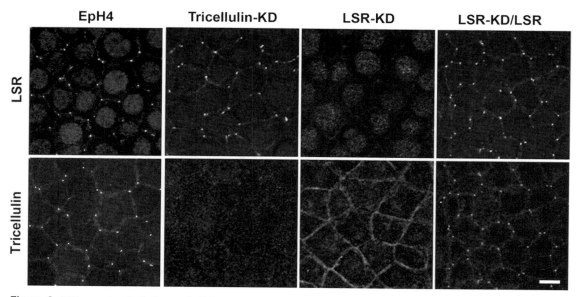

Figure 3. LSR recruits tricellulin to tricellular contacts. Images show double immunofluorescence staining of mouse EpH4 epithelial cells, tricellulin knockdown EpH4 cells (Tricellulin-KD), LSR-knockdown cells (LSR-KD), and LSR-knockdown cells expressing RNAi-resistant HA-tagged LSR (LSR-KD/LSR) with anti-LSR and anti-tricellulin antibodies. Scale bar: 10 μm.

many short TJ strands at tTJs. Perhaps the most intriguing question now is how LSR recognizes tricellular contacts for its own localization. The LSR domain responsible and the factors that recruit LSR to tricellular contacts should be investigated in future studies.

LSR was originally identified as a triglyceride-rich lipoprotein receptor.[17] LSR gene-deficient mice had previously been reported to be embryonic lethal,[18] whereas heterozygous mice show increased fatty acid levels after food intake.[19] However, to date the relationship between these two functions of LSR—in tTJ formation and in lipoprotein uptake—remains elusive. To clarify the role of LSR in tTJ formation *in vivo*, tissue-specific conditional knockout mice for the LSR gene should be generated and analyzed.

Recently, we found that LSR could not be detected in some epithelial tissues by immunofluorescence staining, despite tricellulin localization at tricellular contacts (unpublished observation). Given our conclusion described earlier that LSR recruits tricellulin to tricellular contacts, it seems that other proteins must assume the role of LSR. Two candidates are the LSR-related proteins, ILDR1[20] and ILDR2/c1orf32/Lisch-like.[21,22] Both are membrane proteins containing an Ig-like do-

main and their overall structure shares a primary amino acid sequence with that of LSR. It would be of great interest to examine whether ILDR1 and ILDR2/c1orf32/Lisch-like also share common functional features with LSR in terms of their possible localization at tTJs and recruitment of tricellulin. In particular, ILDR1 is intriguing because it has recently been reported to be a causal gene for autosomal-recessive hearing impairment, in which tricellulin mutations were also identified.[23,24]

Acknowledgments

We thank Dr. Hiroyuki Sasaki (Jikei University School of Medicine) for providing an electron micrograph for Figure 1. M.F. is supported by a NEXT Program from the Japan Society for the Promotion of Science.

Conflicts of interest

The authors declare no conflicts of interest.

References

1. Staehelin, L.A. 1973. Further observations on the fine structure of freeze-cleaved tight junctions. *J. Cell Sci.* **13:** 763–786.
2. Farquhar, M.G. & G.E. Palade. 1963. Junctional complexes in various epithelia. *J. Cell Biol.* **17:** 375–412.

3. Walker, D.C., A. MacKenzie, W.C. Hulbert & J.C. Hogg. 1985. A re-assessment of the tricellular region of epithelial cell tight junctions in trachea of guinea pig. *Acta Anat (Basel).* **122:** 35–38.

4. Wade, J.B. & M.J. Karnovsky. 1974. The structure of the zonula occludens. A single fibril model based on freeze-fracture. *J. Cell Biol.* **60:** 168–180.

5. Staehelin, L.A., T.M. Mukherjee & A.W. Williams. 1969. Freeze-etch appearance of the tight junctions in the epithelium of small and large intestine of mice. Protoplasma. **67:** 165–184.

6. Friend, D.S. & N.B. Gilula. 1972. Variations in tight and gap junctions in mammalian tissues. *J. Cell Biol.* **53:** 758–776.

7. Ikenouchi, J., M. Furuse, K. Furuse, *et al.* 2005. Tricellulin constitutes a novel barrier at tricellular contacts of epithelial cells. *J. Cell Biol.* **171:** 939–945.

8. Krug, S.M., S. Amasheh, J.F. Richter, *et al.* 2009. Tricellulin forms a barrier to macromolecules in tricellular tight junctions without affecting ion permeability. *Mol. Biol. Cell* **20:** 3713–3724.

9. Ikenouchi, J., H. Sasaki, S. Tsukita, *et al.* 2008. Loss of occludin affects tricellular localization of tricellulin. *Mol. Biol. Cell* **19:** 4687–4693.

10. Stevenson, B.R., J.D. Siliciano, M.S. Mooseker & D.A. Goodenough. 1986. Identification of ZO-1: a high molecular weight polypeptide associated with the tight junction (zonula occludens) in a variety of epithelia. *J. Cell Biol.* **103:** 755–766.

11. Citi, S., H. Sabanay, R. Jakes, *et al.* 1988. Cingulin, a new peripheral component of tight junctions. *Nature* **333:** 272–276.

12. Furuse, M., T. Hirase, M. Itoh, *et al.* 1993. Occludin: a novel integral membrane protein localizing at tight junctions. *J. Cell Biol.* **123:** 1777–1788.

13. Martin-Padura, I., S. Lostaglio, M. Schneemann, *et al.* 1998. Junctional adhesion molecule, a novel member of the immunoglobulin superfamily that distributes at intercellular junctions and modulates monocyte transmigration. *J. Cell Biol.* **142:** 117–127.

14. Misawa, K., T. Nosaka, S. Morita, *et al.* 2000. A method to identify cDNAs based on localization of green fluorescent protein fusion products. *Proc. Natl. Acad. Sci. U. S. A.* **97:** 3062–3066.

15. Masuda, S., Y. Oda, H. Sasaki, *et al.* 2011. LSR defines cell corners for tricellular tight junction formation in epithelial cells. *J. Cell Sci.* **124:** 548–555.

16. Stevenson, B.R., J.M. Anderson, D.A. Goodenough & M.S. Mooseker. 1988. Tight junction structure and ZO-1 content are identical in two strains of Madin-Darby canine kidney cells which differ in transepithelial resistance. *J. Cell Biol.* **107:** 2401–2408.

17. Yen, F.T., M. Masson, N. Clossais-Besnard, *et al.* 1999. Molecular cloning of a lipolysis-stimulated remnant receptor expressed in the liver. *J. Biol. Chem.* **274:** 13390–13398.

18. Mesli, S., S. Javorschi, A.M. Berard, *et al.* 2004. Distribution of the lipolysis stimulated receptor in adult and embryonic murine tissues and lethality of LSR-/- embryos at 12.5 to 14.5 days of gestation. *Eur. J. Biochem.* **271:** 3103–3114.

19. Yen, F.T., O. Roitel, L. Bonnard, *et al.* 2008. Lipolysis stimulated lipoprotein receptor: a novel molecular link between hyperlipidemia, weight gain, and atherosclerosis in mice. *J. Biol. Chem.* **283:** 25650–25659.

20. Hauge, H., S. Patzke, J. Delabie & H.C. Aasheim. 2004. Characterization of a novel immunoglobulin-like domain containing receptor. *Biochem. Biophys. Res. Commun.* **323:** 970–978.

21. Roni, V., R. Carpio & B. Wissinger. 2007. Mapping of transcription start sites of human retina expressed genes. *BMC Genomics.* **8:** 42.

22. Dokmanovic-Chouinard, M., W.K. Chung, J.C. Chevre, *et al.* 2008. Positional cloning of "Lisch-Like", a candidate modifier of susceptibility to type 2 diabetes in mice. *PLoS Genet.* **4:** e1000137.

23. Riazuddin, S., Z.M. Ahmed, A.S. Fanning, *et al.* 2006. Tricellulin is a tight-junction protein necessary for hearing. *Am. J. Hum. Genet.* **79:** 1040–1051.

24. Borck, G., A. Ur Rehman, K. Lee, *et al.* 2011. Loss-of-function mutations of ILDR1 cause autosomal-recessive hearing impairment DFNB42. *Am. J. Hum. Genet.* **88:** 127–137.

Ann. N.Y. Acad. Sci. ISSN 0077-8923

ANNALS OF THE NEW YORK ACADEMY OF SCIENCES

Issue: *Barriers and Channels Formed by Tight Junction Proteins*

Overexpression of claudin-5 but not claudin-3 induces formation of *trans*-interaction–dependent multilamellar bodies

Jan Rossa, Dorothea Lorenz, Martina Ringling, Anna Veshnyakova, and Joerg Piontek

Leibniz-Institut für Molekulare Pharmakologie, Berlin, Germany

Address for correspondence: Joerg Piontek, Ph.D., Leibniz-Institut für Molekulare Pharmakologie, Robert-Rössle-Str. 10 13125 Berlin, Germany. piontek@fmp-berlin.de

Tight junctions (TJs) regulate paracellular barriers and claudins (Cld) form the backbone of TJ strands. To elucidate the molecular mechanism of claudin polymer formation, TJs were reconstituted by claudin transfection of TJ-free HEK293 cells. Therewith, typical TJ stands can be found at cell–cell contacts. In addition, overexpression of Cld5-YFP induces formation of huge intracellular multilamellar bodies. In contrast, Cld3 does not induce similar structures. Inhibition of *trans*-interaction of Cld5 by Y148A substitution diminished formation of multilamellar bodies. These results demonstrate claudin subtype-specific oligomerization. Cld3 and Cld5 localize to the plasma membrane differentially. Phosphorylation at T207 of Cld5 was suggested to participate in regulation of Cld5 internalization. However, prevention of potential phosphorylation by T207A substitution did not increase Cld5 amount in the plasma membrane of transfected cells. Taken together, if carefully evaluated, transfection of claudin constructs in nonpolar cells is a powerful strategy to improve understanding of subcellular targeting and assembly of TJ proteins.

Keywords: tight junction; claudins; multilamellar bodies; transfection; electron microscopy

Introduction

Tight junctions (TJs) constitute the paracellular barrier in epithelia and endothelia. They limit and regulate the paracellular permeation of ions, solutes and proteins.[1] TJs appear as an anatomizing network of strands composed of transmembrane proteins (as shown by freeze–fracture electron microscopy) and as fusions of the membranes of two neighboring cells (as shown by transmission electron microscopy).[2] The members of the claudin (Cld) family are tetraspan membrane proteins that constitute the backbone of TJs by *cis*- (side by side) and *trans*- (head to head) interactions.[3,4] The diverse claudin subtypes are expressed in a tissue-specific manner and differ in their barrier properties.[5] They can functionally be divided in barrier-forming, e.g., Cld1, -3, -5, and pore-forming claudins, e.g., Cld2, -10a, 10b, -15.[5] For instance, Cld5 is able to tighten the blood–brain barrier against small molecules up to 800 Da;[6] in contrast, Cld2 and a complex between Cld16 and Cld19 were reported to form pores for cations.[7,8] The ability to form paracellular ion pores and their ion selectivity is mainly determined by residues in the first extracellular loop (ECL1) of claudins.[9,5,10] Based on sequence homology studies, claudins can be grouped as classic claudins (Cld1-10, -14, -15, -17, -19) with high homology within this group and nonclassic claudins (Cld11, -12, -13, -16, -18, -20, -24) with lower homology.[5] In line with the sequence homology, the classic claudins are likely to share a common helix-turn-helix structure of the ECL2. This loop essentially contributes to claudin *trans*-interaction involved in strand formation and paracellular tightening.[3,11–13]

Most epithelial and endothelial cells express an assortment of various claudin subtypes, and it is assumed that these form heteropolymers.[14] The pattern of heterophilic compatibility of claudins is under investigation.[13] To elucidate the molecular mechanism of claudin polymer assembly, we reconstituted TJ strands by claudin transfection of HEK293 (HEK) cells. We focused on Cld3 and Cld5, both of which are assumed to be expressed

doi: 10.1111/j.1749-6632.2012.06546.x

in the blood–brain barrier.[6,15] In this study, we report that overexpression of Cld5-YFP can drive the formation of unique intracellular multilamellar structures that provide information of claudin subtype-specific oligomerization properties.

Material and methods

Cell culture and transfection

HEK cells were maintained in Dulbecco's modified Eagle's medium (DMEM, Invitrogen, Carlsbad, CA) supplemented with 10% fetal calf serum, 100 units/mL streptomycin (Invitrogen, Carlsbad, CA). One day after plating the cells on poly-L-lysine-coated glass coverslips or culture dishes transient transfections of HEK cells were performed with Lipofectamine 2000 (Invitrogen) as described elsewhere.[3]

Mammalian expression vectors and site-directed mutagenesis

Plasmids based on pEYFP-N1 (Clontech) encoding mouse $Cld5_{wt}$-YFP,[16] $Cld5_{Y148A}$-YFP,[3] and $Cld3_{wt}$-YFP[13] have been described previously. The plasmid encoding $Cld5_{T207A}$ was generated by site-directed mutagenesis of pEYFP-N1-$Cld5_{wt}$ by using the primers 5'-GCGCCGC GGCGGCCCGCGGCCAATGGCGATTAC-3' and 5'-GTAATCGCCATTGGCC GCGGGCCGCCGC GGCGC-3'.

Live-cell imaging

Two to three days after transfection, cells were transferred to 1 ml Hank's buffered salt solution (HBSS) with Ca^{2+} and Mg^{2+}. The plasma membrane was visualized either by incubation with 5 μg/mL CellMask™ Deep Red plasma membrane stain (Invitrogen) in HBSS for 15 min or by addition of 20 μL trypan blue, 0.05% in phosphate-buffered saline (PBS) for direct use. Cells were examined with a laser scanning microscope (LSM) 510 META system, using an Axiovert 135 microscope equipped with a PlanNeofluar 100×/1.3 objective (Zeiss, Jena, Germany).[3] LSM 510 software (Zeiss) was used to analyze the claudin enrichment at contacts between two claudin-expressing cells with intensity profiles.

Transmission electron microscopy

For morphology studies, transfected and nontransfected HEK cells grown on culture dishes were fixed with 3% v/v glutaraldehyde, washed, scraped, treated with OsO4 (2%, 1h), and embedded in PolyBed 812 (Polysciences Europe GmbH, Eppelheim, Germany) as previously described.[17] Embedded samples were sectioned (60 nm) on a cryostat (Reichert Ultracut S), stained with uranylacetate and lead citrate, and examined on an electron microscope (Tecnai F20 microscope, FEI, Eindhooven, the Netherlands) at 200 kV using a CCD camera (894 Ultrascan 1000, Gatan Inc., Pleasanton, CA).

Immunogold electron microscopy was performed as previously described.[18] Transfected and nontransfected HEK cells were fixed (0.25% v/v glutaraldehyde, 3% v/v paraformaldehyde), cryosubstituted in a freeze substitution unit (Leica AFS), and embedded in LR-White. Immunolabeling of Cld5-YFP was carried out as previously described[19] on 60 nm sections placed on nickel grids using polyclonal rabbit anti-GFP antiserum 01.[20] and subsequently gold (15 nm) labeled goat antirabbit IgG antibodies (Aurion, Wageningen, the Netherlands). Sections were contrasted with uranylacetate and lead citrate and viewed in on an electron microscope (80 kV Zeiss 902A, Zeiss, Oberkochen, Germany) using a CCD camera (Megaview III, Soft Imaging System, Germany).

Results and discussion

Cellular reconstitution of TJ strands by claudin transfection of TJ-free cells

TJ strands can be reconstituted by transient or stable transfection of TJ-free HEK cells with either Cld5-YFP or Cld3-YFP as described previously.[3,13] At contacts between Cld5-expressing cells, strong enrichment of the YFP-fluorescence indicating *trans*-interaction of Cld5[3] was found (Fig. 1A–F). This leads to formation of TJ strands at cell–cell contacts, which was demonstrated by freeze-fracture EM.[3] These membranous strands are the morphological hallmark of TJs.[2] Similarly, TJ strands could be reconstituted by transfection of Cld3-YFP in HEK cells.[13] Moreover, meshes of reconstituted TJ strands at cell–cell contacts were detected and investigated by super resolution localization microscopy.[21]

Cld5 induces formation of intracellular multilamellar bodies

In addition to the TJ strands at cell–cell contacts, other claudin-including structures were observed. Cld5-YFP and Cld3-YFP signals were also found

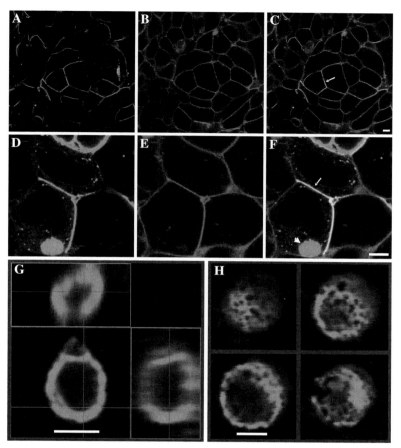

Figure 1. Cld5-YFP induces formation of huge intracellular structures in transiently transfected HEK cells. Three days after transfection or replating stably transfected lines, HEK cells were stained with Cellmask™ (A–C) or trypan blue (D–F) to visualize the plasma membrane. Living cells were analyzed by confocal microscopy. (A–C) In stably transfected HEK cells, Cld5-YFP (green) is strongly enriched (C, merge, arrow) in the plasma membrane (red) at contacts between Cld5-expressing cells. (D–F) In transiently transfected cells, similar enrichment of Cld5-YFP at contacts is found (F, merge, arrow). In addition, huge intracellular structures positive for Cld5-YFP (F, arrowhead) were detected. Z-stacks of high-magnification confocal images revealed hollow spheres with a Cld5-YFP-positive web on the surface (G, *xy* image with *yz* image [red box] and *xz* image [green box]; H series of *xy* images). Bars are 5 μm in C and F and 2 μm in G and H.

in intracellular compartments. These colocalized partly with the endoplasmic reticulum (ER) or were found in small vesicles, probably corresponding to those organelles in which the claudins are either synthesized or transported.[3,13] However, especially for Cld5-YFP, larger and more aggregated intracellular structures were detected (Fig. 1D–F). These structures were prominent after transient transfection but hardly found after stable transfection (Fig. 1A–C).

To investigate these structures in more detail, z-stacks of confocal images were analyzed. The majority of transiently transfected cells had at least

one of these structures, which exhibited a very large diameter of ∼2 μm; z-stacks revealed the structures to be hollow spheres with Cld5-YFP signals on the surface but not the inner core (Fig. 1G). Moreover, the YFP signal was not homogenously spread over the surface but rather showed a mesh-like distribution (Fig. 1H). To our knowledge, so far, similar structures have not been previously described by fluorescence microscopy.

For high-resolution investigation of the morphology of the hollow spheres containing Cld5-YFP, immunogold EM with anti-GFP antibodies was performed. Interestingly, gold particles

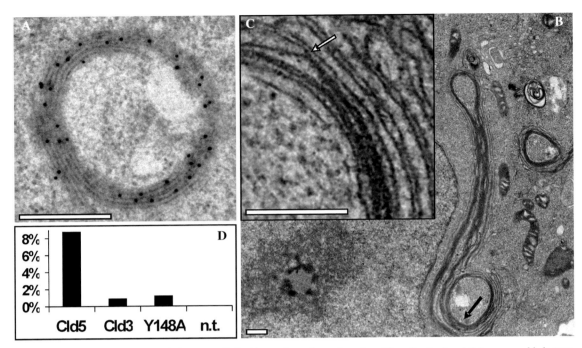

Figure 2. Trans-interaction of Cld5 but not Cdl3 is able to trigger formation of multilamellar bodies. (A) Immunogold electron microscopy of HEK cells transiently transfected with Cld5-YFP using anti-GFP antibodies. The gold particles indicating Cld5-YFP strongly labeled huge hollow spherical multilamellar bodies (MLB) with ~1–2 μm diameter. (B) Using conventional transmission electron microscopy, single spheres (arrow), or two spheres connected by a stack of membranes were detected. (C) Higher magnification revealed areas with collapse of the lumen between two membranes (arrow). (D) Quantification of claudin-YFP–transfected HEK cells revealed that multilamellar bodies were much more frequent for Cld5 than for Cld3 or $Cld5_{Y148A}$ and did not appear in nontransfected (n.t.) cells (given is the percentage of cell sections showing MLB; number of cell sections [from different cells] analyzed: Cld5, $n = 648$; Cld3, $n = 440$; $Cld5_{Y148A}$, $n = 656$; n.t., $n = 583$). Bars are 500 nm (A), 500 nm (B), and 250 nm (C).

indicating Cld5-YFP strongly labeled unique multilamellar bodies (MLB, Fig. 2A). They exhibited a diameter of ~1–2 μm, which was reminiscent to the diameter of the structures identified by confocal microscopy. The inside was of granular appearance, which was not clearly different from the cytosol. The outer part was a spiral of electron-dense (membrane) and electron-light layers (Fig. 2A). The anti-GFP label was strongly associated with the stack of membranes but not with the inside of the spheres. In contrast, similar structures were not found for nontransfected HEK cells. These results clearly demonstrate that Cld5-YFP is able to induce formation of MLB in transiently transfected HEK cells.

Cld3 does not trigger formation of multilamellar bodies

Since the MLB can be identified based on their unique morphology, further analysis was performed without immunogold labeling. In addition, conventional electron microscopy enables better preservation of the membranes due to fixation with 3% glutaraldehyde. Again, in nontransfected HEK cells, no MLB similar to those induced by Cld5 were found (Fig. 2D). For Cld5-YFP–transfected cells, MLB were found in 8.8% of all sectioned cells in the slices, whereas for Cld3-YFP–transfected cells, only 0.7% of the sectioned cells in the slices contained MLB (Fig. 2D). These data show that MLB are formed in a claudin subtype-specific manner.

Trans-*interaction of Cld5 triggers formation of multilamellar bodies*

To investigate the contribution of *trans*-interaction to the formation of MLB, $Cld5_{Y148A}$-YFP–transfected HEK-cells expressing not less Cld5 than $Cld5_{wt}$-YFP–transfected cells were analyzed. Substitution Y148A in the ECL2 strongly inhibits the *trans*-interaction of Cld5.[3] Interestingly, for

Cld5$_{Y148A}$-transfected cells, only 1.2% of the sectioned cells in the slices contained MLB (Fig. 2D). This suggests that formation of MLB depends on *trans*-interaction between Cld5 molecules. The conclusion is further supported by the following morphological observation. The MLB appeared to consist of furled tubules or disks of membranes (Fig. 2B). At some positions, the lumen of the membrane disks was visible, whereas at other positions, the lumen was collapsed (Fig 2C, arrow). This collapse could be explained by luminal *trans*-interaction between Cld5 molecules located in the two opposing membranes.

Similar structures have been reported for untagged occludin, another transmembrane TJ protein. After overexpression of occludin in Sf9 insect cells, Furuse *et al.* found MLB, including many disk-like structures in which the luminal space was collapsed.[22] Probably the oligomerization properties of some transmembrane TJ proteins have the potential to trigger formation of MLB, at least when driven by overexpression.

Trans-interaction of Cld5 can explain the collapse of the lumen and clustering of Cld5 molecules, but it does not explain the stacking of membranes. Therefore, another driving force is required. In the case of occludin, this might be achieved by dimerization of the cytosolic C-terminal domain or by crosslinking via scaffolding proteins.[23,24] In the case of Cld5-YFP, the stacking could be achieved by antiparallel low-affinity dimerization of YFP, which has been shown to support formation of MLB when fused to transmembrane proteins.[25] YFP (GFP, CFP) can dimerize in an antiparallel but not parallel mode. The latter is underlined by the lack of fluorescence resonance energy transfer between CFP- and YFP-fusion proteins of noninteracting Cld5 and corticotrophin-releasing hormone receptor 1.[16]

However, YFP cannot be the major driving force for the formation of MLB, since it has been shown that only particular transmembrane YFP fusion proteins (e.g., calnexin or cytochrome b5) are capable of inducing MLB.[25,26] In line with these reports, we found MLB predominantly for Cld5-YFP but not for Cld3-YFP. Another factor shown to promote MLB formation is a high expression level of MLB-inducing proteins.[25] Consistently, we found MLB after transient overexpression but very little in stable cell lines.

We show that inhibition of *trans*-interaction by the Y148A substitution in the ECL2 strongly reduced the ability of Cld5 to induce MLB. In addition, Cld3-YFP, which does not induce MLB efficiently, shows less *cis*-interaction than Cld5-YFP.[13] Together, these findings indicate that the induction of Cld5-containing MLB depends on claudin subtype-specific oligomerization properties.

Another striking observation was the mesh-like distribution of Cld5-YFP on the surface of the MLB. Furthermore, the fluorescence intensity in the MLB was much stronger than the intensity in TJ strands at cell–cell contacts. The EM images revealed membrane stacks of about 200–400 nm thickness. In confocal microscopy, most signals arising from such a distance cannot be resolved separately and are consequently added up, leading to stronger signals than that of Cld5-YFP at the ~20 nm interface at cell–cell contacts. This explains the much stronger YFP-signals of the MLB. On the other hand, the nonhomogeneous but mesh-like Cld5 signals of the MLB (Fig. 1H) indicate that the Cld5-YFP molecules are clustered in and between the membrane disks. This could be best explained by Cld5-polymers running in parallel in the different membranes of the stack because of crosslinking via YFP.

Biological significance of multilamellar bodies
In stably transfected HEK-cells, MLB were hardly detectable (Fig 1A–C), and in differentiated polar epithelial cells, claudin-containing MLB have not been described so far. In contrast to multivesicular bodies connected to endocytosis,[27] the claudin-containing MLB described here may be a result of overexpression in cells that are not adapted to the presence of claudins. Therefore, it is most useful to analyze the claudins at cell–cell contacts where they do form functional TJ strands similar to those found in polar epithelial cells.[13] Nevertheless, MLB provide additional information about the oligomerization properties of claudins, since they are formed in a subtype-specific and *trans*-interaction–dependent manner. Hence, we propose that claudin-containing MLB could be a tool to improve the molecular understanding of the assembly of transmembrane TJ proteins. In addition, they could be used to enrich and purify claudins expressed in mammalian cells.

MLB were suggested to exist as a physiologically and pathophysiologically relevant ER-derived

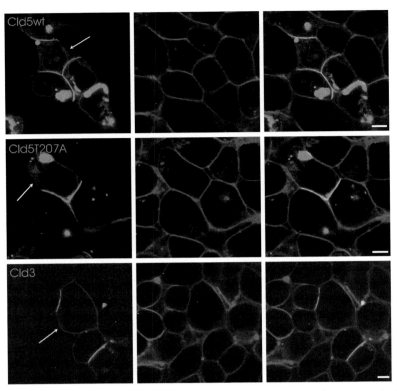

Figure 3. T207A in Cld5 does not change subcellular distribution of the protein. HEK cells were transiently transfected with Cld5$_{wt}$-YFP, Cld5$_{T207A}$-YFP, or Cld3-YFP and three days later stained with Cellmask™ to visualize the plasma membrane (red). Living cells were analyzed by confocal microscopy. All constructs (green) showed enrichment at contacts between claudin-expressing cells. In addition, Cld3$_{wt}$ was clearly detectable in other parts of the plasma membrane (bottom, arrow). This was not the case for Cld5$_{wt}$ nor Cld5$_{T207A}$ (top and middle, arrow). Bar, 5 μm.

compartment in which ER-resident proteins participate in the targeting of misassembled proteins.[26] This was also shown to be relevant for connexins, which are, like claudins, tetraspan junctional proteins.[28] The connexin-50 P88S mutant is linked to hereditary congenital cataracts and targeted to MLB. Similarly, claudins could be misassembled due to mutations linked to hereditary diseases,[5] missing interaction partners, or altered expression level, and targeted to MLB and cause cellular stress.

Cld3 and Cld5 differ in their plasma membrane localization

We have shown that Cld3 and Cld5 colocalize and interact in TJ strands at contacts between claudin-expressing HEK cells.[13] However, in intracellular compartments and in the plasma membrane outside of cell–cell contacts, they weakly colocalize and hardly interact. This suggests that Cld3 and Cld5 are

transported to the plasma membrane and/or internalized independently.

Internalization of claudins is influenced by phosphorylation.[5] Phosphorylation at T207 in the cytoplasmic tail of Cld5 was reported to influence barrier properties[29] and suggested to be involved in destabilization of Cld5 at the plasma membrane.[30] Hence, we speculated that the lower amount of Cld5 in the plasma membrane compared to that of Cld3 could be related to phosphorylation-dependent internalization of Cld5. Consequently, we generated Cld5$_{T207A}$ to prevent potential phosphorylation at T207. However, Cld5$_{T207A}$ showed a similar subcellular distribution like Cld5$_{wt}$. In particular, no increase of the Cld5 signal in the plasma membrane outside of contacts between claudin-expressing cells was observed (Fig. 3). This finding suggests that phosphorylation at T207 is not involved in the subcellular localization of Cld5 in HEK cells. Further structure–function studies using mutagenesis have to be performed to clarify the reason for the difference between the subcellular localization of Cld3 and Cld5.

So far, we have obtained evidence to show that Cld3 and Cld5 have similar capacities for *trans*-interaction.[13] Nevertheless, the dynamic mobile fraction in the plasma membrane is higher for Cld3. In addition, the ability for *cis*-interaction is weaker for Cld3.[13] Both claudins form TJ strands of different morphology. Cld3 forms continuous, protoplasmic face-associated strands, whereas the strands formed by Cld5 are discontinuous and exoplasmic-face associated.[13]

In summary, these data suggest that Cld3 and Cld5 differ in their oligomerization properties, which are assumed to influence the barrier properties of Cld3- and/or Cld5-containing TJs. The exogenous expression of defined claudins in cell systems without endogenous TJs provides a useful tool to investigate the self-assembly properties of transmembrane TJ proteins. Ultimately, this will improve the molecular understanding of paracellular barriers, and the mechanistic information can be used to modulate TJs for therapeutic applications.

Acknowledgments

This work was funded by the Deutsche Forschungsgemeinschaft, DFG PI 837/2-1. We thank Dr. Bevin Gangadharan for critically reading the manuscript.

Conflicts of interest

The authors declare no conflicts of interests.

References

1. Angelow, S., R. Ahlstrom & A.S. Yu. 2008. Biology of Claudins. *Am. J. Physiol Renal Physiol.* **295:** F867–F876.
2. Staehelin, L.A. 1974. Structure and function of intercellular junctions. *Int. Rev. Cytol.* **39:** 191–283.
3. Piontek, J., L. Winkler, H. Wolburg, *et al.* 2008. Formation of tight junction: determinants of homophilic interaction between classic claudins. *FASEB J.* **22:** 146–158.
4. Morita, K., M. Furuse, K. Fujimoto & S. Tsukita. 1999. Claudin multigene family encoding four-transmembrane domain protein components of tight junction strands. *Proc. Natl. Acad. Sci. USA* **96:** 511–516.
5. Krause, G., L. Winkler, S.L. Mueller, *et al.* 2008. Structure and function of claudins. *Biochim. Biophys. Acta.* **1778:** 631–645.
6. Nitta, T., M. Hata, S. Gotoh, *et al.* 2003. Size-selective loosening of the blood-brain barrier in claudin-5-deficient mice. *J. Cell Biol.* **161:** 653–660.
7. Hou, J., A. Renigunta, M. Konrad, *et al.* 2008. Claudin-16 and claudin-19 interact and form a cation-selective tight junction complex. *J. Clin. Invest.* **118:** 619–628.
8. Amasheh, S., N. Meiri, A.H. Gitter, *et al.* 2002. Claudin-2 expression induces cation-selective channels in tight junctions of epithelial cells. *J. Cell Sci.* **115:** 4969–4976.
9. Colegio, O.R., C.M. Van Itallie, H.J. McCrea, *et al.* 2002. Claudins create charge-selective channels in the paracellular pathway between epithelial cells. *Am. J. Physiol Cell Physiol.* **283:** C142–C147.
10. Yu, A.S., M.H. Cheng, S. Angelow, *et al.* 2009. Molecular basis for cation selectivity in claudin-2-based paracellular pores: identification of an electrostatic interaction site. *J. Gen. Physiol.* **133:** 111–127.
11. Piehl, C., J. Piontek, J. Cording, *et al.* 2010. Participation of the second extracellular loop of claudin-5 in paracellular tightening against ions, small and large molecules. *Cell Mol. Life Sci.* **67:** 2131–2140.
12. Zhang, J., J. Piontek, H. Wolburg, *et al.* 2010. Establishment of a neuroepithelial barrier by Claudin5a is essential for zebrafish brain ventricular lumen expansion. *Proc. Natl. Acad. Sci. USA* **107:** 1425–1430.
13. Piontek, J., S. Fritzsche, J. Cording, *et al.* 2011. Elucidating the principles of the molecular organization of heteropolymeric tight junction strands. *Cell Mol. Life Sci.* **68:** 3903–3918.
14. Furuse, M., H. Sasaki & S. Tsukita. 1999. Manner of interaction of heterogeneous claudin species within and between tight junction strands. *J. Cell Biol.* **147:** 891–903.
15. Wolburg, H. & A. Lippoldt. 2002. Tight junctions of the blood-brain barrier: development, composition and regulation. *Vascul. Pharmacol.* **38:** 323–337.
16. Blasig, I.E., L. Winkler, B. Lassowski, *et al.* 2006. On the self-association potential of transmembrane tight junction proteins. *Cell Mol. Life Sci.* **63:** 505–514.
17. Lorenz, D., B. Wiesner, J. Zipper, *et al.* 1998. Mechanism of peptide-induced mast cell degranulation. Translocation and patch-clamp studies. *J. Gen. Physiol.* **112:** 577–591.
18. Henn, V., B. Edemir, E. Stefan, *et al.* 2004. Identification of a novel A-kinase anchoring protein 18 isoform and evidence for its role in the vasopressin-induced aquaporin-2 shuttle in renal principal cells. *J. Biol. Chem.* **279:** 26654–26665.
19. Berryman, M. & R. Rodewald. 1995. Beta 2-microglobulin co-distributes with the heavy chain of the intestinal IgG-Fc receptor throughout the transepithelial transport pathway of the neonatal rat. *J. Cell Sci.* **108**(Pt 6): 2347–2360.
20. Alken, M., C. Rutz, R. Kochl, *et al.* 2005. The signal peptide of the rat corticotropin-releasing factor receptor 1 promotes receptor expression but is not essential for establishing a functional receptor. *Biochem. J.* **390:** 455–464.
21. Kaufmann, R., J. Piontek, F. Grull, *et al.* 2012. Visualization and quantitative analysis of reconstituted tight junctions using localization microscopy. *Plos One* **7:** e31128.
22. Furuse, M., K. Fujimoto, N. Sato, *et al.* 1996. Overexpression of occludin, a tight junction-associated integral membrane protein, induces the formation of intracellular multilamellar bodies bearing tight junction-like structures. *J. Cell Sci.* **109**(Pt 2): 429–435.

23. Muller, S.L., M. Portwich, A. Schmidt, *et al.* 2005. The tight junction protein occludin and the adherens junction protein alpha-catenin share a common interaction mechanism with ZO-1. *J. Biol. Chem.* **280:** 3747–3756.

24. Fanning, A.S., B.P. Little, C. Rahner, *et al.* 2007. The unique-5 and -6 motifs of ZO-1 regulate tight junction strand localization and scaffolding properties. *Mol. Biol. Cell.* **18:** 721–731.

25. Snapp, E.L., R.S. Hegde, M. Francolini, *et al.* 2003. Formation of stacked ER cisternae by low affinity protein interactions. *J. Cell Biol.* **163:** 257–269.

26. Korkhov, V.M., L. Milan-Lobo, B. Zuber, *et al.* 2008. Peptide-based interactions with calnexin target misassembled membrane proteins into endoplasmic reticulum-derived multilamellar bodies. *J. Mol. Biol.* **378:** 337–352.

27. Huotari, J. & A. Helenius. 2011. Endosome maturation. *EMBO J.* **30:** 3481–3500.

28. Lichtenstein, A., G.M. Gaietta, T.J. Deerinck, *et al.* 2009. The cytoplasmic accumulations of the cataract-associated mutant, Connexin50P88S, are long-lived and form in the endoplasmic reticulum. *Exp. Eye Res.* **88:** 600–609.

29. Soma, T., H. Chiba, Y. Kato-Mori, *et al.* 2004. Thr(207) of claudin-5 is involved in size-selective loosening of the endothelial barrier by cyclic AMP. *Exp. Cell Res.* **300:** 202–212.

30. Yamamoto, M., S.H. Ramirez, S. Sato, *et al.* 2008. Phosphorylation of claudin-5 and occludin by rho kinase in brain endothelial cells. *Am. J. Pathol.* **172:** 521–533.

Ann. N.Y. Acad. Sci. ISSN 0077-8923

ANNALS OF THE NEW YORK ACADEMY OF SCIENCES

Issue: *Barriers and Channels Formed by Tight Junction Proteins*

Association between segments of zonula occludens proteins: live-cell FRET and mass spectrometric analysis

Christine Rueckert,* Victor Castro,* Corinna Gagell, Sebastian Dabrowski, Michael Schümann, Eberhard Krause, Ingolf E. Blasig, and Reiner F. Haseloff

Leibniz-Institut für Molekulare Pharmakologie, Berlin-Buch, Germany

Address for correspondence: Ingolf Blasig, Leibniz-Institut für Molekulare Pharmakologie, Robert-Rössle-Str. 10, 13125 Berlin-Buch, Germany. iblasig@fmp-berlin.de

The tight junction protein ZO-1 (zonula occludens protein 1) has recruiting/scaffolding functions in the junctional complex of epithelial and endothelial cells. Homodimerization was proposed to be crucial for ZO-1 function. Here, we investigated the ability of ZO-1 domains to mediate self-interaction in living cells. We expressed ZO-1 truncation mutants as fusions with derivatives of green fluorescent protein in tight junction–free HEK-293 cells and determined self-association by means of fluorescence resonance energy transfer measurements using live-cell imaging. We show that both an SH3-hinge-GuK fusion protein and the PDZ2 domain self-associate in our test system. The recombinant PDZ2 domain also binds to ZO-1 and ZO-2 in tight junction–forming HT29/B6 cell lysates, as demonstrated by coprecipitation. Both interaction types are of relevance for the function of ZO-1 in the regulation of the junctional complex in polar cells.

Keywords: zonula occludens proteins; tight junctions; protein–protein interaction; PDZ domains; live-cell imaging

Introduction

Zonula occludens 1 (ZO-1) is a cell junction-associated intracellular phosphoprotein of ∼225 kDa, which belongs to the MAGuK (membrane-associated guanylate kinase) protein family.[1–3] Its composition is characterized by three PDZ (PSD95-DlgA-ZO-1 homology), an SH3 (Src homology 3), and a GuK (guanylate kinase-like) domain followed by an acidic stretch, which are arranged in this order and linked by the variable regions U1–6.[4] This multidomain protein interacts with numerous partners, such as the tight junctional transmembranal proteins occludin and tricellulin,[5–8] claudins,[9] JAM (junctional adhesion molecule),[10] the adherens junction protein α-catenin,[11,12] the MAGuK proteins ZO-2 and ZO-3,[13–15] and cytoskeletal elements.[16,17] ZO-1 was also reported to interact with the transcription factor ZONAB (ZO-1–associated nucleic acid binding protein),[18,19] the heat shock protein Apg-2[20] and proteins involved in cellular signaling, for example, G-protein α-subunits,[21–23] ZO-1 associated kinase,[3] phosphatase DEP-1,[24] and AF-6.[25] ZO-1 was proposed to function in the formation of cell junctions by spatially and temporally organizing proteins of the junctional complexes.[26–28] It was suggested that self-association of ZO-1 is required to build a scaffold that allows specific binding and accumulation of proteins to promote cellular ultrastructures at the sites of formation of adherens and tight junctions.[14,29,30] Analysis of ZO-1 constructs showed that ZO-1 dimerization is a prerequisite for claudin polymerization at the plasma membrane.[26] Dimerization was achieved in these experiments by additional molecular modification of ZO-1 constructs with an inducible dimerization tag, leaving the identification of dimerization-mediating ZO-1 regions an open problem. To elucidate the molecular

---

*These authors contributed equally to this work.

doi: 10.1111/j.1749-6632.2012.06571.x
Ann. N.Y. Acad. Sci. 1257 (2012) 67–76 © 2012 New York Academy of Sciences.

mechanisms of the ZO-1 self-interaction, *in vitro* and *in silico* studies revealed two ZO-1 regions that may play a role in this process. The PDZ2 domain is necessary and sufficient for dimer formation of insect cell-expressed ZO-1 deletion mutants.[31] In an X-ray crystallization study, the PDZ2 domain expressed in *E. coli* was found to form a dimer.[32] By using a molecular modeling approach, we suggested that a dimeric structural arrangement of the SH3-hinge-GuK unit (ShG) of ZO-1 is a precondition for its interaction with occludin.[29] This prediction was supported by the observation that the isolated ZO-1 ShG expressed in *E. coli* forms oligomeric complexes. However, neither the ShG nor the PDZ2 domain was shown to dimerize in cells. In general, it has to be considered difficult to unequivocally determine the involved regions in ZO-1 dimerization: experiments in a cellular background cannot directly rule out additional molecules mediating ZO-1 self-association by providing a link between two or more ZO-1 molecules or by facilitating dimerization-prone ZO-1 conformations. On the other hand, *in vitro* experiments on isolated ZO-1 molecules expressed in nonmammalian organisms may result in effects observed on proteins with compromised folding that does not reflect the situation inside the cell.

Here, we used the human embryonic kidney cell line HEK-293 as an experimental system that allows expression of ZO-1 constructs in a cellular background lacking functional tight junction proteins, thus minimizing their influence as potential mediators for the studied interactions. We asked which of the suggested ZO-1 regions self-interacts under physiological conditions as they exist inside the cell. To this end we generated constructs that allowed expression of the PDZ2 domain and the ShG unit of ZO-1 as fusions with the fluorescence proteins CFP (cyan fluorescent protein) and YFP (yellow fluorescent protein) and used them for investigating their self-association in transfected HEK-293 cells by FRET (fluorescence resonance energy transfer) analysis in a live-cell imaging approach. The microscopic investigations were supplemented by an affinity-based approach to identify ZO proteins associating with recombinant PDZ domains of ZO-1. Our results demonstrate that both of the isolated binding domains associate in a homophilic manner within the cell.

Material and methods

Construction of the ZO-1 molecule expression vectors

DNA coding for ZO-1 polypeptide chains was inserted into the mammalian expression vectors pECFP and pEYFP (BD Biosciences Clontech) in frame with the CFP/YFP coding sequences and using the SalI and XmaI restriction sites within the multiple cloning site of the vectors. The fluorescent protein was located at the respective N-terminus of the construct. The template used for PCR amplification of ZO-1–derived inserts was the cDNA clone pSPORT-ZO1 DKFZp686A1195QZ (RZPD, Berlin, Germany) originating from adult human retina. The primer pairs for ZO-1 insert amplification were 5′-ACTATGTCGACATGGCTAAACCTACTAAAGTC-3′ (forward), 5′-CTATACCCGGGCAAGATCAGGG ACATTCAATAGCG-3′ (reverse) for amino acids 170–262 (PDZ2 domain), and 5′-ACTATGTC GACATGGGAGATTCTTTCTATATTAGAACC-3′ (forward), 5′-CTATACCCGGGCCGCCTTTCCCT CGGAAACCCATACC-3′ (reverse) for amino acids 504–793 (ShG unit). The ZO-1 construct plasmids were amplified in and purified from *E. coli* (strain DH5α) with a Qiagen® Plasmid Mini Kit (Qiagen, the Netherlands), sequence verified and used for transfection of cultured mammalian cells.

Cell culture and transfection

HEK-293 and MDCK-II (Madin Darby canine kidney cell line) were grown in Dulbecco's modified Eagle's Medium (DMEM; 3.7 g/L NaHCO₃, 1 g/L D-glucose; Biochrom, Berlin, Germany), supplemented with 10% FCS (fetal calf serum; PAA Laboratories GmbH, Linz, Austria), 2 mM L-glutamine, 5 g/mL penicillin, and 5 U/mL streptomycin (Biochrom AG, Berlin, Germany), at 37 °C and 5% CO₂. For the affinity-based isolation of binding proteins, stable isotope labeling with amino acids in cell culture (SILAC) was used to discriminate unspecifically binding proteins. HT29/B6 (untransfected human colon adenocarcinoma grade II cell line; kindly provided by M. Fromm, Berlin) were grown in RPMI-1640 medium lacking arginine and lysine (Perbio Science, Bonn, Germany) supplemented with Arg/Lys-¹³C₆- (PDZ pull-down) or Arg/Lys-¹²C₆ (control). For transfection, HEK-293 cells were grown to ∼90% confluence (in 30 mm dishes) and incubated under low serum

(5% FCS) and antibiotic-free conditions for two hours. A total of 10 µg lipofectamine™ 2000 (Invitrogen, Carlsbad, CA) and 12 µg plasmid DNA in 500 µL Optimem® (Invitrogen/Gibco) were added, and the cells were incubated for another six hours before the medium was replaced by supplemented DMEM. The same procedure was applied to MDCK-II cells grown on 12 mm glass coverslips, with the transfection solution containing 1 µg lipofectamine™ 2000 and 0.4 µg plasmid DNA in 100 µL Optimem®. HEK cells were split 36 h after transfection to poly-L-lysine (Sigma Aldrich, St. Louis, MO) coated glass coverslips. Cells were used for FRET analysis or immunofluorescence experiments 48 h after transfection.

Immunoblotting

Transfected HEK-293 cells were scraped from coverslips, washed in PBS, pelleted, and lysed by sonication in a protein loading buffer (2% SDS, 100 mM dithiotreitol, 10% (v/v) glycerol, 30 mM Tris, pH 6.8) followed by incubation for 10 minutes at 95 °C. After centrifugation, the cell debris-free supernatant proteins were separated by SDS-PAGE and blotted to a nitrocellulose membrane (Hybond™-ECL; Amersham Biosciences Europe, UK). The membrane was blocked over night at 4 °C with 5% (w/v) powdered milk in TBST (25 mM Tris, pH 7.4, 300 mM NaCl, 0.01% (v/v) Tween 20) and incubated first with 10 µg/mL mouse, anti-GFP (green fluorescent protein) antibody (clone JL-8, Clontech, also recognizes CFP and YFP) in TBST for 1.5 h and, then, with 0.2 µg/mL horse radish peroxidase (HRP)-conjugated goat, antimouse antibody for 0.75 h at room temperature. HRP activity was detected by LumiAnalyzer measurements of chemoluminescence catalyzed by ECL Western blotting detection reagents I and II (Amersham Biosciences).

Cell fixation and imaging

Transfected MDCK-II cells on glass coverslips were washed in PBS, fixed with 2.5% (w/v) paraformaldehyde (PFA), 100 mM sodium cacodylate, 100 mM sucrose, remaining PFA was quenched in 100 mM glycine in PBS, washed again, and mounted on microscopical slides in Immu-Mount (Thermo Electron Corporation, Waltham, MA). The fixed cells were analyzed by laser scanning microscopy with an LSM 510, objective 100×/1.3 oil, ∞/0.17, Plan-Neofluar (Carl Zeiss Jena GmbH, Germany). YFP was excited with an Ar⁺ laser at 514 nm, fluores-

cence was filtered (530–600 nm) and detected with a PMT-CCD array. Image acquisition was controlled by LSMib software (Carl Zeiss Jena GmbH, Germany).

Acceptor photobleaching FRET measurements and live-cell imaging

Two days after transfection, HEK-293 cells were transferred to DMEM (without phenol red, with 5 g/mL penicillin, 5 U/mL streptomycin, 10 mM Hepes), and FRET analysis was performed in the cell membrane (CFP/YFP fusions of PDZ2, ShG) or cytosol (CFP, YFP alone), using an LSM 510 microscope. CFP and YFP were excited at 458 nm and 514 nm and detected at 463–495 nm and 527–634 nm, respectively. Photobleaching of YFP was performed at 514 nm with 75% Ar⁺ laser intensity. Image data were acquired and processed using LSMib software (Carl Zeiss Jena, Germany). FRET efficiency (E_F) was calculated as $E_F = (I_a - I_b)/I_a$ (I_a and I_b, donor fluorescence intensity after and before acceptor photobleaching).[33]

Recombinant protein expression and affinity chromatography

PDZ domains were created by PCR using Advantage 2 Polymerase Mix (DB Biosciences, Heidelberg, Germany) based on the human ZO-1 template described above (DKFZp686A1195QZ). Deletion mutants were generated by the cloning system pTriEx-6 3C/LIC (Novagen, Darmstadt, Germany) resulting in constructs with an N-terminal Strep tag II. The primer pairs were 5′-CAGGGACCCGGTATATGGGAACAACATAC AG-3′ (forward), 5′-GGCACCAGAGCGTTTTACT TCTTCCTTC-3′ (reverse) for amino acids 18–110 (PDZ1), 5′-CAGGGACCCGGTCCTGCTAAACCT ACTAAAG-3′ (forward), and 5′-GGCACCAGA GCGTTTTATTCATCTCTTTGAAC-3′ (reverse) for amino acids 181–264 (PDZ2). The plasmids were amplified in and purified from *E. coli* (strain Nova Blue GigaSingle; Novagen) with a Mini Kit (Qiagen), sequence verified and used for transfection in *E. coli* (BL21). After induction, bacteria were grown to an optical density of 0.7±0.1 (at 600 nm) and were harvested by centrifugation for 10 min at 4000×*g*. The pellets were resuspended in a 20 mM Tris/HCl lysis buffer (pH 7.5) containing 100 mM NaCl, 5% glycerol, 1% Triton X-100, 0.1 mM phenylmethylsulphonyl fluoride, 1 mM ethylenediaminetetraacetic acid (EDTA) and protease inhibitor cocktail (Sigma-Aldrich, Taufkirchen, Germany), and underwent two passages in an EmulsiFlex-C3 homogenizer (Avestin Europe GmbH, Mannheim, Germany). The insoluble cell debris was removed by centrifugation for 30 min at $30000 \times g$. The expression of the PDZ domains was checked qualitatively and quantitatively by mass spectrometry and SDS-PAGE (Coomassie staining), respectively. Purification of the *Strep*-tagged recombinant proteins and affinity chromatographic isolation of binding proteins were accomplished using *Strep*-Tactin-Sepharose (IBA, Goettingen, Germany) according to the manufacturer's instructions. Briefly, 1 mL of sepharose was incubated (2 h, 4 °C) with a lysate of *E. coli* (not transfected with fluorescent proteins) containing either *Strep*-tagged PDZ domains. After removal of unbound bacterial protein, 2 mL (corresponding to 10 mg of ^{13}C-labeled protein) of epithelial cell lysate (lysis buffer as described) were incubated for two hours at 4 °C. Unbound epithelial proteins were removed in three washing steps, and elution was performed using 3×0.2 mL of elution buffer (100 mM Tris/HCl, pH 8.0, 150 mM NaCl, 1 mM EDTA, 2.5 mM desthiobiotin). The same procedure (without adding PDZ domains) was performed with the control sample using unlabeled epithelial proteins. The complete experiment (pull-down and subsequent mass spectrometry) was conducted three times, two times using rat epithelial cells, and once using human HT29/B6 cells; experiments with human and rat cells gave qualitatively similar results.

Mass spectrometry

For mass spectrometry (MS) analysis, reduced (dithiothreitol) and alkylated (iodacetamide) proteins were separated using a Tris glycine gradient 4–20% gel (Invitrogen); gel bands were cut into 40 slices of equal size. Tryptic in-gel digestion of proteins and nano-LC-MS/MS experiments were performed as described previously.[34] Briefly, tryptic peptides were separated by a reversed-phase capillary liquid chromatography system (Eksigent 2D nanoflow LC; Axel Semrau GmbH, Germany) connected to an LTQ-Orbitrap XL mass spectrometer (Thermo Scientific). Mass spectra were acquired in a data-dependent mode with one MS survey scan (with a resolution of 60,000) in the Orbitrap followed by MS/MS scans of the five most intense

precursor ions in the LTQ. The MS survey range was m/z 350–1,500. The dynamic exclusion time (for precursor ions) was set to 120 sec and automatic gain control was set to 3×10^6 and 20,000 for Orbitrap-MS and LTQ-MS/MS scans, respectively. Identification and quantification of proteins were carried out with version 1.0.13.13 of the Max-Quant software package.[35] Generated peak lists (∗.msm files) were submitted to a MASCOT search engine (version 2.2, Matrix Science Ltd., Boston, MA) and searched against the IPI human protein data base (version 3.52). The mass tolerance of precursor and sequence ions was set to 7 ppm and 0.35 Da, respectively. Methionine oxidation was used as variable modification, alkylation (carbamidomethyl) as fixed modification. False discovery rates were <1% based on matches to reversed sequences in the concatenated target-decoy data base. Proteins were considered for further analysis if, at least, two sequenced peptides could be quantified.

Results

ZO-1 molecule expression in HEK-293 cells

Cleared lysates of HEK-293 cells, cotransfected with pEYFP/pECFP constructs of ZO-1 PDZ2 or ZO-1 ShG, were analyzed by SDS-PAGE/Western blot for expression of the ZO-1 regions as fusions with YFP/CFP. Immunostaining with anti-GFP resulted in detection of protein bands with apparent molecular weights as expected for the fusions of ZO-1 ShG (61 kDa) and ZO-1 PDZ2 (38 kDa). In addition, minor signals were detected for smaller molecules, presumably originating from fusion free CFP/YFP (Fig. 1A). This result indicated that most of the fluorescence signals observed in transfected HEK-293 cells originated from the ZO-1 fusion constructs. The distribution of the fusion proteins in HEK-293 cells was analyzed based on their fluorescence by laser scanning microscopy. Both constructs showed localization throughout the cell, including the nuclei where the observed fluorescence signals were slightly lower than in the cytoplasm (Fig. 1B, upper part).

ZO-1 molecule localization in tight junction-forming MDCK-II cells

MDCK-II cells were transfected with pEYFP constructs of ShG or PDZ2. Transfection efficiency and expression levels were not sufficient for Western blot analysis but allowed imaging of fusion protein

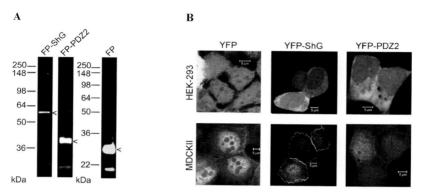

Figure 1. Expression of the PDZ2 domain and ShG unit (SH3 domain, hinge region, GuK domain) of ZO-1 transfected in cells. (A) Lysates of HEK-293 cells cotransfected with plasmid DNA coding for YFP-ShG and CFP-ShG (FP-ShG), YFP-PDZ2 and PDZ2-CFP (FP-PDZ2), CFP and YFP (FP) were analyzed by SDS-PAGE and Western blot immunostained against GFP (antibody also detects CFP and YFP). The observed apparent molecular weights correspond to the theoretical values of the fusions (arrow heads): 61 kDa for ZO-1 ShG and 38 kDa for ZO-1 PDZ2. (B) Subcellular distribution of ZO-1 constructs in transfected HEK-293 and MDCK-II-cells, respectively. Scale bars, 5 μm. Representative results of at least three experiments are shown. C/Y/GFP, cyan/yellow/green fluorescence protein; HEK, human embryonic kidney; MDCK, Madin Darby canine kidney.

localization in single cells by laser scanning microscopy. While the YFP-PDZ2 fusion was localized throughout the cells including nucleus and cell membrane, the YFP-ShG fusion was particularly localized at cell margins (Fig. 1B, lower part). This differential distribution might point to a preferred incorporation of the ShG region into protein complexes at the plasma membrane of the tight junction-expressing MDCK-II cells.

FRET analysis of ZO-1 molecules in HEK-293 cells

HEK-293 cells were cotransfected to express ZO-1 ShG or ZO-1 PDZ2 as fusion proteins with CFP or YFP in a single cell. The living cells were tested two days after transfection by FRET analysis for the ability of the expressed fusion proteins to self-associate. Transfected HEK-293 cells expressing a CFP-YFP tandem construct served as positive control (courtesy of B. Wiesner). These cells expressed the FRET pair as one connected polypeptide chain resulting in strong FRET signals due to the close spatial arrangement of the two fluorescent proteins. HEK-293 cells coexpressing untagged CFP and YFP molecules in a single cell were used as negative control. Analysis of the cells in the acceptor photobleaching assay showed FRET signals for both ZO-1 ShG and ZO-1 PDZ2 constructs, which were lower than those measured for the positive control, but significantly higher than the signals of the negative control (Fig. 2).

Mass spectrometric analysis of eluates from recombinant PDZ pull-downs

Further investigation directed at the interactions of ZO proteins were accomplished by MS analysis of eluates obtained from affinity-based isolation of ZO-1 binding proteins. The experimental approach is shown in Figure 3 (upper part). Epithelial cells (HT29/B6) were cultivated in arginine/lysine-deficient culture medium supplemented with either L-arginine-$^{13}C_6$ and L-lysine-$^{13}C_6$ or corresponding $^{12}C_6$ amino acids. After performing the coprecipitation, eluates were mixed, proteins were separated by SDS gel electrophoresis and subjected to in-gel digestion. Mass spectrometry identified peptides originating from ZO proteins; the sequence coverage is given in Figure 3 (lower part).

In both pull-down experiments, peptides coming from both *E. coli*-expressed recombinant PDZ domains (possibly, but to a minor extent, also originating from the control pull-down) were identified, indicated by a heavy-to-light ratio $R \ll 1$ (Table 1). In contrast to PDZ1, eluate fractions obtained from the PDZ2 domain contained fragments of both ZO-1 and ZO-2 derived from HT29/B6. Strong enrichment is indicated by identification of numerous ^{13}C-containing peptides with $R \gg 1$ (ZO-1, 12 peptides, cumulative Mascot® score 643; ZO-2, 16 peptides, cumulative Mascot® score 633) pointing to homotypic and heterotypic (ZO-2) interactions when ZO-1 PDZ2 was used as a bait.

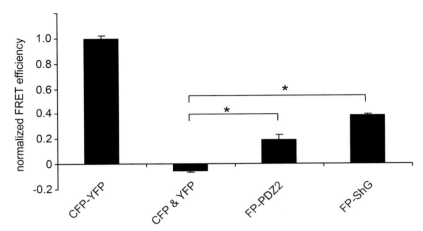

Figure 2. Fluorescence resonance energy transfer (FRET) analysis of the self-interaction of the PDZ2 domain and ShG unit (SH3 domain, hinge region, GuK domain) of ZO-1 in transfected HEK-293 cells. HEK-293 cells were (co)transfected with plasmid DNA coding for the positive control tandem construct CFP-YFP, CFP and YFP as single molecules (negative control), CFP- and YFP fusions of ShG (FP-ShG), or of PDZ2 (FP-PDZ2), respectively, and were analyzed in an acceptor photobleaching FRET assay using live-cell imaging. FP-ShG and FP-PDZ2 gave rise to FRET signals significantly higher than those measured for the negative control. Data were collected from 20 cells per experiment. *significant differences, Mann–Whitney U-test, $P < 0.001$; Mean $\pm$ SEM, $n = 4$ independent experiments.

Discussion

Here, we investigated the self-association properties of ZO-1 that were reported to be important for ZO-1 function in the tight junction organization. The objective was to gather information on the molecular region of ZO-1 that mediates dimerization in living cells. The experimental model is based on the expression of those ZO-1 regions that were hypothesized to be involved in homodimerization[29,31] in HEK-293 cells, that is, in the absence of functional tight junctions. We tested the PDZ2 domain and the ShG unit of ZO-1 for their intrinsic ability to self-associate when expressed as YFP/CFP fusions by employing a FRET assay in living cells. Significant FRET signals were observed for both constructs suggesting their self-association in cells that do not have tight junctions. This demonstrates that the dimerization properties proposed for ZO-1 PDZ2 and -ShG based on *in vitro* experiments[31,32] and molecular modeling[29] are also valid in a cellular environment. Consequently, both domains exhibit an intrinsic potential to mediate ZO-1 self-association in cells that do not have tight junctions. However, these findings do not finally elucidate the specific molecular conditions under which these ZO-1 regions act as dimerization mediators in tight junction formation. The presence of ZO-1 interaction partners may have an impact on the conformation and/or availability of certain ZO-1 domains or modules, which could regulate their dimerization functions in a spatial and/or temporal fashion.

In epithelial cells with mature tight junctions, ZO-1 localizes to junctions.[36] In other situations, such as epithelial/endothelial cells under low extracellular Ca^{2+}, ATP depletion, or without contacts to neighboring cells or in nonpolar cells, ZO-1 can be found in the cytoplasm, for example, colocalizing with actin complexes,[11,16,17] or even in the nucleus where it was suggested to fulfill regulatory functions in conjunction with a transcription factor.[18,19] In cells that lack tight junctions, ZO-1 was reported to interact with catenins in the absence of E-cadherin[11,12] making it part of the adherence junction protein complex. Given the huge diversity of the ZO-1 interaction partners ranging from cytoplasmic as well as membrane proteins of different subcellular structures, it remains difficult to come up with a clear picture of the structural and regulatory functions of ZO-1. This explains why so many efforts are made to first characterize the interaction and interdependencies of ZO-1 with other molecules in different cellular situations. Several studies suggest that ZO-1 functions may also depend on its phosphorylation state and on structural features determined by intramolecular interactions.[24]

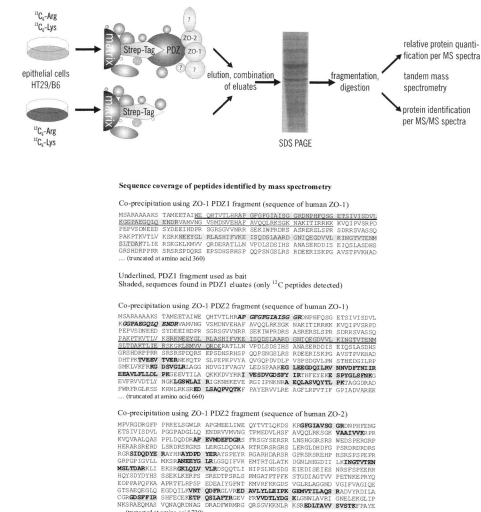

Figure 3. Affinity chromatographic isolation and mass spectrometric identification of binding partners of ZO-1 PDZ domains 1 and 2 (upper part); and sequence coverage of peptides detected in the eluate fractions (lower part). Homogenates of the epithelial cell line HT29/B6 were incubated with recombinant PDZ domains of ZO-1. Associated proteins were eluted, separated by SDS-PAGE and identified by mass spectrometry.

Especially the latter has to be kept in mind when interpreting data obtained with truncated versions of ZO-1. Umeda *et al.*[26] showed that the formation of ZO-1 homodimers has a strong impact on its function in tight junction strand organization in cells. The authors suggest a pivotal role of the ZO-1 SH3/GuK domain-mediated homodimerization but not membrane localization in this process.[26] However, *in vitro* studies of ZO-1 truncation mutants expressed in insect cells assign ZO-1 homodimerization activity to its PDZ2 domain but not its SH3/GuK unit lacking the hinge region.[31] These contradictory findings point to a role of the specific experimental system and construction of the

Table 1. Peptides detected in eluates of ZO-1 PDZ1/PDZ2 coprecipitation fractions

ZO-1/PDZ1 – ZO-1$_l$	ZO-1/PDZ2 – ZO-1$_h$	ZO-1/PDZ2 – ZO-1$_l$	ZO-1/PDZ2 – ZO-2$_h$
APGFGFGIAISGGR	AEQLASVQYTLPK	APGFGFGIAISGGR	AFEVMDEFDGR
DGNIQEGDVVLK	APGFGFGIAISGGR	DGNIQEGDVVLK	ANEEYGLR
DNPHFQSGETSIVI SDVLK	EDLSAQPVQTK	DGNIQEGDVVLKING TVTENMSLTDAK	AYDPDYER
EISQDSLAAR	EEAVLFLLDLPK	EISQDSLAAR	EDAVLYLLEIPK
GGPAEGQLQENDR	EGLEEGDQILR	EISQDSLAARDGNIQE GDVVLK	EDLTAVVSVSTK
INGTVTENMSLTDAK	ESPYGLSFNK	GGPAEGQLQENDR	ETPQSLAFTR
LASHIFVK	GGPAEGQLQENDR	INGTVTENMSLTDAK	GDSFFIR
NEEYGLR	IVESDVGDSFYIR	INGTVTENMSLTDAKT LIER	GEMVTILAQSR
	KGDSVGLR	LASHIFVK	GFGIAVSGGR
	LGSWLAIR	LASHIFVKEISQDSLAAR	GKLQLVVLR
	TVEEVTVER	LKMVVQR	INGTVTENMSLTDAR
	VNNVDFTNIIR	NEEYGLR	LQLVVLR
		SRKNEEYGLR	SIDQDYER
			VAAIVVK
			VNTQDFR
			VVDTLYDGK
R = 0.057 ± 0.019	R = 14.3 ± 2.1	R = 0.064 ± 0.016	R = 21.1 ± 2.2

Indices *h* and *l* indicate whether the respective heavy or light peptide was sequenced for MS identification.
R, mean intensity ratio of MS signals of detected peptides (I_{C13}/I_{C12}); $R \ll 1$ indicates peptides originating from *E. coli*-expressed recombinant PDZ domains used as a bait or from control coprecipitation.

ZO-1 molecules under investigation. We found earlier that the hinge region between the SH3 and the GuK domain plays a major role in protein–protein interactions of ZO-1.[37] Consequently, the incomplete structure may explain the different findings.[31]

In this study, we compared the distribution of ZO-1 constructs in cells without (HEK-293) and with tight junctions (MDCK-II). In the absence of tight junctions, both the PDZ2 and the ShG constructs were distributed throughout the cell (Fig. 1B). In the tight junction-forming cell line MDCK-II, YFP-ShG localizes to the plasma membrane whereas the PDZ2 fusion construct is distributed throughout the cell with a slight enrichment at the plasma membrane, resembling that of the YFP control (Fig. 1B). These observations are in agreement with Umeda *et al.* who localized a construct without the ShG (comprising the three ZO-1 PDZ domains) in the cytoplasm and the construct comprising the ShG in addition to the three PDZ domains at the cell boundaries.[26]

The affinity chromatographic isolation of the binding partners of the PDZ domains revealed, based on the strong enrichment of both ZO-1 and ZO-2 in the eluate fractions of ZO-1 PDZ2, that the binding activity of this domain is not restricted to homodimerization. Although the binding epitope cannot be located based on the pull-down assay data, the binding is likely to be favored by the similarity of the PDZ2 domains of ZO-1 and ZO-2 (similarity index, 84.8%; identity index, 68.4% according to http://www.ebi.ac.uk/Tools/psa/emboss_needle/). In contrast, there was no indication found for an association of ZO-1 PDZ1 with ZO-1.

This and other studies aiming at assignment of ZO-1 functions to its domains and modules underline the fact that the controversial findings may originate from the complex interdependency of the multidomain structure of ZO-1 and its regulation by a pool of molecules not defined yet. With this report, we contribute the finding that at least two regions of ZO-1, the ShG unit and the PDZ2 domain, can

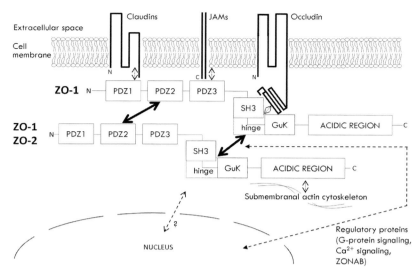

Figure 4. Structural organization and interaction possibilities of protein binding segments of ZO-1. The upper part shows which of the protein binding parts (squares) of ZO-1 interact with transmembrane tight junction proteins, such as claudins, junctional adhesion molecules (JAMs) or occludin. The middle part depicts the association areas between ZO proteins. The lower part indicates a proposed interplay with the nucleus. The first PDZ domain (PDZ1) recruits claudins forming the backbone and determines the tightness of the tight junctions. PDZ2 is involved in the homologous association of ZO-1 as well as heterologous binding of ZO-2, while PDZ3 interacts with JAMs. The SH3-hinge-GuK unit contains a multiple binding site for occludin, the adherens-junction molecule α-catenin, G-proteins, and the transcription factor ZONAB. It constitutes a domain with paramount regulatory relevance that, in addition, has the ability to interact with itself, thus being able to influence the homologous ZO-1 dimerization/polymerization upon the activation of different signaling cascades. These homologous and heterologous associations may affect the subcellular localization of ZO-1. Furthermore, the binding of ZO-1 to ZO-2 can also influence the cellular localization and behavior of the latter.

account for homotypic, the latter also heterotypic, interactions of ZO-1 (Fig. 4, lower part).

Acknowledgment

This work was supported by DFG BL308/6-4, 7-4, and 9-1.

Conflicts of interest

The authors declare no conflicts of interest.

References

1. Stevenson, B.R. *et al.* 1986. Identification of ZO-1: a high molecular weight polypeptide associated with the tight junction (zonula occludens) in a variety of epithelia. *J. Cell Biol.* **103:** 755–766.

2. te Velthuis, A.J., J.F. Admiraal & C.P. Bagowski. 2007. Molecular evolution of the MAGUK family in metazoan genomes. *BMC Evol. Biol.* **7:** 129.

3. Balda, M.S., J.M. Anderson & K. Matter. 1996. The SH3 domain of the tight junction protein ZO-1 binds to a serine protein kinase that phosphorylates a region C-terminal to this domain. *FEBS Lett.* **399:** 326–332.

4. Fanning, A.S. *et al.* 2007. The unique-5 and -6 motifs of ZO-1 regulate tight junction strand localization and scaffolding properties. *Mol. Biol. Cell* **18:** 721–731.

5. Schmidt, A. *et al.* 2004. Occludin binds to the SH3-hinge-GuK unit of zonula occludens protein 1: potential mechanism of tight junction regulation. *Cell Mol. Life Sci.* **61:** 1354–1365.

6. Furuse, M. *et al.* 1994. Direct association of occludin with ZO-1 and its possible involvement in the localization of occludin at tight junctions. *J. Cell Biol.* **127:** 1617–1626.

7. Raleigh, D.R. *et al.* 2010. Tight junction-associated MARVEL proteins marveld3, tricellulin, and occludin have distinct but overlapping functions. *Mol. Biol. Cell.* **21:** 1200–1213.

8. Riazuddin, S. *et al.* 2006. Tricellulin is a tight-junction protein necessary for hearing. *Am. J. Hum. Genet.* **79:** 1040–1051.

9. Itoh, M. *et al.* 1999. Direct binding of three tight junction-associated MAGUKs, ZO-1, ZO-2, and ZO-3, with the COOH termini of claudins. *J. Cell Biol.* **147:** 1351–1363.

10. Ebnet, K. *et al.* 2000. Junctional adhesion molecule interacts with the PDZ domaincontaining proteins AF-6 and ZO-1. *J. Biol. Chem.* **275:** 27979–27988.

11. Itoh, M. *et al.* 1997. Involvement of ZO-1 in cadherin-based cell adhesion through its direct binding to alpha catenin and actin filaments. *J. Cell Biol.* **138:** 181–192.

12. Imamura, Y. *et al.* 1999. Functional domains of alpha-catenin required for the strong state of cadherin-based cell adhesion. *J. Cell Biol.* **144:** 1311–1322.

13. Haskins, J. *et al.* 1998. ZO-3, a novel member of the MAGUK protein family found at the tight junction, interacts with ZO-1 and occludin. *J. Cell Biol.* **141:** 199–208.

14. Wittchen, E.S., J. Haskins & B.R. Stevenson. 1999. Protein interactions at the tight junction. Actin has multiple binding partners, and ZO-1 forms independent complexes with ZO-2 and ZO-3. *J. Biol. Chem.* **274:** 35179–35185.

15. Gumbiner, B., T. Lowenkopf & D. Apatira. 1991. Identification of a 160-kDa polypeptide that binds to the tight junction protein ZO-1. *Proc. Natl. Acad. Sci. USA* **88:** 3460-3464.

16. Fanning, A.S., T.Y. Ma & J.M. Anderson. 2002. Isolation and functional characterization of the actin binding region in the tight junction protein ZO-1. *FASEB J.* **16:** 1835–1837.

17. Katsube, T. *et al.* 1998. Cortactin associates with the cell–cell junction protein ZO-1 in both Drosophila and mouse. *J. Biol. Chem.* **273:** 29672–29677.

18. Balda, M.S., M.D. Garrett & K. Matter. 2003. The ZO-1-associated Y-box factor ZONAB regulates epithelial cell proliferation and cell density. *J. Cell Biol.* **160:** 423–432.

19. Balda, M.S. & K. Matter. 2000. The tight junction protein ZO-1 and an interacting transcription factor regulate ErbB-2 expression. *EMBO J.* **19:** 2024–2033.

20. Tsapara, A., K. Matter & M.S. Balda. 2006. The heat-shock protein Apg-2 binds to the tight junction protein ZO-1 and regulates transcriptional activity of ZONAB. *Mol. Biol. Cell* **17:** 1322–1330.

21. Meyer, T.N. *et al.* 2003. Galpha12 regulates epithelial cell junctions through Src tyrosine kinases. *Am. J. Physiol. Cell Physiol.* **285:** C1281–C1293.

22. Meyer, T.N., C. Schwesinger & B.M. Denker. 2002. Zonula occludens-1 is a scaffolding protein for signaling molecules. Galpha (12) directly binds to the Src homology 3 domain and regulates paracellular permeability in epithelial cells. *J. Biol. Chem.* **277:** 24855–24858.

23. Denker, B.M. *et al.* 1996. Involvement of a heterotrimeric G protein alpha subunit in tight junction biogenesis. *J. Biol. Chem.* **271:** 25750–25753.

24. Sallee, J.L. & K. Burridge. 2009. Density-enhanced phosphatase 1 regulates phosphorylation of tight junction proteins and enhances barrier function of epithelial cells. *J. Biol. Chem.* **284:** 14997–15006.

25. Yamamoto, T. *et al.* 1997. The Ras target AF-6 interacts with ZO-1 and serves as a peripheral component of tight junctions in epithelial cells. *J. Cell Biol.* **139:** 785–795.

26. Umeda, K. *et al.* 2006. ZO-1 and ZO-2 independently determine where claudins are polymerized in tight-junction strand formation. *Cell* **126:** 741–754.

27. Umeda, K. *et al.* 2004. Establishment and characterization of cultured epithelial cells lacking expression of ZO-1. *J. Biol. Chem.* **279:** 44785–44794.

28. Furuse, M. 2010. Molecular basis of the core structure of tight junctions. *Cold Spring Harb. Perspect Biol.* **2:** a002907.

29. Muller, S.L. *et al.* 2005. The tight junction protein occludin and the adherens junction protein alpha-catenin share a common interaction mechanism with ZO-1. *J. Biol. Chem.* **280:** 3747–3756.

30. Mitic, L.L. & J.M. Anderson. 1998. Molecular architecture of tight junctions. *Annu. Rev. Physiol.* **60:** 121–142.

31. Utepbergenov, D.I., A.S. Fanning & J.M. Anderson. 2006. Dimerization of the scaffolding protein ZO-1 through the second PDZ domain. *J. Biol. Chem.* **281:** 24671–24677.

32. Fanning, A.S. *et al.* 2007. Domain swapping within PDZ2 is responsible for dimerization of ZO proteins. *J. Biol. Chem.* **282:** 37710–37716.

33. Piontek, J. *et al.* 2008. Formation of tight junction: determinants of homophilic interaction between classic claudins. *FASEB J.* **22:** 146–158.

34. Lange, S. *et al.* 2010. Identification of phosphorylation-dependent interaction partners of the adapter protein ADAP using quantitative mass spectrometry: SILAC vs (18)O-labeling. *J. Proteome Res.* **9:** 4113–4122.

35. Cox J. & Mann M. 2008. MaxQuant enables high peptide identification rates, individualized p.p.b.-range mass accuracies and proteome-wide protein quantification. *Nature Biotechnol.* **26:** 1367–1372.

36. Fanning, A.S. & J.M. Anderson. 2009. Zonula occludens-1 and -2 are cytosolic scaffolds that regulate the assembly of cellular junctions. *Ann. N. Y. Acad. Sci.* **1165:** 113–120.

37. Singh Bal, M. *et al.* 2012. The hinge region of the scaffolding protein of cell contacts, zonula occludens protein 1, regulates interacting with various signaling proteins. *J. Cell Biochem.* **113:** 934–945.

Ann. N.Y. Acad. Sci. ISSN 0077-8923

ANNALS OF THE NEW YORK ACADEMY OF SCIENCES
Issue: *Barriers and Channels Formed by Tight Junction Proteins*

Dynamic properties of the tight junction barrier

Christopher R. Weber

Department of Pathology, the University of Chicago, Chicago, Illinois

Address for correspondence: Christopher R. Weber, the University of Chicago, 5841 S. Maryland Avenue, Chicago, IL 60637.
christopher.weber@uchospitals.edu

A principal role of tight junctions is to seal the apical intercellular space and limit paracellular flux of ions and molecules. Despite the fact that tight junctions form heavily cross-linked structures, functional studies have fostered the hypothesis that the tight junction barrier is dynamic and defined by opening and closing events. However, it has been impossible to directly measure tight junction barrier function with sufficient resolution to detect such events. Nevertheless, recent electrophysiological and sieving studies have provided tremendous insight into the presence of at least two pathways of trans-tight junction flux: a high-capacity ion-selective "pore" pathway and a low-capacity "leak" pathway that allows the passage of macromolecules. Furthermore, it is now known that the tight junction molecular structure is highly dynamic and that dynamics are correlated with barrier function. Taken together, these data support a dynamic model of tight junction conductance and suggest that regulation of tight junction openings and closings may provide sensitive means of barrier regulation.

Keywords: tight junction; FRAP; claudin; occludin; ZO-1

Fluid-filled structures throughout the body are lined by sheets of specialized epithelial or endothelial cells. One of the principal roles for these cells is to regulate luminal fluid composition through absorption and secretion of ions, organic molecules, and water. While transporters in the apical and basolateral cell membranes of these cells are critical to the active transport of ions against concentration gradients, solute would easily diffuse back if it were not for the paracellular barrier established by tight junctions. Now, almost 50 years after tight junctions were first identified at the apical intercellular space, we understand that the tight junction barrier is more than a simple seal of the paracellular space. Rather, tight junctions are dynamic structures with a capacity to finely regulate the paracellular passage of ions and molecules via different size- and charge-selective pathways.

Tight junction ultrastructure provides theoretical basis of tight junction "pores"

Tight junction ultrastructure, demonstrated by freeze-fracture scanning electron microscopy, provides insight into the organization of the tight junction barrier. Lipid bilayers fracture along hydrophobic planes and lateral views of the intercellular space reveal multiple anastomosing stands at the tight junction. The number of strands varies between cell type and organ, and strand number correlates roughly with barrier function.[1] For example, leaky epithelia, like the proximal convoluted tubule and gallbladder, have, on average, one or two strands, but tight epithelia, such as urinary bladder, have five or more tight junction strands. A simple model of the tight junction barrier would be to consider each strand as a resistor. Since resistors sum in series, overall transepithelial electrical resistance (TER) should be directly proportional (or conductance inversely proportional) to the number of strands. However, this turns out not to be the case. Rather, TER is a logarithmic function of the number of tight junction strands.[2] One possible explanation to account for such nonlinear behavior is that strands are populated by "pores" that can open and close to regulate tight junction ion conductance.[2] To date, it has not been possible to measure tight junction function with high enough resolution to detect such openings and closings, primarily due to limitations in the

doi: 10.1111/j.1749-6632.2012.06528.x

ability to electrically isolate a single tight junction area. However, despite the lack of local tight junction recordings, multiple studies have demonstrated that alterations in the molecular composition of the tight junction may have profound effects on pore-like tight junction barrier, even though differences at the ultrastructural level are not detected. One large family of tight junction proteins, the claudins, has been hypothesized to provide the structural basis of the paracellular pore.

Members of the claudin family are a major component of tight junction strands, and, when expressed in fibroblasts lacking tight junctions, claudins can direct formation of tight junction strand-like structures between adjacent cells.[3] These interactions have been proposed to occur through homotypical and heterotypical interactions of claudin extracellular domains, and are believed to form the basis of small ion permeability. For example, overexpression of claudin-2 in high-resistance MDCK I cells, which express little claudin-2, induces an 85% decrease in overall electrical resistance, and an 8.5-fold increase in Na^+ permeability.[4] Similarly, there is an 82% decrease in Na^+ permeability and a fourfold increase in TER in Caco-2 cells after knockdown of claudin-2.[5] Furthermore, expression of other claudins, such as claudin-4 or -5, has the capacity to decrease Na^+ permeability,[6,7] or in the case of claudin-10a, induce anion selectivity.[8] Multiple studies have demonstrated that the first extracellular loop of claudins is critical in

defining the ion selectivity. For example, expressing chimeric claudins where one loop one of claudin-2 is swapped with that of claudin-4 makes claudin-2 behave like claudin-4, while swapping the same extracellular loop of claudin-4 with that of claudin-2 makes claudin-4 behave more like claudin-2. These changes occur without any alterations in tight junction strand ultrastructure.[9] Since claudins from one cell interact with claudins on adjacent cells,[10–12] and more than one claudin is expressed in a single cell, it is possible that different claudins can cooperate to form a broad range of pores with unique sieving properties. For example, claudin-16 and claudin-19 can interact with one another and cooperatively form sodium-selective paracellular pores in the thick ascending loop of Henle that play an important role in generating transepithelial ion gradients. These gradients are in turn essential to driving reabsorption of divalent cations Mg^{2+} and Ca^{2+} (see Ref. 13). Indeed, inherited mutations of either claudin-16 or claudin-19 result in a syndrome associated with hypomagnesemia and hypercalciuria.[14] Thus, regulation of claudin expression provides a possible mechanism to finely regulate pore-like tight junction behavior. However, in addition to ions, it is well recognized that noncharged molecules with varying sizes, such as serum albumin and various sized dextrans, can pass through the tight junction. Since these molecules would be too large to pass through ion-selective pores, an additional pathway to allow larger molecules to pass must exist.

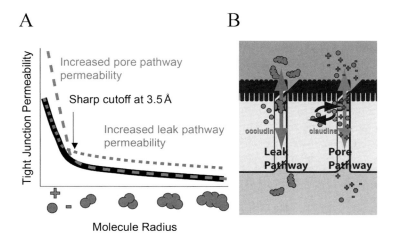

Figure 1. Size dependence of paracellular flux (solid black line) indicates the presence of two pathways. (A) A high-capacity pore pathway (dashed green line) is responsible for *trans*-tight junction flux of small ions and molecules, whereas a leaky pathway (dashed red line) allows paracellular flux of larger molecules (after Refs. 14, 16, and 17). (B) Claudins primarily regulate the pore pathway, while occludin appears to be important to leak pathway regulation (after Ref. 29).[15,17,28]

Multiple pathways of *trans*-tight conductance

In order to assess the differential permeability of the tight junction to small and large molecules (i.e., tight junction size selectivity), there are two commonly used techniques. The first technique assesses the transepithelial flux of series polyethylene glycol oligomers ranging in radius from 2.8 Å to 7.0 Å,[15,16] and a second approach relies on electrical reversal potential measurements after apical or basolateral substitution of Na^+, with different sized cations ranging in size from 1.0 Å to 3.6 Å.[17] Using either method, plotting tracer permeability as a function of size, reveals a bimodal relationship between size and permeability in a variety of epithelial cells. A high-capacity size-restrictive component has a sharp radius cut-off to molecules with radii larger than about 3.5 Å and a low-capacity pathway has no apparent upper cutoff, up to at least 7.5 Å (Fig. 1A). These two pathways have been termed *pore* and *leak pathways*, respectively, and data support that the structural basis of these two pathways is distinct. Specifically, expression of claudin-2 increases only pore flux,

whereas occludin and tricellulin appear to be important to leak pathway regulation (Fig. 1B).[15,17–19] Furthermore, multiple studies have demonstrated that these two pathways are differentially regulated as stimuli affecting tight junction permeability to small ions and larger macromolecules do not always correlate.

One good example of selective pore regulation is that which occurs following activation of apical sodium-glucose cotransport (SGLT1). In the presence of high apical glucose, there is reduction in TER along with an increase in the flux of the small molecule mannitol (radius = 3.6 Å), yet there is no change in the flux of larger inulin ($r = 11.5$Å).[20] These pore and leak pathways may also be differentially regulated via pathological stimuli such as inflammatory cytokines IL-13 and TNF-α. IL-13 specifically upregulates expression of claudin-2 (Fig. 2A and B), results in a decrease in TER (Fig. 2C), and increases the relative permeability of Na^+ to Cl^- (Fig. 2D).[21–23] However, IL-13 does not induce increased flux of 4 kDa FITC-dextran (Fig. 2E).[23] In contrast,

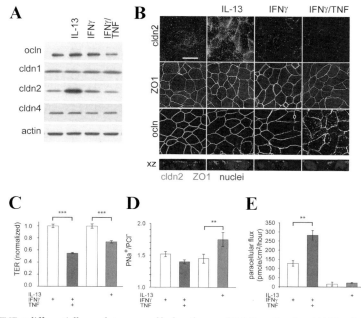

Figure 2. IL-13 and TNF-α differentially regulate pore and leak pathways of tight junction flux. (A) Claudin-2 protein was increased by 308%, and occludin protein was reduced by 60% after respective treatments with IL-13 or TNF-α. (B) IL-13 specifically increased claudin-2 localization at the tight junction without affecting occludin or ZO-1. In contrast, TNF-α caused occludin internalization without affecting ZO-1 or claudin-2. Bar, 10 μm. (C) TNF-α (red; 2.5 ng/mL) and IL-13 (green; 0.1 ng/mL) reduced TER of T84 monolayers cells at 4 and 12 h, respectively. (D) IL-13 (green), but not TNF-α (red), increased PNa$^+$/PCl$^-$ of T84 monolayers. ($^{**}P \leq 0.01$; $^{***}P \leq 0.001$, SEM). (E) TNF-α (red), but not IL-13 (green), increased the paracellular permeability of 4 kDa dextran (from Ref. 23).

TNF-α, induces TER loss with a corresponding increase in macromolecular flux (Fig. 2C–D), representative of a leak pathway.[23,24] Consistent with what is known about the molecular basis of the leak pathway, occludin internalization is associated with TNF-induced barrier dysfunction (Fig. 2A and B). Since TNF-α does not induce a change in PNa^+/PCl^- (Fig. 2D). TER loss corresponds with equivalent increases in tight junction permeability to both Na^+ and Cl^-.[23] Since such noncharge-selective increases in Na^+ and Cl^- permeability would be expected to bring PNa^+/PCl^- closer to 1.0, the fact that this ratio is only minimally reduced supports that tight junction "leak" is a low-capacity pathway, contributing at most ~20% to overall ion conductance, even in the presence of TNF-α.

Thus, these data support the presence of distinct and differentially regulated high-capacity pore and low-capacity leak pathways across the tight junction. However, despite our knowledge of multiple distinct pathways, current assessments of barrier function reflect average measurements of epithelial barrier and do not provide insight on a local,

molecular level. Nonetheless, recent studies of fluorescently tagged tight junction protein dynamics after local photobleaching (i.e., FRAP (fluorescence recovery after photobleaching) studies) reveal that global tight junction function does in fact correlate with tight junction dynamics, at a molecular level.

Tight junction dynamics correlate with tight junction function

The principle behind FRAP experiments is that the time course of fluorescence recovery, after transient photobleaching of a fluorescently tagged tight junction protein, provides insight into the stability of molecular interactions. By comparing the mobilities of different tight junction proteins, it is possible to infer interactions at a molecular level. For example, the FRAP kinetics of ZO-1, occludin, and claudin-1 are each unique, and recovery occurs from different pools of unbleached proteins. Experimental data, recapitulated in simulations (Fig. 3A–E), show that occludin exchanges within the tight junction and with non-tight junction lateral membrane pools. In contrast, claudin-1 is distinguished by a much larger immobile fraction, and ZO-1 does not

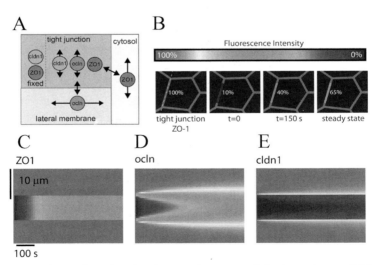

Figure 3. Tight junction proteins are dynamic and undergo continuous remodeling in steady state. (A) Computer models were established based on fluorescence recovery after photobleaching of fluorescently labeled claudin-1, occludin, or ZO-1. The model reflects distinct dynamic behavior and exchange of these proteins within three cellular compartments: tight junction, lateral membrane, and cytosol. Claudin-1 is largely anchored at the tight junction, while occludin exchanges with a small lateral membrane pool (20% of total). ZO-1 is immobile at the tight junction and exchanges with intracellular pools (from Ref. 30). (B) A laser is used to transiently photobleach a region of tight junction–expressing EGFP-labeled tight junction proteins (here demonstrated for ZO-1 before and after a laser pulse at $t = 0$). (C–E) Kymographs demonstrate time course of fluorescence recovery in the tight junction region indicated by the dashed line in Figure 3C (simulation parameters from Ref. 31). (C) ZO-1 recovery occurs entirely from intracellular pools. (D) Occludin recovery is fast and occurs from the tight junction and lateral membrane. (E) Claudin-1 recovery is slow and occurs from a mobile pool of claudin in the adjacent area of the tight junction.[31]

diffuse as an intact tight junction complex along the plasma membrane, but rather exchanges with intracellular pools.

Furthermore, the dynamics of ZO-1 exchange between the intracellular pools and the tight junction are highly dependent on interactions with the actin cytoskeleton.[25] Thus, these data support that tight junctions undergo continuous remodeling at steady state, and it will be important to understand how these dynamics correlate with barrier regulation. One hypothesis is that stimuli that influence tight junction mobility will tend to stabilize the tight junction within particular conformations, that is, either in open or closed states.

An example of such dynamic tight junction regulation involves CK2 phosphorylation of occludin in its C-terminal tail, which causes the mobilities of occludin, claudin-1, claudin-2, and ZO-1 converge.[5] These data, combined with complementary binding assays, support the formation of a protein complex and favors reductions in small ion permeability, consistent with pore pathway regulation. Furthermore, inhibition of phosphorylation prevents IL-13–induced barrier dysfunction, consistent with the known role of IL-13 in increasing claudin-2 pores. Thus, in addition to the role for occludin in pathologic leak pathway regulation, there exists a possible role for occludin in the regulation of the pore pathway, via modulation of claudin-2 pore formation. These data suggest that dynamic pore-like behavior relies both on claudin expression and also the stability of claudin interactions with other tight junction proteins. It is presently unclear if claudin-2 dynamics may provide the basis of tight junction pore openings and closings, as proposed by Claude,[2] since it is presently not possible to measure the tight junction barrier with similar molecular resolution. Understanding tight junction pore behavior requires understanding of pore number, pore size, charge selectivity, as well as the frequency and duration of pore openings. Nevertheless, some predictions, based on macroscopic tight junction behavior, have been possible about how altering the open and closing probabilities of claudin pores would be expected to influence barrier function.

Modeling the tight junction function

In order to better understand how dynamic pore properties could potentially play an important role in barrier regulation, it will be important to first consider a simple static model of pore function. A reductionist model of pore function ignores strand-to-strand and regional variations in tight junction composition as well as potential unique properties of different claudins, and rather will focus on claudin-2. Claudin-2 is one of the best defined pore-forming claudins, and is known to regulate pore behavior in multiple cell lines *in vitro* and *in vivo*.[5,17,23,26] Charge-altering amino acid substitutions and cysteine scanning mutagenesis approaches have helped to identify critical amino acids within the first extracellular loop of claudin-2 that are responsible for pore-like behavior when expressed in MDCK cells.[11] Such experiments have enabled Yu *et al.* to model the claudin-2 selectivity filter as a cylinder with 70pS conductance.[17] The simplest possible static model of the tight junction barrier is therefore one that considers each tight junction strand as a resistor populated by a fixed number of such claudin-2 pores.

In epithelia-expressing claudin-2, there would need to be approximately four 70 pS claudin pores per 100 nm of tight junction strands in order to account for measured TER values.[17] Also, in order to obtain a reasonable conductance in the absence of claudin-2, a relatively small claudin-2–independent tight junction conductance of 2 pS per 100 nm will be assumed here. In a simple static model, claudin-2 conductance is inversely proportional to the number of tight junction strands, ranging from 70 pS/100 nm for four strands and increasing to 282 pS/100 nm for one strand (Fig. 4A). Furthermore, in this static model, conductance is directly proportional to the number of claudin-2 pores. However, as described earlier, when conductance is assessed across a range of epithelial with different numbers of tight junction strands, conductance is not inversely correlated with strand number, which has been an argument for the presence of tight junction pores.[2] In order to assess how opening and closing of pores could potentially provide a means to regulate the tight junction barrier, the static model will be compared to a dynamic model.

Since the ultrastructural appearance of a tight junction strand is continuous, it is conceivable that there may be many more claudins in a strand than was predicted in the above static model. Therefore, a starting point for a dynamic model will be to assume that only 10% of claudins form conductive pores, which requires the presence of 53 claudin-2

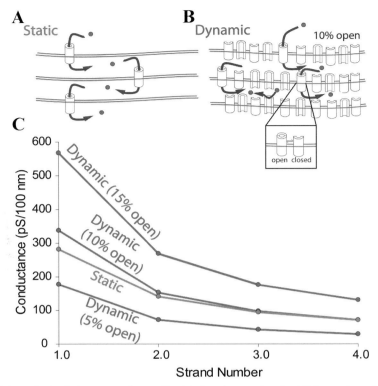

Figure 4. Static and dynamic simulations are used to predict tight junction pore barrier. (A) In a static model, each tight junction strand has a low conductance (2 pS per 100 nm) and is populated by four 70 pS claudin pores per 100 nm of tight junction.[17] (B) In a dynamic model, only a percentage of claudins form conductive pores, and pores randomly flicker between open and closed states with defined probabilities. The percentage of open pores is defined by the probability of a closed claudin moving to an open state and an open claudin moving to a closed state. (C) In a dynamic model (red), conductance is higher than would be predicted for a simple inverse relationship with strand number (as is true for the static model, green), and conductance may be quite sensitive to the percentage of open claudin-2 pores.

proteins per 100 nm of tight junction strand to match the known global epithelial conductance. Also, in order to maintain this relationship in steady state, the probability of an open pore moving to a closed state is necessarily 10-fold higher than the probability of a closed pore moving to an open state. If one looks at the average conductance of a population of pores in steady-state using time as a spatially averaged simulation, the frequency and rate of transition between pore states averages out, and it is possible to determine the relationship between conductance and strand number. As suggested by Claude, this relationship no longer follows a simple inverse relationship. For a low number of strands, conductance is higher than would be predicted from an analogous static model (Fig. 4C). Intuitively, this can be explained by the fact that synchronized openings on multiple strands are more likely to occur as

strand number decreases. It also follows that barrier function may be exquisitely sensitive to the number of claudin pores per strand, which is in contrast to what one would expect for a static model, where conductance is directly proportional to the number of pores per strand. For example, in the dynamic model, decreasing all claudins (open and closed) by 75% results in an 89% decrease in conductance, while increasing claudins by 75% results in a 92% increase in overall conductance. In contrast, for a static model, changing the number of pores by 75% in either direction would result in a corresponding 75% change in barrier function. Thus, a dynamic model may be more sensitive to changes in expression level, and may partially explain the apparent nonlinear relationship between claudin-2 expression and measured barrier function (unpublished personal observation).

In addition to the above differences between a static and dynamic model, an important feature of the dynamic model is that it provides two additional parameters, pore open and closed probability, which may be sensitive to small perturbations. This is best illustrated by the profound effects on barrier function, which occur when the percentage of open pores is varied, while keeping the total number of claudins constant (Fig. 4C). For example, starting with 10% of claudins in the open pore state, a decrease in the pore-forming probability by 50% decreases overall conductance by 60%, and increasing pore openings by 50% increases conductance by 84%. Thus, a dynamic model can better describe known tight junction properties and may provide a potential sensitive means of barrier regulation, even though it is presently not possible to experimentally detect pore openings and closing with molecular resolution.

Tight junction leak is explained by a rare high conductance pathway

Much less is known about the biophysical basis of tight junction leak flux. One possible explanation for tight junction leak is that it is spatially restricted to only a small portion of total linear tight junction. For example, tricellulin has been shown to form a barrier to macromolecules in tricellular tight junctions without contributing increased ion permeability.[19] However, a second possibility is that the leak pathway is dynamic like the pore pathway. Leak may be explained by infrequent high conductance tight junction openings that are large enough to allow macromolecules to pass but rare enough to maintain overall tight junction ion selectivity when studied on a global scale. Alternatively, dynamic leak openings may be associated with transient breaks or separations of a tight junction strands. Such breaks would allow diffusion of molecules across a strand and thus diffusion across the entire tight junction may occur in a stepwise fashion followed by resealing of the first strand and breakage of a new strand. This may occur pathophysiologically during injury, but may also occur normally during cell extrusion, during cell proliferation, or even as a result of normal tight junction strand rearrangements.[27] A leak event may be rare, relative to the probability of pore openings, and the overall effect on tight junction conductance would depend on the original number of tight junction strands. The impact of such an event may be equivalent to moving to the left along one of the

conductance curves in Figure 4C, while keeping the pore open probability constant. It is intriguing to consider that heavily branched tight junction structures may provide a functional advantage by limiting lateral diffusion between tight junction strands. Functionally, this means that a high conductance leak event would only be expected to affect ∼500 nm of linear tight junction.

Summary and future directions

Tight junctions were once thought to be static structures with the ability to seal off the intercellular space. This hypothesis was bolstered by the fact that tight junctions are heavily cross-linked structures. However, predictions based on early ultrastructural studies have suggested that this cannot be correct, and many have proposed the existence of a dynamic pore pathway. Recent electrophysiological sieving studies support the existence of such an ion-selective pore pathway, and are also consistent with an additional low-capacity leak pathway, but it has been impossible to study these tight junction functions at a molecular level. Additional molecular insight into tight junction function can be obtained from high-resolution FRAP studies, and there is a strong correlation between the mobility of different tight junction proteins and epithelial barrier function. Taken together, these data support a dynamic model of the tight junction barrier and also suggest that regulation of tight junction openings and closings may provide a sensitive means to modulate barrier function without changing protein expression. However, single tight junction opening and closing events have never been measured, and our ability to detect single openings and closings will require novel high resolution techniques. Improved insight into these pathways will likely facilitate the development of more specific pharmacologic modulators designed to target and modulate specific aspects of barrier function in the future.

Conflicts of interest

The author declares no conflicts of interest.

References

1. Claude, P. & D.A. Goodenough. 1973. Fracture faces of zonulae occludentes from "tight" and "leaky" epithelia. *J. Cell. Biol.* **58:** 390–400.
2. Claude, P. 1978. Morphological factors influencing transepithelial permeability: a model for the resistance of the zonula occludens. *J. Membr. Biol.* **39:** 219–232.

3. Furuse, M., H. Sasaki, K. Fujimoto & S. Tsukita. 1998. A single gene product, claudin-1 or -2, reconstitutes tight junction strands and recruits occludin in fibroblasts. *J. Cell. Biol.* **143:** 391–401.

4. Amasheh, S., N. Meiri, A.H. Gitter, *et al.* 2002. Claudin-2 expression induces cation-selective channels in tight junctions of epithelial cells. *J. Cell. Sci.* **115:** 4969–4976.

5. Raleigh, D.R., D.M. Boe, D. Yu, *et al.* 2011. Occludin S408 phosphorylation regulates tight junction protein interactions and barrier function. *J. Cell. Biol.* **193:** 565–582.

6. Van Itallie, C., C. Rahner & J.M. Anderson. 2001. Regulated expression of claudin-4 decreases paracellular conductance through a selective decrease in sodium permeability. *J. Clin. Invest.* **107:** 1319–1327.

7. Wen, H., D.D. Watry, M.C. Marcondes & H.S. Fox. 2004. Selective decrease in paracellular conductance of tight junctions: role of the first extracellular domain of claudin-5. *Mol. Cell. Biol.* **24:** 8408–8417.

8. Van Itallie, C.M., S. Rogan, A. Yu, L.S. Vidal, J. Holmes & J.M. Anderson. 2006. Two splice variants of claudin-10 in the kidney create paracellular pores with different ion selectivities. *Am. J. Physiol. Renal Physiol.* **291:** F1288–F1299.

9. Colegio, O.R., C.V. Itallie, C. Rahner & J.M. Anderson. 2003. Claudin extracellular domains determine paracellular charge selectivity and resistance but not tight junction fibril architecture. *Am. J. Physiol. Cell. Physiol.* **284:** C1346–C1354.

10. Mitic, L.L., V.M. Unger & J.M. Anderson. 2003. Expression, solubilization, and biochemical characterization of the tight junction transmembrane protein claudin-4. *Protein Science : a publication of the Protein Society* **12:** 218–227.

11. Angelow, S. & A.S. Yu. 2009. Structure-function studies of claudin extracellular domains by cysteine-scanning mutagenesis. *J. Biol. Chem.* **284:** 29205–29217.

12. Van Itallie, C.M., L.L. Mitic & J.M. Anderson. 2011. Claudin-2 forms homodimers and is a component of a high molecular weight protein complex. *J. Biol. Chem.* **286:** 3442–3450.

13. Hou, J., A. Renigunta, M. Konrad, *et al.* 2008. Claudin-16 and claudin-19 interact and form a cation-selective tight junction complex. *J. Clin. Invest.* **118:** 619–628.

14. Hou, J., A. Renigunta, A.S. Gomes, *et al.* 2009. Claudin-16 and claudin-19 interaction is required for their assembly into tight junctions and for renal reabsorption of magnesium. *Proc. Natl. Acad. Sci. USA* **106:** 15350–15355.

15. Van Itallie, C.M., J. Holmes, A. Bridges, *et al.* 2008. The density of small tight junction pores varies among cell types and is increased by expression of claudin-2. *J. Cell. Sci.* **121:** 298–305.

16. Watson, C.J., M. Rowland & G. Warhurst. 2001. Functional modeling of tight junctions in intestinal cell monolayers using polyethylene glycol oligomers. *Am. J. Physiol. Cell. Physiol.* **281:** C388–C397.

17. Yu, A.S., M.H. Cheng, S. Angelow, *et al.* 2009. Molecular basis for cation selectivity in claudin-2-based paracellular pores: identification of an electrostatic interaction site. *J. Gen. Physiol.* **133:** 111–127.

18. Yu, A.S., K.M. McCarthy, S.A. Francis, *et al.* 2005. Knockdown of occludin expression leads to diverse phenotypic alterations in epithelial cells. *Am. J. Physiol. Cell. Physiol.* **288:** C1231–C1241.

19. Krug, S.M., S. Amasheh, J.F. Richter, *et al.* 2009. Tricellulin forms a barrier to macromolecules in tricellular tight junctions without affecting ion permeability. *Mol. Biol. Cell.* **20:** 3713–3724.

20. Turner, J.R., B.K. Rill, S.L. Carlson, *et al.* 1997. Physiological regulation of epithelial tight junctions is associated with myosin light-chain phosphorylation. *Am. J. Physiol.* **273:** C1378–C1385.

21. Prasad, S., R. Mingrino, K. Kaukinen, *et al.* 2005. Inflammatory processes have differential effects on claudins 2, 3, and 4 in colonic epithelial cells. *Lab. Invest.* **85:** 1139–1162.

22. Heller, F., P. Florian, C. Bojarski, *et al.* 2005. Interleukin-13 Is the key effector Th2 cytokine in ulcerative colitis that affects epithelial tight junctions, apoptosis, and cell restitution. *Gastroenterology* **129:** 550–564.

23. Weber, C.R., D.R. Raleigh, L. Su, *et al.* 2010. Epithelial myosin light chain kinase activation induces mucosal interleukin-13 expression to alter tight junction ion selectivity. *J. Biol. Chem.* **285:** 12037–12046.

24. Wang, F., W.V. Graham, Y. Wang, *et al.* 2005. Interferongamma and tumor necrosis factor-alpha synergize to induce intestinal epithelial barrier dysfunction by upregulating myosin light chain kinase expression. *Am. J. Pathol.* **166:** 409–419.

25. Yu, D., A.M. Marchiando, C.R. Weber, *et al.* 2010. MLCK-dependent exchange and actin binding region-dependent anchoring of ZO-1 regulate tight junction barrier function. *Proc. Natl. Acad. Sci. USA* **107:** 8237–8241.

26. Furuse, M., K. Furuse, H. Sasaki & S. Tsukita. 2001. Conversion of zonulae occludentes from tight to leaky strand type by introducing claudin-2 intoMadin-Darby canine kidney I cells. *J. Cell. Biol.* **153:** 263–272.

27. Marcial, M.A. & J.L. Madara. 1987. Analysis of absorptive cell occluding junction structure-function relationships in a state of enhanced junctional permeability. *Lab. Invest.* **56:** 424–434.

28. Van Itallie, C.M., A.S. Fanning, A. Bridges & J.M. Anderson. 2009. ZO-1 stabilizes the tight junction solute barrier through coupling to the perijunctional cytoskeleton. *Mol. Biol. Cell.* **20:** 3930–3940.

29. Turner, J.R. 2006. Molecular basis of epithelial barrier regulation: from basic mechanisms to clinical application. *Am. J. Pathol.* **169:** 1901–1909.

30. Shen, L., C.R. Weber, D.R. Raleigh, D. Yu & J.R. Turner. 2011. Tight junction pore and leak pathways: a dynamic duo. *Annu. Rev. Physiol.* **73:** 283–309.

31. Shen, L., C.R. Weber & J.R. Turner. 2008. The tight junction protein complex undergoes rapid and continuous molecular remodeling at steady state. *J. Cell. Biol.* **181:** 683–695.

Ann. N.Y. Acad. Sci. ISSN 0077-8923

ANNALS OF THE NEW YORK ACADEMY OF SCIENCES

Issue: *Barriers and Channels Formed by Tight Junction Proteins*

Regulation of tight junctions in human normal pancreatic duct epithelial cells and cancer cells

Takashi Kojima and Norimasa Sawada

Department of Pathology, Sapporo Medical University School of Medicine, Sapporo, Japan

Address for correspondence: Takashi Kojima, Ph.D., Department of Pathology, Sapporo Medical University School of Medicine, S1. W17, Sapporo 060-8556, Japan. ktakashi@sapmed.ac.jp

To investigate the regulation of tight junction molecules in normal human pancreatic duct epithelial (HPDE) cells and pancreatic cancer cells, we introduced the human telomerase reverse transcriptase (hTERT) gene into HPDE cells in primary culture and compared them to pancreatic cancer cell lines. The hTERT-transfected HPDE cells were positive for PDE markers and expressed claudin-1, claudin-4, claudin-7, and claudin-18, occludin, tricellulin, marvelD3, JAM-A, zonula occludens (ZO)-1, and ZO-2. The tight junction molecules, including claudin-4 and claudin-18 of normal HPDE cells, were in part regulated via a protein kinase C signal pathway by transcriptional control. In addition, claudin-18 in normal HPDE cells and pancreatic cancer cells was markedly induced by a PKC activator, and claudin-18 in pancreatic cancer cells was also modified by DNA methylation. In the marvel family of normal HPDE cells and pancreatic cancer cells, tricellulin was upregulated via a c-Jun N-terminal kinase pathway, and marvelD3 was downregulated during Snail-induced epithelial–mesenchymal transition.

Keywords: tight junctions; normal human pancreatic duct epithelial cells; pancreatic cancer cells; claudins; tricellulin; marvelD3

Introduction

Pancreatic cancer, which has a strong invasive capacity with frequent metastasis and recurrence, is known to be one of the most malignant human diseases. It is the fourth leading cause of cancer death in the United States, and its death rate has not decreased over the past few decades.[1] Thus, there is an urgent need to develop novel diagnostic and therapeutic strategies to reduce the mortality of pancreatic cancer patients. Recently, it was found that in several human cancers, including pancreatic cancer, some tight junction protein claudins are abnormally regulated and therefore are promising molecular targets for diagnosis and therapy.[2,3]

The tight junction, the most apically located of the intercellular junctional complexes, inhibits solute and water flow through the paracellular space (termed the *barrier* function).[4,5] It also separates the apical from the basolateral cell surface domains to establish cell polarity (termed the *fence function*).[6,7]

Recent evidence suggests that tight junctions also participate in signal transduction mechanisms that regulate epit'.elial cell proliferation, gene expression, differentiation, and morphogenesis.[8]

Tight junctions are formed by not only the integral membrane proteins claudins, occludin, tricellulin, and junctional adhesion molecules (JAMs), but also many peripheral membrane proteins, including the scaffold PDZ-expression proteins zonula occludens (ZO)-1, ZO-2, ZO-3, multi-PDZ domain protein-1 (MUPP1), membrane-associated guanylate kinase with inverted orientation-1 (MAGI)-1, MAGI-2, MAGI-3, cell polarity molecules atypical PKC isotype-specific interacting protein/PAR-3, PAR-6, PALS-1, and PALS-1-associated tight junction (PATJ), as well as the non-PDZ–expressing proteins cingulin, symplekin, ZONAB, GEF-H1, aPKC, PP2A, Rab3b, Rab13, PTEN, and 7H6.[9–12] These tight junction proteins are regulated by various cytokines and growth factors via distinct signal transduction pathways.[13,14]

doi: 10.1111/j.1749-6632.2012.06579.x

Table 1. Changes of tight junction proteins and barrier function in normal human pancreatic duct epithelial cells and cancer cells

Cell type	Treatment	Tight junction proteins	Barrier function	Ref.
hTERT-HPDE	Fetal bovine serum	CLDN-1, -4, -7 ↑; OCDN↑; ZO-1, -2↑	Upregulation	26
	TPA	CLDN-1, -4, -7, -18 ↑; OCDN↑; ZO-1, -2↑	—	26
	Anisomycin	TRIC↑	Upregulation	27
	IL-1β/TNF-α/IL-1α	TRIC↑	—	27
HPAC/HPAF-II	TPA	CLDN-1↓; CLDN-18↑	Downregulation	28, 29
	5-aza-cdR	CLDN-18↑	—	28
HPAC	Anisomycin	TRIC↑	Upregulation	27
	IL-1β/TNF-α/IL-1α	TRIC↑	—	27
	siRNA-TRIC	TRIC↓	Downregulation	27
	Hypoxia (2% O_2)	CLDN-1↓; TRIC↓; marvelD3↓	—	30
	TGF-β	CLDN-1↓; marvelD3↓	—	30
	siRNA-marvelD3	marvelD3↓	Downregulation	30
	siRNA-FOXA2	TRIC↓; marvelD3↓	—	30

NOTE: HPAC, HPAF-II: well-differentiated human pancreatic cancer cell lines. hTERT-HPDE, hTERT-transfected human pancreatic duct epithelial cells; IL, Interleukin; TNF, tumor necrosis factor; TPA, 12-O-tetradecanoylphorbol 13-acetate.

The claudin family, which consists of at least 24 members, is solely responsible for forming tight junction strands and has four transmembrane domains and two extracellular loops.[10] The first extracellular loop is the coreceptor of the hepatitis C virus[15] and influences the paracellular charge selectivity,[16] and the second extracellular loop is the receptor of *Clostridium perfringens* enterotoxin (CPE).[17] In pancreatic cancer, claudin-4 and claudin-18 are highly expressed[18,3] and are diagnostic or therapeutic targets of monoclonal antibodies against their extracellular loops.[19,20] In addition, as claudin-4 is also a high-affinity receptor of CPE,[21] full-length CPE with a direct cytotoxic effect and the C-terminal receptor-binding domain of CPE without a cytotoxic effect are employed for selective treatment and drug delivery against claudin-4 expressing tumors.[18,22] However, the regulatory mechanisms of claudin-based tight junctions remain unknown even in normal human pancreatic duct epithelial (HPDE) cells.

Both occludin and tricellulin (marvelD2) contain the tetra-spanning MARVEL (MAL and related proteins for vesicle trafficking and membrane link) domain that is present in proteins involved in membrane apposition and concentrated in cholesterol-rich microdomains.[23] The novel tight junction protein marvelD3 contains a conserved MARVEL domain like occludin and tricellulin.[24,25] However, little is yet known about how tricellulin and marvelD3 are regulated in HPDE cells.

In this study, we introduced the human telomerase reverse transcriptase (hTERT) gene into HPDE cells in primary culture and compared them to pancreatic cancer cell lines. The changes of tight junction molecules and barrier function in the hTERT-HPDE cells and pancreatic cancer cell lines HPAF-II and HPAC were observed after various treatments (Table 1).

Tight junction molecules of hTERT-HPDE cells

The introduction of the catalytic subunit of human telomerase, hTERT into human somatic cells, such as fibroblasts and retinal pigment epithelial cells, typically extends their life span without altering the growth requirements, disturbance of the cell-cycle checkpoints, tumorigenicity, or chromosomal abnormality.[31,32] We also established hTERT-transfected human nasal epithelial cells with an extended life span.[33]

To investigate the regulation of tight junction molecules, including claudins in normal HPDE cells, we introduced the hTERT gene into HPDE

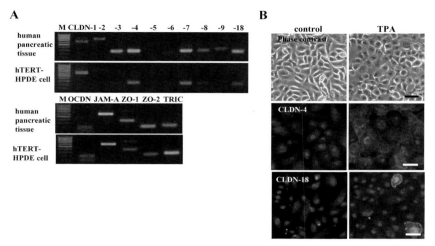

Figure 1. (A) RT-PCR for tight junction molecules in hTERT-HPDE cells. mRNAs of claudin-1, claudin-4, claudin-7, and claudin-18, occludin, JAM-A, ZO-1, ZO-2, and tricellulin are detected in hTERT-HPDE cells. M: 100-bp ladder DNA marker. (B) Phase-contrast images and immunocytochemistry for claudin-4 and claudin-18 in hTERT-HPDE cells at 24 h after treatment with 100 nM 12-O-tetradecanoylphorbol 13-acetate (TPA). Black bar, 80 μm; white bar, 20 μm. CLDN, claudin; OCDN, occludin; TRIC, tricellulin.

cells in primary culture.[26] The hTERT-transfected HPDE (hTERT-HPDE) cells were positive for PDE markers, such as CK7, CK19, and carbonic anhydrase isozyme 2 (CA-II), and expressed epithelial tight junction molecules claudin-1, claudin-4, claudin-7, and claudin-18, occludin, tricellulin, marvelD3, JAM-A, ZO-1, and ZO-2 (Fig. 1A and B, and 3B).[26]

Regulation of tight junction molecules in hTERT-HPDE cells at the transcriptional level via a protein kinase C signal pathway

Protein kinase C (PKC) is a family of serine-threonine kinases known to regulate epithelial barrier function via tight junctions.[34,35] PKC has been shown to induce both assembly and disassembly of tight junctions depending on the cell type and conditions of activation.[35–37] To confirm whether the PKC signal pathway was closely associated with the regulation of tight junctions in HPDE cells, hTERT-HPDE cells were treated with the PKC activator 12-O-tetradecanoylphorbol 13-acetate (TPA). Treatment with TPA-enhanced expression of most tight junction proteins.[26] In addition, claudin-4 and claudin-18 were markedly induced by TPA at the transcriptional level together with an increase of tight junction strands (Fig. 1B).[26] The upregulation of tight junction proteins by TPA was inhibited

completely by a pan-PKC inhibitor (GF109203X). A PKCθ inhibitor (myristoylated PKCθ pseudo-substrate peptide inhibitor) prevented upregulation of claudin-18 by TPA; a PKC-α inhibitor (Gö6976) prevented upregulation of claudin-4 and claudin-18 by TPA; and a PKC-δ inhibitor (rottlerin) prevented upregulation of claudin-7, claudin-18, occludin, ZO-1, and ZO-2 by TPA (Fig. 2A and B).[26]

By GeneChip analysis, upregulation of transcription factor ELF3 was observed in both fetal bovine serum- and TPA-treated cells. ELF3 belongs to the ELF (E74-like factor) subfamily of the ETS transcription factors, but it is distinguished from most ETS family members by its expression pattern, which is specific in epithelial tissues of the lung, liver, kidney, pancreas, prostate, small intestine, and colon mucosa.[38] ELF3 controls intestinal epithelial differentiation.[39] It is reported that the expression of claudin-7 in epithelial structures in synovial sarcoma is regulated by ELF3.[40] In hTERT-HPDE cells, ELF3 mRNA is increased by TPA and a pan-PKC inhibitor prevents upregulation of ELF3 mRNA by TPA.[26] When knockdown of ELF3 was caused by siRNAs in hTERT-HPDE cells, upregulation of claudin-7 by TPA was inhibited. These results suggested that claudin-7 in normal HPDE cells might be regulated via a PKC-δ/ELF-3 pathway (Fig. 2A).

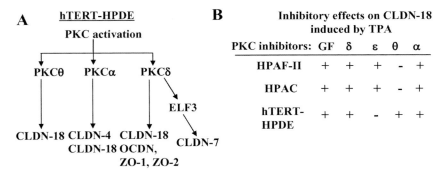

Figure 2. (A) Diagram showing regulation of tight junction molecules in hTERT-HPDE cells via PKC isoforms. (B) Inhibitory effects of GF109203X (GF) as a panPKC inhibitor, Gö6976 as a PKCα inhibitor, rottlerin as a PKCδ inhibitor, PKCε inhibitor, and PKCθ inhibitor for claudin-18 induced by TPA in HPAF-II cells, HPAC cells, and hTERT-HPDE cells. CLDN, claudin.

Claudin-18 in pancreatic cancer cells regulated at the transcriptional level via specific PKC signal pathways and modified by DNA methylation

Claudin-18 has two alternatively spliced variants, claudin-18a1 and claudin-18a2, which are highly expressed in the lung and stomach, respectively.[41] Furthermore, a PKC/MAPK/AP-1–dependent pathway regulates claudin-18a2 expression in TPA-stimulated gastric cancer cells.[41] Claudin-18a2 is activated in a wide range of human malignant tumors, including gastric, esophageal, pancreatic, lung, and ovarian cancers, and can be specifically targeted by monoclonal antibodies.[19] Claudin-18 is highly expressed in pancreatic intraepithelial neoplasia, including precursor PanIN lesions and pancreatic duct carcinoma, and serves as a diagnostic marker.[3]

PKC is commonly dysregulated in cancers of the prostate, breast, colon, pancreas, liver, and kidney.[42] It is also reported that levels of PKCα, PKCβ1, and PKCδ are higher in pancreatic cancer, whereas that of PKCε is higher in normal tissue.[43] Claudin-18 mRNA, indicated as claudin-18a2, was markedly induced by TPA and in well- and moderately differentiated human pancreatic cancer cell lines HPAF-II and HPAC and hTERT-HPDE cells, the protein was also strongly increased.[28] The upregulation of claudin-18 by TPA in human pancreatic cancer cell lines was prevented by inhibitors of PKCδ, PKCε, and PKCα (rottlerin, PKCε translocation inhibitor peptide, Gö6976), whereas the upregulation of claudin-18 by TPA in hTERT-HPDE cells was prevented by inhibitors of PKCδ, PKCθ, and PKCα (rottlerin, myristoylated PKCθ pseudosubstrate peptide inhibitor, Gö6976) (Fig. 2B).

On the other hand, in the development and progression of cancer, tumor suppressor genes may be silenced by mechanisms such as methylation. Silencing of the expression of some claudins in several human cancers is correlated with promoter hypermethylation. These include claudin-7 in breast cancer, claudin-4 in bladder cancer, and claudin-11 in gastric cancer.[44–46] Furthermore, a CpG island is present within the promoter region of the claudin-18a2 gene, and treatment with the demethylating agent 5-aza-CdR restores the expression in primary cultures prepared from gastric cancer cell line SNUI.[19] A CpG island has been identified within the coding sequence of the claudin-18 gene and in HPAF-II and HPAC, and treatment with the demethylating agent 5-azadeoxycytidine enhances upregulation of claudin-18 by TPA.[28]

Regulation of tricellular tight junctions in HPDE cells via a JNK pathway

The tricellular tight junction forms at the convergence of bicellular tight junctions where three epithelial cells meet in polarized epithelia, and it is required for the maintenance of the transepithelial barrier.[12] More recently, tricellulin was identified as the first marker of the tricellular tight junction in epithelial cells, and its loss affects the organization of the tricellular tight junction and the barrier function of epithelial cells.[12] Knockdown of occludin causes mislocalization of tricellulin to bicellular tight

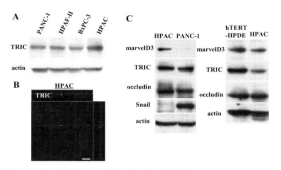

Figure 3. (A) Western blotting for TRIC in pancreatic cancer cell lines PANC-1, HPAF-II, BXPC-3, and HPAC. (B) A confocal laser scanning microscopic image for TRIC in HPAC cells. Bar: 10 μm. (C) Western blotting for marvelD3, TRIC, occludin, and snail in pancreatic cancer cell lines HPAC and PANC-1, and normal pancreatic duct epithelial hTERT-HPDE cells. TRIC, tricellulin.

junctions.[47] It has also been reported that tricellulin forms a barrier in tricellular tight junctions effective only for macromolecules and in bicellular tight junctions for solutes of all sizes.[48] We recently reported expression and localization of tricellulin in human nasal epithelial cells *in vivo* and *in vitro*.[49] Furthermore, tricellulin mRNA was also detected in human monocytic cell line THP-1.[50]

On the other hand, c-Jun N-terminal kinase (JNK) activation is essential for disassembly of adherens and tight junctions in human keratinocytes and colonic epithelial cells.[51,52] Recently, inhibition of JNK activity was shown to enhance epithelial barrier function through differential modulation of claudin expression in murine mammary epithelial cells.[53]

To investigate whether tricellulin is regulated via a JNK pathway, human pancreatic HPAC cells, highly expressed at tricellular contacts (Fig. 3A and B), were exposed to various stimuli such as the JNK activators anisomycin and the proinflammatory cytokines IL-1β, tumor necrosis factor (TNF)-α, and interleukin (IL)-1α.[27] Tricellulin expression and the barrier function were upregulated together with the activity of phospho-Rac1/cdc42 and phospho-JNK by treatment with the JNK activator anisomycin, and suppressed not only by inhibitors of JNK and PKC but also by siRNAs of tricellulin.[27] Tricellulin expression was induced by treatment with the proinflammatory cytokines IL-1β, TNF-α, and IL-1α, whereas the changes were inhibited by a JNK inhibitor.[27] Furthermore, in nor-

mal HPDE cells that were in hTERT-transfected primary culture, the responses of tricellulin expression to the various stimuli were similar to those in HPAC cells.[27] Tricellulin expression in tricellular tight junctions is strongly regulated together with the barrier function via the JNK transduction pathway. These findings suggest that JNK may be involved in the regulation of tricellular tight junctions, including tricellulin expression and the barrier function, during normal remodeling of epithelial cells, and may prevent disruption of the epithelial barrier in inflammation and other disorders in pancreatic duct epithelial cells.

Regulation of marvelD3 in HPDE cells during Snail-induced epithelial–mesenchymal transition

The transcription factor Snail has a key role in epithelial–mesenchymal transition (EMT) during development and in tumor progression via negative regulation of adherens and tight junctions, such as E-cadherin, claudins, and occludin.[54–57] EMT is characterized by a loss of cell–cell contact and apicobasal polarity, which are hallmarks of dysfunction of the tight junction fence.[58,59] The repression of tricellulin is also related to Snail-induced EMT in human gastric carcinoma.[60]

We first investigated marvelD3 expression in well- and poorly differentiated human pancreatic cancer cell lines and normal pancreatic duct epithelial cells into which the hTERT gene was introduced in primary culture. Furthermore, to investigate the changes of marvelD3 in well-differentiated pancreatic cancer HPAC cells during Snail-induced EMT, we used models of EMT *in vitro* subjected to hypoxia,[61,62] TGF-β treatment,[63,64] and knockdown of forkhead box transcription factor A2 (FOXA2) by siRNAs.[65,66]

MarvelD3 was transcriptionally downregulated in poorly differentiated pancreatic cancer cells.[30] It was also downregulated during Snail-induced EMT of pancreatic cancer cells in which Snail was highly expressed and the fence function downregulated, whereas it was maintained in well-differentiated human pancreatic cancer cells and normal pancreatic duct epithelial cells (Fig. 3C).[30] Depletion of marvelD3 by siRNAs in HPAC cells resulted in downregulation of barrier functions indicated as a decrease in transepithelial electric resistance and an increase of permeability to fluorescent dextran tracers, whereas it did not affect the fence function of tight junctions (Table 1).[30] These results suggested that marvelD3 is transcriptionally downregulated in Snail-induced EMT during the progression of pancreatic cancer.

Conclusion

By using hTERT-HPDE cells, we found that the expression of tight junction molecules, including claudins, and barrier functions in normal HPDE cells were regulated via a PKC signal pathway. The regulation of claudin-4 and claudin-18, which are highly expressed in the pancreatic cancers, in part differed from that of normal HPDE cells with regard to the types of PKC isoforms. Furthermore, claudin-18 in pancreatic cancer cells, but not normal HPDE cells, was modified by DNA methylation. The regulation of claudin-4 and claudin-18 in normal HPDE cells is essential as biomarkers for diagnosis and molecular therapeutic targets in pancreatic cancer. Thus, it is important that the regulation of MARVEL family members, including tricellulin and marvelD3, be further investigated and compared to that of the claudin family in human pancreatic cancer cells and normal pancreatic duct epithelial cells.

hTERT-HPDE cells have good growth potential and a long life span, and provide us with an indispensable and stable model for studying the regulation of tight junctions in normal HPDE cells.

Acknowledgments

This work was supported by the Suhara Memorial Foundation; the Pancreas Research Foundation of Japan; Grants-in-Aid from the National Project "Knowledge Cluster Initiative" (2nd stage, "Sapporo Biocluster Bio-S") Program for developing a supporting system for upgrading education and research; the Ministry of Education, Culture, Sports, Science, and Technology; and the Ministry of Health, Labour, and Welfare of Japan.

Conflicts of interest

The authors declare no conflicts of interest.

References

1. Jemal, A. *et al*. 2010. Cancer statistis. *CA Cancer J. Clin.* **60:** 277–300.
2. Michl, P. *et al*. 2003. Claudin-4 expression decreases invasiveness and metastatic potential of pancreatic cancer. *Cancer Res.* **63:** 6265–6271.
3. Karanjawala, Z.E. *et al*. 2008. New markers of pancreatic cancer identified through differential gene expression analyses: claudin 18 and annexin A8. *Am. J. Surg. Pathol.* **32:** 188–196.
4. Gumbiner, B.M. 1993. Breaking through the tight junction barrier. *J. Cell Biol.* **123:** 1631–1633.
5. Schneeberger, E.E & R.D. Lynch. 1992. Structure, function, and regulation of cellular tight junctions. *Am. J. Physiol.* **262:** L647–L661.
6. Van Meer, G. & K. Simons. 1986. The function of tight junctions in maintaining differences in lipid composition between the apical and the basolateral cell surface domains of MDCK cells. *EMBO J.* **5:** 1455–1464.
7. Cereijido, M. *et al*. 1998. Role of tight junctions in establishing and maintaining cell polarity. *Annu. Rev. Physiol.* **60:** 161–177.
8. Matter, K. & M.S. Balda. 2003. Signalling to and from tight junctions. *Nat. Rev. Mol. Cell Biol.* **4:** 225–237.
9. Schneeberger, E.E. & R.D. Lynch. 2004. The tight junction: a multifunctional complex. *Am. J. Physiol. Cell Physiol.* **286:** C1213–C1228.
10. Tsukita, S. *et al*. 2001. Multifunctional strands in tight junctions. *Nat. Rev. Mol. Cell Biol.* **2:** 285–293.
11. Sawada, N. *et al*. 2003. Tight junctions and human diseases. *Med. Electron. Microsc.* **36:** 147–156.
12. Ikenouchi, J. *et al*. 2005. Tricellulin constitutes a novel barrier at tricellular contacts of epithelial cells. *J. Cell Biol.* **171:** 939–945.
13. González-Mariscal, L. *et al*. 2008. Crosstalk of tight junction components with signaling pathways. *Biochim. Biophys. Acta.* **1778:** 729–756.
14. Kojima, T. *et al*. 2009. Tight junction proteins and signal transduction pathways in hepatocytes. *Histol. Histopathol.* **24:** 1463–1472.
15. Evans, M.J. *et al*. 2007. Claudin-1 is a hepatitis C virus co-receptor required for a late step in entry. *Nature* **446:** 801–805.
16. Van Itallie, C.M. & J.M. Anderson. 2006. Claudins and epithelial paracellular transport. *Annu. Rev. Physiol.* **68:** 403–429.
17. Fujita, K. *et al*. 2000. *Clostridium perfringens* enterotoxin binds to the second extracellular loop of claudin-3, a tight junction integral membrane protein. *FEBS Lett.* **476:** 258–261.
18. Michl, P. *et al*. 2001. Claudin-4: a new target for pancreatic cancer treatment using *Clostridium Perfringens* enterotoxin. *Gastroenterology* **121:** 678–684.
19. Sahin, U. *et al*. 2008. Claudin-18 splice variant 2 is a pan-cancer target suitable for therapeutic antibody development. *Clin. Cancer Res.* **14:** 7624–7634.
20. Suzuki, M. *et al*. 2009. Therapeutic antitumor efficacy of monoclonal antibody against claudin-4 for pancreatic and ovarian cancers. *Cancer Sci.* **100:** 1623–1630.
21. Katahira, J. *et al*. 1997. *Clostridium perfringens* enterotoxin utilizes two structurally related membrane proteins as functional receptors *in vivo*. *J. Biol. Chem.* **272:** 26652–26658.
22. Saeki, R. *et al*. 2009. A novel tumor-targeted therapy using a claudin-4-targeting molecule. *Mol. Pharmacol.* **76:** 918–926.

23. Sanchez-Pulido, L. *et al.* 2002. MARVEL: a conserved domain involved in membrane apposition events. *Trends Biochem. Sci.* **27**: 599–601.

24. Steed, E. *et al.* 2009. Identification of MarvelD3 as a tight junction-associated transmembrane protein of the occludin family. *BMC Cell Biol.* **10**: 95.

25. Raleigh, D.R. *et al.* 2010. Tight junction-associated MARVEL proteins marveld3, tricellulin, and occludin have distinct but overlapping functions. *Mol. Biol. Cell* **21**: 1200–1213.

26. Yamaguchi, H. *et al.* 2010. Transcriptional control of tight junction proteins via a protein kinase C signal pathway in human telomerase reverse transcriptase-transfected human pancreatic duct epithelial cells. *Am. J. Pathol.* **177**: 698–712.

27. Kojima, T. *et al.* 2010. c-Jun N-terminal kinase is largely involved in the regulation of tricellular tight junctions via tricellulin in human pancreatic duct epithelial cells. *J. Cell. Physiol.* **225**: 720–733.

28. Ito, T. *et al.* 2011. Transcriptional regulation of claudin-18 via specific protein kinase C signaling pathways and modification of DNA-methylation in human pancreatic cancer cells. *J. Cell. Biochem.* **112**: 1761–1772.

29. Kyuno, D. *et al.* 2011. Protein kinase Cα inhibitor enhances sensitivity of human pancreatic cancer HPAC cells to Clostridium perfringes enterotoxin via claudin-4. *Cell Tissue Res.* **346**: 369–381.

30. Kojima, T. *et al.* 2011. Downregulation of tight junction-associated MARVEL protein marvelD3 during epithelial-mesenchymal transition in human pancreatic cancer cells. *Exp. Cell Res.* **317**: 2288–2298.

31. Bodnar, A.G. *et al.* 1998. Extension of life-span by introduction of telomerase into normal human cells. *Science* **279**: 349–352.

32. Vaziri, H. & S. Benchimol. 1998. Reconstitution of telomerase activity in normal human cells leads to elongation of telomeres and extended replicative life span. *Curr. Biol.* **8**: 279–282.

33. Kurose, M. *et al.* 2007. Induction of claudins in passaged hTERT-transfected human nasal epithelial cells with an extended life span. *Cell Tissue Res.* **330**: 63–74.

34. Balda, M.S. *et al.* 1993. Assembly of the tight junction: the role of diacylglycerol. *J. Cell Biol.* **123**: 293–302.

35. Andreeva, A.Y. *et al.* 2006. Assembly of tight junction is regulated by the antagonism of conventional and novel protein kinase C isoforms. *Int. J. Biochem. Cell Biol.* **38**: 222–233.

36. Ellis, B. *et al.* 1992. Cellular variability in the development of tight junctions after activation of protein kinase C. *Am. J. Physiol.* **263**: F293–F300.

37. Sjö, A. *et al.* 2003. Distinct effects of protein kinase C on the barrier function at different developmental stages. *Biosci. Rep.* **23**: 87–102

38. Tymms, M.J. *et al.* 1997. A novel epithelial-expressed ETS gene, *ELF3*: human and murine cDNA sequences, murine genomic organization, human mapping to 1q32.2 and expression in tissues and cancer. *Oncogene* **15**: 2449–2462.

39. Jedlicka, P. & A. Gutierrez-Hartmann. 2008. Ets transcription factors in intestinal morphogenesis, homeostasis and disease. *Histol. Histopathol.* **23**: 1417–1424.

40. Kohno, Y. *et al.* 2006. Expression of claudin7 is tightly associated with epithelial structures in synovial sarcomas and regulated by an Ets family transcription factor, ELF3. *J. Biol. Chem.* **281**: 38941–38950.

41. Yano, K. *et al.* 2008. Transcriptional activation of the human claudin-18 gene promoter through two AP-1 motifs in PMA-stimulated MKN45 gastric cancer cells. *Am. J. Physiol. Gastrointest. Liver Physiol.* **294**: G336–G343.

42. Ali, A.S. *et al.* 2009. Exploitation of protein kinase C: a useful target for cancer therapy. *Cancer Treat Rev.* **35**: 1–8.

43. El-Rayes, B.F. *et al.* 2008. Protein kinase C: a target for therapy in pancreatic cancer. *Pancreas* **36**: 346–352.

44. Kominsky, S.L. *et al.* 2003. Loss of the tight junction protein claudin-7 correlates with histological grade in both ductal carcinoma in situ and invasive ductal carcinoma of the breast. *Oncogene* **22**: 2021–2033.

45. Boireau, S. *et al.* 2007. DNA-methylation-dependent alterations of claudin-4 expression in human bladder carcinoma. *Carcinogenesis* **28**: 246–258.

46. Agarwal, R. *et al.* 2009. Silencing of claudin-11 is associated with increased invasiveness of gastric cancer cells. *PLoS One* **4**: e8002.

47. Ikenouchi, J. *et al.* 2008. Loss of occludin affects tricellular localization of tricellulin. *Mol. Biol. Cell.* **19**: 4687–493.

48. Krug, S.M. *et al.* 2009. Tricellulin forms a barrier to macromolecules in tricellular tight junctions without affecting ion permeability. *Mol. Biol. Cell* **20**: 3713–3724.

49. Ohkuni, T. *et al.* 2009. Expression and localization of tricellulin in human nasal epithelial cells *in vivo* and *in vitro*. *Med. Mol. Morphol.* **42**: 204–211.

50. Ogasawara, N. *et al.* 2009. Induction of JAM-A during differentiation of human THP-1 dendritic cells. *Biochem. Biophys. Res. Commun.* **389**: 543–549.

51. Naydenov, N.G. *et al.* 2009. c-Jun N-terminal kinase mediates disassembly of apical junctions in model intestinal epithelia. *Cell Cycle* **8**: 2110–2121.

52. Lee, M.H. *et al.* 2009. JNK phosphorylates beta-catenin and regulates adherens junctions. *FASEB J.* **23**: 3874–3883.

53. Carrozzino, F. *et al.* 2009. Inhibition of basal p38 or JNK activity enhances epithelial barrier function through differential modulation of claudin expression. *Am. J. Physiol. Cell. Physiol.* **297**: C775–C787.

54. Nieto, M.A. 2002. The snail superfamily of zinc-finger transcription factors. *Nat. Rev. Mol. Cell Biol.* **3**: 155–166.

55. Batlle, E. 2000. The transcription factor snail is a repressor of E-cadherin gene expression in epithelial tumour cells. *Nat. Cell Biol.* **2**: 84–89.

56. Cano, A. *et al.* 2000. The transcription factor snail controls epithelial–mesenchymal transitions by repressing E-cadherin expression. *Nat. Cell Biol.* **2**: 76–83.

57. Ikenouchi, J. *et al.* 2003. Regulation of tight junctions during the epithelium-mesenchyme transition: direct repression of the gene expression of claudins/occludin by Snail. *J. Cell Sci.* **116**: 1959–1967.

58. Balda, M.S. *et al.* 1996. Functional dissociation of paracellular permeability and transepithelial electrical resistance and disruption of the apical-basolateral intramembrane

diffusion barrier by expression of a mutant tight junction membrane protein. *J. Cell Biol.* **134:** 1031–1049.

59. Lee, D.B. *et al.* 2006. Tight junction biology and kidney dysfunction. *Am. J. Physiol. Renal Physiol.* **290:** F20–F34.

60. Masuda, R. *et al.* 2010. Negative regulation of the tight junction protein tricellulin by snail-induced epithelial-mesenchymal transition in gastric carcinoma cells. *Pathobiology* **77:** 106–113.

61. Hotz, B. *et al.* 2007. Epithelial to mesenchymal transition: expression of the regulators snail, slug, and twist in pancreatic cancer. *Clin. Cancer Res.* **13:** 4769–4776.

62. Cannito, S. *et al.* 2008. Redox mechanisms switch on hypoxia-dependent epithelial-mesenchymal transition in cancer cells. *Carcinogenesis* **29:** 2267–2278.

63. Medici, D. *et al.* 2006. Cooperation between snail and LEF-1 transcription factors is essential for TGF-beta1-induced epithelial-mesenchymal transition. *Mol. Biol. Cell* **17:** 1871–1879.

64. Kojima, T. *et al.* 2008. Transforming growth factor-beta induces epithelial to mesenchymal transition by down-regulation of claudin-1 expression and the fence function in adult rat hepatocytes. *Liver Int.* **28:** 534–545.

65. Song, Y. *et al.* 2010. Loss of FOXA1/2 is essential for the epithelial-to-mesenchymal transition in pancreatic cancer. *Cancer Res.* **70:** 2115–2125.

66. Tang, Y. *et al.* 2011. FOXA2 functions as a suppressor of tumor metastasis by inhibition of epithelial-to-mesenchymal transition in human lung cancers. *Cell Res.* **21:** 316–326.

Ann. N.Y. Acad. Sci. ISSN 0077-8923

ANNALS OF THE NEW YORK ACADEMY OF SCIENCES

Issue: *Barriers and Channels Formed by Tight Junction Proteins*

The role for protein tyrosine phosphatase nonreceptor type 2 in regulating autophagosome formation

Michael Scharl[1,2,3] and Gerhard Rogler[2,3]

[1]Department of Oncology, [2]Division of Gastroenterology and Hepatology, University Hospital Zurich, Zurich, Switzerland. [3]Zurich Center for Integrative Human Physiology, University of Zurich, Zurich, Switzerland

Address for correspondence: Michael Scharl, M.D., Division of Gastroenterology and Hepatology, University Hospital Zurich, Rämistrasse 100, 8091 Zurich, Switzerland. michael.scharl@usz.ch

Genome-wide association studies have identified single nucleotide polymorphisms within the gene locus encoding protein tyrosine phosphatase nonreceptor type 2 (PTPN2) as a risk factor for the development of chronic inflammatory diseases, such as inflammatory bowel disease (IBD), type 1 diabetes, and rheumatoid arthritis. IBD is characterized by a breakdown of the intestinal epithelial barrier function leading to an overwhelming and uncontrolled immune response to bacterial antigens. Recent studies demonstrated that PTPN2 regulates cytokine-induced signaling pathways, epithelial barrier function, and cytokine secretion in human intestinal cells. Dysfunction of PTPN2 is also associated with impaired autophagosome formation and defective bacterial handling in intestinal cells. All of these cellular functions have been demonstrated to play a crucial role in the pathogenesis of IBD. The genetic variations within the PTPN2 gene may result in altered protein function, thereby essentially contributing to the onset and perpetuation of chronic inflammatory conditions in the intestine.

Keywords: PTPN2; Crohn's disease; autophagy; bacteria

Introduction

Inflammatory bowel disease (IBD) is mainly represented by its major subtypes, Crohn's disease (CD), and ulcerative colitis (UC). Although UC is restricted to the colon and reflects a continuous inflammation confined to the mucosal layer, CD is characterized by a discontinuous, granulomatuous, and transmural inflammation that can occur anywhere in the gastrointestinal tract. Genetic, immunological, and bacterial factors are believed to contribute essentially to the pathogenesis of IBD. In particular, an epithelial barrier defect, coupled with a dysfunctional immune response of the innate as well as the acquired immune system to commensal flora, results either in excessive upregulation or impaired downregulation of inflammatory events. This combination of pathogenetic events finally drives the development of chronic intestinal inflammation.[1] The predispositions seem to be genetically determined; genome-wide association studies (GWAS) have identified variations in 99 gene loci that have been associated with IBD.[2]

The surface of the intestinal tract is covered by intestinal epithelial cells (IEC) that form a barrier against luminal antigens and toxins and regulate essential immune responses. Of note, an epithelial defect causing increased epithelial barrier permeability may also contribute significantly to the major clinical symptom of IBD, diarrhea, due to leak flux of electrolytes and water into the lumen.[3] Bacteria and bacterial components play a crucial role for the onset and perpetuation of chronic intestinal inflammation.[1] Thus, the appropriate response to bacterial stimuli plays a key role for the maintenance of intestinal homeostasis. Many of the identified risk genes have been demonstrated to be critically involved in bacterial recognition, induction of antimicrobial factors, activation, and modulation of innate, as well as adaptive immune responses, and the maintenance of intestinal epithelial barrier function.

Autophagy

A recent GWAS confirmed a strong association of genetic variations in autophagy genes, such as

doi: 10.1111/j.1749-6632.2012.06578.x

autophagy-related 16-like 1 (ATG16L1), and the gene encoding immunity-related GTPase family M (*IRGM*), with CD but not with UC. Studies have demonstrated that autophagy is a fundamental process for bulk degradation of cytoplasmic compartments, damaged organelles, and/or misfolded proteins. In autophagosomes, these structures are sequestered into double-membrane–enclosed vesicles and delivered to lysosomes for final degradation.[4–6] Autophagy is activated by stress conditions, such as starvation or hypoxia, and dysfunction of autophagy has been implicated in numerous pathologies, such as cancer or neurodegeneration.[7] Autophagy is also critically involved in host defense against intracellular pathogens, such as *Listeria monocytogenes* (LM) or *Salmonella typhimurium*.[8–10] This observation further supports the hypothesis that abnormal immune responses to luminal bacteria or bacterial antigens play an essential role in the manifestation of CD.

Previous data have demonstrated that presence of genetic variation within autophagy genes results in defective bacterial handling, prolonged intracellular survival of pathogenic bacteria, and an uncontrolled inflammatory situation. The molecular target of rapamycine (mTOR) plays a central role in autophagosome formation;[11] for example, increased mTOR activation prevents autophagy, and prolonged activation of autophagy results in the activation of mTOR as part of a negative feedback mechanism finally inhibiting autophagy.[12] Following inhibition of mTOR, autophagosome formation is mediated by two highly conserved protein conjugation systems. The initial step is the activation of beclin-1, after which autophagosome assembly involves ATG12–ATG5 conjugation, which is catalyzed by ATG7 and ATG10. The resulting ATG5–ATG12 conjugate is stabilized by a noncovalent complex with ATG16L1. In addition to ATG7 and ATG3, this complex mediates the conversion of LC3B-I to LC3B-II, by lipidation with phosphatidyl-ethanolamine, into what finally establishes the formation of functional autophagosomes.[13]

Protein tyrosine phosphatase nonreceptor type 2

Variations within the gene locus encoding protein tyrosine phosphatase nonreceptor type 2 (PTPN2) have recently been associated with a number of chronic inflammatory diseases, such as CD, UC, type 1 diabetes (T1D), and rheumatoid arthritis (RA).[14,15] PTPN2 exists in two different splice forms, nuclear 45 kDa, and a cytoplasmic 48 kDa protein variant. Upon activation, the nuclear variant, which seems to be more important for the catalytic activity of PTPN2 than the 48 kDa variant, can exit the nucleus and accumulate in the cytoplasm.[16] Recent findings suggest that PTPN2 regulates growth factor receptor signaling, such as epidermal growth factor receptor (EGFR) and vascular endothelial growth factor receptor, a number of proinflammatory mediators, such as signal transducers and activators of transcription 1 and 3, as well as mitogen-activated protein kinases (MAPK) p38, c-Jun N-terminal kinase, and extracellular signal-regulated kinase 1/2 (ERK1/2).[17–22] Recently, PTPN2 has been implicated in the regulation of endoplasmatic reticulum stress and in T cell receptor signaling.[23,24]

Studies with PTPN2-deficient (PTPN2$^{-/-}$) mice have confirmed that PTPN2 is a key negative regulator of cytokine signaling. PTPN2$^{-/-}$ mice have elevated serum levels of interferon-gamma (IFN-γ) and tumor necrosis factor (TNF-α), as well as elevated production of nitric oxide. Interestingly, they also present with increased sensitivity to the bacterial wall component lipopolysaccharide, and die within five weeks of birth from severe diarrhea, anemia, and progressive systemic inflammatory disease.[25,26] The observations that PTPN2 is involved in both regulating proinflammatory signaling cascades and T cell receptor signaling suggest an important role for PTPN2 in the pathogenesis of autoinflammatory diseases, such as RA and T1D.[27] In particular, PTPN2 has been demonstrated to regulate insulin signaling by inactivating the insulin receptor.[20] On a functional level, loss of PTPN2 promotes cytokine-induced pancreatic β cell apoptosis[28] and regulates gluconeogenesis in the liver.[29] These observations provide the rationale for the association of dysfunction of PTPN2 and the onset of T1D.

PTPN2 and intestinal inflammation

Recent data have demonstrated that PTPN2 regulates IFN-γ–induced signaling and effects in cell models of chronic intestinal inflammation. Treatment of human T$_{84}$ IEC with IFN-γ increases PTPN2 mRNA and protein levels, elevates enzymatic PTPN2 activity, and causes cytoplasmic accumulation of PTPN2. Similar to those in fibroblasts, these effects are mediated via the cellular energy sensor adenosine-monophosphate activated protein kinase. Knockdown studies using PTPN2-specific siRNA constructs revealed that dephosphorylation of STAT1 and 3 is dependent on PTPN2 in these cells. As a functional consequence, loss of PTPN2 accentuates the epithelial barrier defect caused by IFN-γ, that is, PTPN2-deficiency enables IFN-γ to increase the expression of the pore-forming protein claudin-2.[30] In PTPN2-deficient human THP-1 monocytic cells, IFN-γ–induced activity of p38 MAPK and secretion of monocyte chemoattractant protein 1 and interleukin (IL)-6 were enhanced.[31] Further studies demonstrated that TNF-α induces PTPN2 protein and mRNA levels in T_{84} IEC via an NF-κB–dependent mechanism. PTPN2 in turn regulates TNF-α–induced ERK and p38 MAPK activity, as well as IL-6 and IL-8 secretion.[32] In EGF-treated T_{84} cells, PTPN2 regulates EGFR activity and attenuates the inhibitory effect of EGF on chloride secretion.[33] Of special interest is the observation that PTPN2 mRNA and protein levels are elevated in colonic biopsies of CD patients with active disease, compared to healthy controls.[30,32] Of note, heterozygous PTPN2 knockout mice develop a more severe colitis following administration of dextran sulphate sodium than do PTPN2 wild-type (WT) mice.[34] These findings strongly suggest that PTPN2 is crucial for maintaining intestinal homeostasis. In addition, loss of PTPN2 is associated with a number of events that have been shown to play an important role in IBD pathogenesis. These observations indicate how PTPN2 dysfunction could essentially contribute to the onset and perpetuation of chronic intestinal inflammation.

PTPN2 regulates activation and expression of autophagy-associated proteins

As mentioned earlier, PTPN2 is activated by administration of TNF-α or IFN-γ in IEC, and PTPN2 regulates EGFR signaling in T_{84} IEC.[30,32,33] Further, the EGFR-phosphatidyl-inositol 3-kinase (PI3K)-AKT pathway is known to regulate mTOR activation. Increased EGFR-PI3K-AKT signaling results in elevated mTOR activity and, ultimately, inhibition of autophagosome formation.[35] Therefore, the question arose whether PTPN2 would be able to regulate autophagy via modulating the EGFR-PI3K-AKT-

mTOR pathway. This was investigated in a recent study using PTPN2-deficient T_{84} IEC treated with TNF-α and IFN-γ.[36] Cytokine cotreatment significantly enhanced PTPN2 protein levels after 24-hour treatment and PTPN2 was shown to regulate activation of EGFR- and EGFR-dependent pathways. Of particular interest, the cytokine mix only revealed a limited effect on EGFR phosphorylation and on the association of the regulatory p110-subunit of PI3K to EGFR (a well-established marker for PI3K activity) in PTPN2-competent cells. However, loss of PTPN2 resulted in increased EGFR phosphorylation and in elevated p110 association to EGF, followed by elevated activation of AKT and mTOR (Fig. 1A).[36] These findings demonstrate that functional PTPN2 is necessary to control the activity of the autophagy inhibitor mTOR in human IEC, suggesting that PTPN2 might also be involved in autophagosome formation in these cells. Further, PTPN2 not only regulated activity of autophagy-related molecules but also the expression of autophagy-related genes. For example, loss of PTPN2 prevented the cytokine-induced increase in protein levels of the autophagy genes, beclin-1, ATG7, ATG5–ATG12 conjugates, and ATG16L1 in human T_{84} IEC.[36] These data provide further evidence for a pathophysiological role for PTPN2 in IBD, because PTPN2 is involved in regulating the activation of autophagy and the expression of autophagy-related proteins and dysfunction of PTPN2 might contribute to aberrant autophagosome formation in IEC as observed in IBD.

PTPN2 dysfunction results in abnormal autophagosome formation

Interestingly, recent data demonstrated that PTPN2 not only regulates the cytokine-induced activation and expression of autophagy-related molecules, but is also involved in the regulation of autophagosome formation in T_{84} IEC.[36] In particular, loss of PTPN2 in human IEC completely abrogated the increase in LC3B-II protein levels in response to TNF-α/IFN-γ cotreatment. Of note, LC3B-II protein level is a well-established marker for autophagosome formation.[37] By immunofluorescence analysis of LC3B-stained T_{84} cells, cytokine-treated PTPN2-competent cells contained more LC3B-positive (LC3B+) subcellular structures than untreated control cells, indicating increased autophagosome formation. In contrast, PTPN2-deficient cells showed only a small number of LC3B+ vesicles overall and TNF-α/IFN-γ

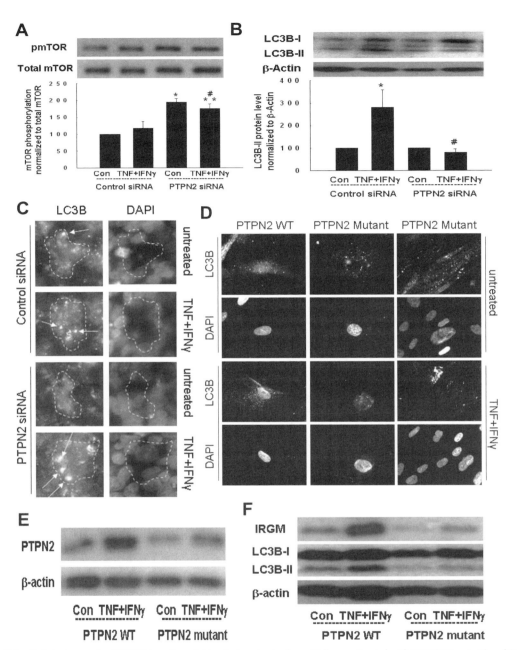

Figure 1. Deficient or variant PTPN2 impairs cytokine-induced autophagy. Cells were treated with TNF (100 ng/mL) and IFN-γ (100 ng/mL) for 24 hours. T_{84} cells were transfected with either nonspecific or PTPN2-specific siRNA. Representative Western blots show protein levels of (A) phosphor-mTOR (Ser^{2448}) and total mTOR ($n = 3$) and (B) LC3B, as well as of the respective loading control, β-actin. Histograms represent the densitometric analysis ($n = 3$ each). (C) Each panel shows representative images of LC3B- (on the left) and DAPI-stained (nuclear staining, on the right) cells ($n = 3$ per condition). Representative cells are circled by a dotted line and white arrows point onto LC3B$^+$ vesicles, indicative for autophagosomes. Magnification: 100-fold. (D) Each panel demonstrates representative images of primary intestinal PTPN2-WT or variant CLPF from CD patients. Magnification: 200-fold. Representative Western blots show protein levels of (E) PTPN2 and (F) IRGM or LC3B, respectively, as well as the corresponding loading control, β-actin in PTPN2-WT or variant CLPF from CD patients. Asterisks indicate significant differences from the respective control ($^*P < 0.05$, $^{**}P < 0.01$). $\#P < 0.05$ versus 24-h cytokine treatment of control siRNA cells. Data in this figure are from Ref. 36.

cotreatment caused the formation of fewer, but larger LC3B$^+$ vacuoles that were localized close to cell borders.[36]

The appearance of such abnormal, large autophagic vacuoles has been regarded as a marker of an ineffective formation of dysfunctional autophagosomes due to a defective autophagy process in these cells.[38] Overall, these observations indicated that PTPN2 activity is crucial for regulating autophagosome formation IEC.

Because these data have been obtained in human cell lines featuring an siRNA-induced PTPN2 knockdown, we further aimed to study the relevance of the CD-associated PTPN2 variant, rs2542151, with respect to autophagosome formation. Using primary colonic lamina propria fibroblasts (CLPF) isolated from CD patients featuring either the PTPN2-WT or the PTPN2-variant gene, it could be demonstrated that presence of the disease-associated PTPN2 variant exerts similar effects to siRNA-induced loss of PTPN2 expression.[36] In particular, by LC3B-immunofluorescence studies, increased numbers of LC3B$^+$ vesicles in PTPN2-WT cells following cytokine-treatment could be detected. In contrast, and similar to PTPN2 knockdown cells, a smaller amount of LC3B$^+$ vesicles was observed in PTPN2-variant CLPF. Further, the detectable LC3B$^+$ spots in variant CLPF had a similar appearance as those in PTPN2 knockdown cells, being large and bulky, indicative of defective autophagosome formation.[36] Presence of the CD-associated PTPN2 variant affected PTPN2 protein levels in CLPF: basal levels of PTPN2 protein were lower in PTPN2-variant CLPF, when compared to PTPN2-WT fibroblasts. Although PTPN2 protein levels were increased in PTPN2-WT cells in response to TNF-α/IFN-γ cotreatment, in PTPN2-variant CLPF, treatment with IFN-γ and TNF-α did not increase PTPN2 protein levels at all.[36] This observation demonstrates that presence of the disease-associated variants within the PTPN2 gene significantly affects PTPN2 expression and might thereby critically contribute to dysfunction of the gene product.

Looking at autophagy molecules in PTPN2-WT and variant CLPF, further significant differences were observed: the cytokine-induced increase in IRGM protein levels in PTPN2-WT cells was absent in cytokine-treated PTPN2-variant cells. In accordance with these observations, cytokine treatment elevated LC3B-II protein levels in PTPN2-WT CLPF, but had no effect in PTPN2-variant cells (Fig. 1B–F).[36] Taken together, these observations demonstrate that presence of the CD-associated variant, rs2542151, within the PTPN2 gene causes abnormal autophagosome formation in the setting of chronic inflammatory conditions. Further, they support the hypothesis that the disease-related PTPN2 variant can be regarded as a loss-of-function variation of the resulting gene product.

PTPN2 regulates autophagy in response to *L. monocytogenes*

Recent data strongly suggest that a defective handling of luminal and/or invading bacteria critically contributes to the onset of IBD. Dysfunction of PTPN2 results in impaired autophagosome formation, and impaired autophagy has been described to result in defective handling of invading bacteria. To further define the role for PTPN2 in regulating autophagy, and to further unravel a possible role for PTPN2 in IBD pathogenesis, it is essential to investigate whether loss of PTPN2 would result in ineffective clearance of invading bacteria due to the altered autophagosome formation.

Recent data indicated that treatment of PTPN2-competent cells with green fluorescent protein (GFP)–labeled *L. monocytogenes* activates autophagy.[36] This was demonstrated by increased numbers of LC3B$^+$ vesicles, indicative for autophagosome formation, and only a small number of visible GFP$^+$ dots, indicative for *L. monocytogenes*. However, PTPN2-deficient T$_{84}$ IEC revealed a strong GFP-staining, suggestive for the presence of a large number of intracellular bacteria, but only a small number of LC3B$^+$ dots in response to invading *L. monocytogenes*.[36]

These data could be confirmed on a protein level, showing decreased levels of LC3B-II in *L. monocytogenes*–treated PTPN2 knockdown cells. Interestingly, LC3B-II levels were clearly increased in PTPN2-WT CLPF upon stimulation with GFP-labeled *L. monocytogenes*, an effect that was not detectable in PTPN2-mutant cells.[36] These findings could be fully confirmed in LC3B-immunofluorescence studies. The presence of the CD-associated PTPN2 variant caused similar effects on autophagosome formation and the number of invading bacteria as does PTPN2 deficiency.[36] Correlating with observations in T$_{84}$ cells,

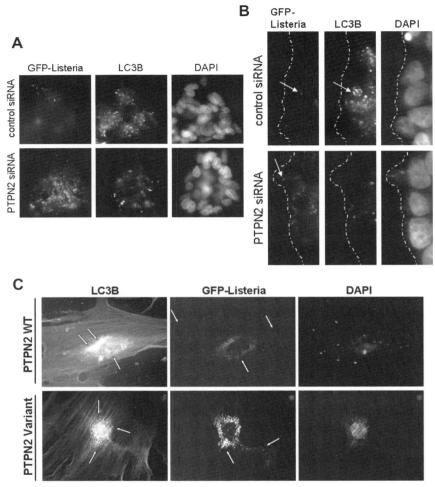

Figure 2. Deficient or mutant PTPN2 results in defective handling of intracellular *L. monocytogenes* (LM). (A, B) Each panel demonstrates representative images of T_{84} cells treated with GFP-labeled LM (MOI 0.1) for 24 hours. Images show GFP-labeled LM (left), LC3B- (middle), or DAPI-stained (nuclear staining, right) cells (*n* = 3 per condition). Dotted lines indicate apical cell surfaces and white arrows point onto GFP-labeled LM or LC3B+ dots, indicative for autophagosomes, respectively. Magnification: (A) 100-fold or (B) 200-fold. (C) Representative images of CLPF from CD patients treated with GFP-labeled LM (MOI 0.1) for 24 hours. Images show LC3B- (left), GFP-labeled LM (middle), or DAPI-stained (nuclear staining, right) cells (*n* = 3 per condition). White arrows point onto GFP-labeled LM or LC3B+ dots, indicative for autophagosomes, respectively. Magnification: 200-fold. Data in this figure are from Ref. 36.

treatment of PTPN2-WT cells with *L. monocytogenes* caused a strong increase in LC3B+ dots, which is indicative of autophagosome formation.[36] In contrast, only a limited number of GFP+ dots, indicative of *Listeriae*, could be observed. In PTPN2-variant CLPF, LC3B staining was clearly less detectable, while a strong GFP+ staining, suggestive of a large number of intracellular bacteria, could be observed (Fig. 2).[36] These data demonstrate that loss or genetically caused dysfunction of PTPN2 results in impaired autophagosome formation and defective handling of invading bacteria, and suggest how presence of the CD-associated PTPN2 variant within intestinal cells could critically contribute to the onset of IBD.

Association of PTPN2 SNP rs1893217 with CD

Recent data demonstrated that presence of the CD-associated variation rs2542151 within the PTPN2 gene locus results in aberrant PTPN2 protein expression and dysfunction of the phosphatase, resulting in impaired autophagosome formation and inadequate bacterial handling. However, the disease-associated SNP rs2542151 (A > C) on chromosome 18p11 is located about 30,000 bp upstream of the transcriptional start site of the PTPN2 gene within the PTPN2 locus region.[14] Therefore, it remains questionable whether a genetic variation at such a position might be solely responsible for the observed effects. However, consistent associations of SNP rs2542151 have been demonstrated with the onset of chronic inflammatory diseases, such as CD, UC, RA, and T1D.[14,15] As demonstrated by Todd *et al.*, a further SNP within the PTPN2 gene locus, namely SNP rs1893217 (A > G), is located within intron 7 of the PTPN2 gene, exists in linkage disequilibrium (LD; $r^2 = 1$) with SNP rs2542151 and shows an even stronger association with T1D as rs2542151. In addition, rs1893217 is independently associated with disease.[39]

A recent study investigated a possible association of the PTPN2 variant rs1893217 with CD. 663 non-IBD control patients and 343 CD patients (both

male and female patients) from a mixed Swiss, German, and Polish cohort were genotyped and analyzed with respect to a possible association of the SNPs rs2542151 and rs1893217 with CD. In the CD patient group there were significantly more homozygous rarer variant carriers detected in both SNP positions than in the non-IBD group ($P = 0.018$). However, no significant differences could be observed between the groups in the proportion of heterozygous carriers. Of note, both of the SNPs were in 100% linkage disequilibrium, the control group was in Hardy–Weinberg equilibrium and both of the SNPs featured the same allele frequency.[40] This new association of SNP rs1893217 with CD is of great interest, especially due to its location in an intronic region of the PTPN2 gene.

PTPN2 regulates muramyl-dipeptide (MDP)-induced autophagosome formation

Because the previously identified SNP rs2542151 is located about 30,000 bp upstream of the transcriptional start site of the PTPN2 gene, the functional relevance of the SNP is questionable. In contrast, the newly associated PTPN2 SNP rs1893217 is

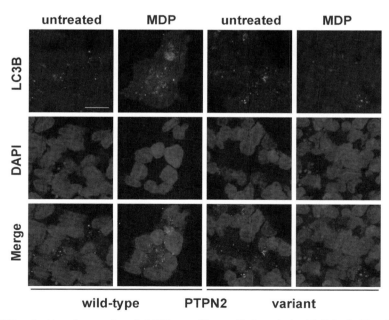

Figure 3. The PTPN2 variant impairs autophagy in MDP-treated human T_{84} intestinal epithelial cells. T_{84} cells, respectively, were transduced with either a mock-control vector or the PTPN2-variant rs1893217-containing vector. Cells were stimulated with MDP (100 ng/mL) for 24 hours. Each panel shows representative images of LC3B- and DAPI-stained (nuclear staining) cells. Each image is representative for three similar experiments per condition. Merged images are presented later. LC3B$^+$ vesicles are indicative for autophagosomes. Size bar indicates 10 μm. Magnification: 100-fold. Data in this figure are from Ref. 40.

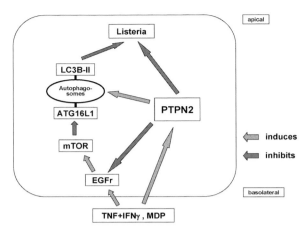

Figure 4. Schematic summary. TNF-α and IFN-γ, as well as MDP induce PTPN2 protein in IEC. PTPN2 inhibits cytokine-induced EGFR and, likely indirectly, mTOR activity, thereby promoting autophagy. Loss of PTPN2 results in increased mTOR activation causing impaired autophagy. Defective autophagy correlates with increased infection rates of invasive *L. monocytogenes*. Data in this figure are from Ref. 36.

located within intron 7 of the PTPN2 gene, and is more likely to affect PTPN2 function, for example, it could result in aberrant splicing. This was studied in subsequent experiments using the bacterial wall component MDP as a stimulus for autophagy induction.[40] MDP is the ligand of nucleotide-oligomerization domain 2 (NOD2), another CD risk gene that initiates cellular antibacterial responses by activating the innate immune system upon activation.[41–43] Autophagosome formation can be induced by MDP, and the presence of CD-associated NOD2-variants results in impaired autophagy.[9,10] The functional relevance of the rs1893217 SNP within the PTPN2 gene was studied by introducing either a mock or the PTPN2 variant containing viral vector constructs into T$_{84}$ cells.[40] Using this approach, more LC3B[+] signal was detected by immunofluorescence in 24-h MDP-treated PTPN2-WT (mock-transfected) IEC than in untreated IEC, indicating increased autophagosome formation.[40] In contrast, the PTPN2 variant carrying IEC revealed only a small number of LC3B[+] vesicles overall and MDP-treatment caused the formation of fewer, but larger, LC3B[+] vacuoles that were localized close to the cell borders (Fig. 3).[40] These abnormal structures might also indicate an ineffective formation of autophagosomes and/or formation of dysfunctional autophagosomes, resulting in a defective autophagy process in these cells. These observations indicate that presence

of rs1893217 results in dysfunction of the PTPN2 gene and suggest how genetic variations within the PTPN2 gene could contribute to the onset of disease.

Summary

Recent studies have demonstrated that PTPN2 regulates the expression of autophagy-related genes, as well as the formation of autophagosomes in human IEC and primary human CLPF. Knockdown of PTPN2 by specific siRNA constructs, or the presence of the CD-associated variants within the PTPN2 gene, most likely cause the formation of dysfunctional autophagosomes, which can result in defective bacterial clearance from intestinal cells. A dysregulated immune response to bacteria or bacterial components has been demonstrated to play a critical role for the onset and perpetuation of chronic intestinal inflammation. Furthermore, these observations fit well with previous findings highlighting a protective role for PTPN2 on the intestinal mucosa by maintaining intestinal epithelial barrier function and by limiting the effects of proinflammatory cytokines in intestinal mucosa cells.

L. monocytogenes accumulate in PTPN2-deficient or PTPN2-variant cells with aberrant autophagosome formation (Fig. 4). This observation is in accordance with the hypothesis that a defective cellular response to invading bacteria plays a crucial role for the pathogenesis of chronic intestinal inflammation. First, these cells are not able to adequately respond to invading pathogens, which cause prolonged bacterial survival within cells and, consequently, a state of cell activation. Second, the cells are not able to control the pathogen-induced inflammation because dysfunction of inflammation-"silencing" factors, such as PTPN2, renders them incapable of limiting the inflammatory reaction. These observations highlight the role for PTPN2 in maintaining cell homeostasis, and support the hypothesis that overexpression of a dysfunctional PTPN2 gene variant could make an important contribution to the onset and perpetuation of chronic inflammatory conditions as observed in many diseases.

Conflicts of interest

The authors declare no conflicts of interest.

References

1. Xavier, R.J. & D.K. Podolsky. 2007. Unravelling the pathogenesis of inflammatory bowel disease. *Nature* **448**: 427–434.
2. Lees, C.W. *et al.* 2011. New IBD genetics: common pathways with other diseases. *Gut* **60**: 1739–1753.
3. Hollander, D. 1999. Intestinal permeability, leaky gut, and intestinal disorders. *Curr. Gastroenterol. Rep.* **1**: 410–416.
4. Levine, B. & V. Deretic. 2007. Unveiling the roles of autophagy in innate and adaptive immunity. *Nat. Rev. Immunol.* **7**: 767–777.
5. Ohsumi, Y. 2001. Molecular dissection of autophagy: two ubiquitin-like systems. *Nat. Rev. Mol. Cell Biol.* **2**: 211–216.
6. Mizushima, N. *et al.* 2008. Autophagy fights disease through cellular self-digestion. *Nature* **451**: 1069–1075.
7. Levine, B. & G. Kroemer. 2008. Autophagy in the pathogenesis of disease. *Cell* **132**: 27–42.
8. Birmingham, C.L. *et al.* 2007. Listeria monocytogenes evades killing by autophagy during colonization of host cells. *Autophagy* **3**: 442–451.
9. Travassos, L.H. *et al.* 2010. Nod1 and Nod2 direct autophagy by recruiting ATG16L1 to the plasma membrane at the site of bacterial entry. *Nat. Immunol.* **11**: 55–62.
10. Cooney, R. *et al.* 2010. NOD2 stimulation induces autophagy in dendritic cells influencing bacterial handling and antigen presentation. *Nat. Med.* **16**: 90–97.
11. Ravikumar, B. *et al.* 2004. Inhibition of mTOR induces autophagy and reduces toxicity of polyglutamine expansions in fly and mouse models of Huntington disease. *Nat. Genet.* **36**: 585–595.
12. Yu, L. *et al.* 2010. Termination of autophagy and reformation of lysosomes regulated by mTOR. *Nature* **465**: 942–946.
13. Hanada, T. *et al.* 2007. The Atg12-Atg5 conjugate has a novel E3-like activity for protein lipidation in autophagy. *J. Biol. Chem.* **282**: 37298–37302.
14. The Wellcome Trust Case Control Consortium. 2007. Genome-wide association study of 14,000 cases of seven common diseases and 3,000 shared controls. *Nature* **447**: 661–678.
15. Franke, A. *et al.* 2008. Replication of signals from recent studies of Crohn's disease identifies previously unknown disease loci for ulcerative colitis. *Nat. Genet.* **40**: 713–715.
16. Lam, M.H. *et al.* 2001. Cellular stress regulates the nucleocytoplasmic distribution of the protein-tyrosine phosphatase TCPTP. *J. Biol. Chem.* **276**: 37700–37707.
17. Tiganis, T. *et al.* 1998. Epidermal growth factor receptor and the adaptor protein p52Shc are specific substrates of T cell protein tyrosine phosphatase. *Mol. Cell Biol.* **18**: 1622–1634.
18. Van Vliet, C. *et al.* 2005. Selective regulation of tumor necrosis factor-induced Erk signaling by Src family kinases and the T cell protein tyrosine phosphatase. *Nat. Immunol.* **6**: 253–260.
19. Mattila, E. *et al.* 2008. The protein tyrosine phosphatase TCPTP controls VEGFR2 signaling. *J. Cell Sci.* **121**: 3570–3580.
20. Galic, S. *et al.* 2003. Regulation of insulin receptor signaling by the protein tyrosine phosphatase TCPTP. *Mol. Cell Biol.* **23**: 2096–2108.
21. Ten Hoeve, J. *et al.* 2002. Identification of a nuclear Stat1 protein tyrosine phosphatase. *Mol. Cell Biol.* **22**: 5662–5668.
22. Yamamoto, T. *et al.* 2002. The nuclear isoform of protein-tyrosine phosphatase TC-PTP regulates interleukin-6-mediated signaling pathway through STAT3 dephosphorylation. *Biochem. Biophys. Res. Commun.* **297**: 811–817.
23. Bettaieb, A. *et al.* 2011. Differential regulation of endoplasmic reticulum stress by protein tyrosine phosphatase 1B and T cell protein tyrosine phosphatase. *J. Biol. Chem.* **286**: 9225–9235.
24. Wiede, F. *et al.* 2011. T cell protein tyrosine phosphatase attenuates T cell signaling to maintain tolerance in mice. *J. Clin. Invest.* **121**: 4758–4774.
25. Heinonen, K.M. *et al.* 2004. T cell protein tyrosine phosphatase deletion results in progressive systemic inflammatory disease. *Blood* **103**: 3457–3464.
26. You-Ten, K.E. *et al.* 1997. Impaired bone marrow microenvironment and immune function in T cell protein tyrosine phosphatase-deficient mice. *J. Exp. Med.* **186**: 683–693.
27. Zikherman, J. & A. Weiss. 2011. Unraveling the functional implications of GWAS: how T cell protein tyrosine phosphatase drives autoimmune disease. *J. Clin. Invest.* **121**: 4618–4621.
28. Moore F. *et al.* 2009. PTPN2, a candidate gene for type 1 diabetes, modulates interferon-gamma-induced pancreatic beta-cell apoptosis. *Diabetes* **58**: 1283–1291.
29. Fukushima, A. *et al.* 2010. T cell protein tyrosine phosphatase attenuates STAT3 and insulin signaling in the liver to regulate gluconeogenesis. *Diabetes* **59**: 1906–1914.
30. Scharl, M. *et al.* 2009. Protection of epithelial barrier function by the Crohn's disease associated gene protein tyrosine phosphatase n2. *Gastroenterology* **137**: 2030–2040, e2035.
31. Scharl, M., P. Hruz & D.F. McCole. 2010. Protein tyrosine phosphatase non-receptor Type 2 regulates IFN-gamma-induced cytokine signaling in THP-1 monocytes. *Inflamm. Bowel Dis.* **16**: 2055–2064.
32. Scharl, M. *et al.* 2011. Protein tyrosine phosphatase N2 regulates TNF{alpha}-induced signaling and cytokine secretion in human intestinal epithelial cells. *Gut* **60**: 189–197.

33. Scharl, M., I. Rudenko & D.F. McCole. 2010. Loss of protein tyrosine phosphatase N2 potentiates epidermal growth factor suppression of intestinal epithelial chloride secretion. *Am. J. Physiol. Gastrointest. Liver Physiol.* **299:** G935–G945.

34. Hassan, S.W. *et al.* 2010. Increased susceptibility to dextran sulfate sodium induced colitis in the T cell protein tyrosine phosphatase heterozygous mouse. *PLoS One* **5:** e8868.

35. Galbaugh, T. *et al.* 2006. EGF-induced activation of Akt results in mTOR-dependent p70S6 kinase phosphorylation and inhibition of HC11 cell lactogenic differentiation. *BMC Cell Biol.* **7:** 34.

36. Scharl, M. *et al.* 2011. Protein tyrosine phosphatase nonreceptor type 2 regulates autophagosome formation in human intestinal cells. *Inflamm. Bowel Dis.* doi: 10.1002/ibd.21891. [Epub ahead of print: Oct. 10].

37. Klionsky, D.J. *et al.* 2008. Guidelines for the use and interpretation of assays for monitoring autophagy in higher eukaryotes. *Autophagy* **4:** 151–175.

38. Maiuri, M.C. *et al.* 2007. Self-eating and self-killing: crosstalk between autophagy and apoptosis. *Nat. Rev. Mol. Cell Biol.* **8:** 741–752.

39. Todd, J.A. *et al.* 2007. Robust associations of four new chromosome regions from genome-wide analyses of type 1 diabetes. *Nat. Genet.* **39:** 857–864.

40. Scharl, M. *et al.* 2011. Crohn's disease-associated polymorphism within the PTPN2 gene affects muramyl-dipeptide-induced cytokine secretion and autophagy. *Inflamm. Bowel Dis.* **18:** 900–912.

41. Girardin, S.E. *et al.* 2003. Nod2 is a general sensor of peptidoglycan through muramyl dipeptide (MDP) detection. *J. Biol. Chem.* **278:** 8869–8872.

42. Inohara, N. *et al.* 2003. Host recognition of bacterial muramyl dipeptide mediated through NOD2. Implications for Crohn's disease. *J. Biol. Chem.* **278:** 5509–5512.

43. Hisamatsu, T. *et al.* 2003. CARD15/NOD2 functions as an antibacterial factor in human intestinal epithelial cells. *Gastroenterology* **124:** 993–1000.

Conflicts of interest

The authors declare no conflicts of interest.

References

1. Feltkamp, C.A. & A.W. Van der Waerden. 1983. Junction formation between cultured normal rat hepatocytes. An ultrastructural study on the presence of cholesterol and the structure of developing tight-junction strands. *J. Cell Sci.* **63:** 271–286.

2. Nusrat, A. *et al.* 2000. Tight junctions are membrane microdomains. *J. Cell Sci.* **113:** 1771–1781.

3. Lajoie, P. *et al.* 2009. Lattices, rafts, and scaffolds: domain regulation of receptor signaling at the plasma membrane. *J. Cell Biol.* **185:** 381–385.

4. Lynch, R.D. *et al.* 2007. Cholesterol depletion alters detergent-specific solubility profiles of selected tight junction proteins and the phosphorylation of occludin. *Exp. Cell Res.* **313:** 2597–2610.

5. Lambert, D., C.A. O'Neill & P.J. Padfield. 2005. Depletion of Caco-2 cell cholesterol disrupts barrier function by altering the detergent solubility and distribution of specific tight-junction proteins. *Biochem. J.* **387:** 553–560.

6. Schubert, W. *et al.* 2002. Microvascular hyperpermeability in caveolin-1 (-/-) knock-out mice. Treatment with a specific nitric-oxide synthase inhibitor, L-NAME, restores normal microvascular permeability in Cav-1 null mice. *J. Biol. Chem.* **277:** 40091–40098.

7. Razani, B. *et al.* 2002. Caveolin-1-deficient mice are lean, resistant to diet-induced obesity, and show hypertriglyceridemia with adipocyte abnormalities. *J. Biol. Chem.* **277:** 8635–8647.

8. Shen, L. & J.R. Turner 2005. Actin depolymerization disrupts tight junctions via caveolae-mediated endocytosis. *Mol. Biol. Cell.* **16:** 3919–3936.

9. Marchiando, A.M. *et al.* 2010. Caveolin-1-dependent occludin endocytosis is required for TNF-induced tight junction regulation in vivo. *J. Cell Biol.* **189:** 111–126.

10. Furuse, M. *et al.* 1994. Direct association of occludin with ZO-1 and its possible involvement in the localization of occludin at tight junctions. *J. Cell Biol.* **127:** 1617–1626.

11. Fanning, A.S. & J.M. Anderson. 2009. Zonula occludens-1 and -2 are cytosolic scaffolds that regulate the assembly of cellular junctions. *Ann. N. Y. Acad. Sci.* **1165:** 113–120.

12. Shen, L., C.R. Weber & Turner, J.R. 2008. The tight junction protein complex undergoes rapid and continuous molecular remodeling at steady state. *J. Cell Biol.* **181:** 683–695.

13. Raleigh, D.R. *et al.* 2011. Occludin S408 phosphorylation regulates tight junction protein interactions and barrier function. *J. Cell Biol.* **193:** 565–582.

14. Elias, B.C. *et al.* 2009. Phosphorylation of Tyr-398 and Tyr-402 in occludin prevents its interaction with ZO-1 and destabilizes its assembly at the tight junctions. *J. Biol. Chem.* **284:** 1559–1569.

15. Van Itallie, C.M., L.L. Mitic & J.M. Anderson. 2011. Claudin-2 forms homodimers and is a component of a high molecular weight protein complex. *J. Biol. Chem.* **286:** 3442–3450.

16. Colegio, O.R. *et al.* 2003. Claudin extracellular domains determine paracellular charge selectivity and resistance but not tight junction fibril architecture. *Am. J. Physiol. Cell Physiol.* **284:** C1346–1354.

17. Van Itallie, C.M., O.R. Colegio & J.M. Anderson. 2004. The cytoplasmic tails of claudins can influence tight junction barrier properties through effects on protein stability. *J. Memb. Biol.* **199:** 29–38.

18. Couet, J. *et al* 1997. Identification of Peptide and Protein Ligands for the Caveolin-scaffolding Domain. *J. Biol. Chem.* **272:** 6525–6533.

19. Bruewer, M. *et al.* 2005. Interferon-gamma induces internalization of epithelial tight junction proteins via a macropinocytosis-like process. *Faseb. J.* **19:** 923–933

20. Ikari, A. *et al.* 2011. Epidermal growth factor increases clathrin-dependent endocytosis and degradation of claudin-2 protein in MDCK II cells. *J. Cell. Physiol.* **226:** 2448–2456.

Ann. N.Y. Acad. Sci. ISSN 0077-8923

ANNALS OF THE NEW YORK ACADEMY OF SCIENCES

Issue: *Barriers and Channels Formed by Tight Junction Proteins*

Regulation of epithelial barrier function by the inflammatory bowel disease candidate gene, *PTPN2*

Declan F. McCole

Department of Medicine, University of California, San Diego, School of Medicine, La Jolla, California

Address for correspondence: Declan F. McCole, Ph.D., Division of Gastroenterology, University of California, San Diego, 9500 Gilman Drive, La Jolla, CA 92093-0063. dmccole@ucsd.edu

Protein tyrosine phosphatase nonreceptor type 2 (*PTPN2*) has been identified as an inflammatory bowel disease (IBD) candidate gene. However, the mechanism through which mutations in the *PTPN2* gene contribute to the pathogenesis of IBD has not been identified. *PTPN2* acts as a negative regulator of signaling induced by the proinflammatory cytokine, interferon-gamma (IFN-γ). IFN-γ is known not only to play an important role in the pathogenesis of Crohn's disease (CD), but also to increase permeability of the intestinal epithelial barrier. We have shown that *PTPN2* protects epithelial barrier function by restricting the capacity of IFN-γ to increase epithelial permeability and prevent induction of expression of the pore-forming protein, claudin-2. These data identify an important functional role for *PTPN2* as a protector of the intestinal epithelial barrier and provide clues as to how *PTPN2* mutations may contribute to the pathophysiology of CD.

Keywords: IFN-γ; STAT proteins; claudin-2; Crohn's disease

Introduction

The protein product of the protein tyrosine phosphatase nonreceptor type 2 (*PTPN2*) gene, referred to as T cell protein tyrosine phosphatase (TCPTP), is a ubiquitously expressed classical phosphatase that was originally cloned in T lymphocytes.[1,2] TCPTP exists in two different splice forms: a 45 kD (TC-45) isoform that contains a nuclear localization signal and can shuttle between the nucleus and the cytoplasm, and a cytoplasmic 48 kD (TC-48) variant that is targeted to the endoplasmic reticulum.[3–5] Several important tyrosine kinase receptors and signaling intermediates are recognized substrates known to be dephosphorylated by TCPTP. These include the epidermal growth factor receptor, the insulin receptor, and the signal transducers and activators of transcription (STAT) 1 and STAT3.[3,4,6–9] Thus, TCPTP acts as an important negative regulator of a number of signaling events, including signaling mediated by the inflammatory cytokines, interferon-gamma (IFN-γ), and interleukin (IL)-6.[8,10]

Inflammatory bowel disease (IBD), comprising Crohn's disease (CD) and ulcerative colitis (UC), is a chronic intestinal inflammatory condition that arises through a combination of genetic, immunological, bacterial, and environmental factors.[11] There is increasing evidence of a major role for an epithelial barrier defect, coincident with an inappropriate immune response to commensal bacteria, as a critical contributor to the development of chronic intestinal inflammation in the pathogenesis of IBD.[12] In addition to allowing increased access of bacteria and bacterial products to the lamina propria and triggering an immune cell response, decreased epithelial barrier function may also have a role in the loss of fluid and electrolytes into the lumen and thus contribute directly to the major clinical symptom of CD, diarrhea. With respect to their pathologies, both CD and UC are characterized by specific cytokine profiles with CD in particular being associated with elevated levels of IFN-γ.[12,13] IFN-γ is the major effector cytokine of Th1- and, possibly, Th17-mediated immune responses.[14] With respect to its involvement in IBD pathogenesis, in addition

doi: 10.1111/j.1749-6632.2012.06522.x

to its role(s) in tissue destruction, IFN-γ can also induce more discreet effects on intestinal epithelial cells by reducing epithelial barrier function via signaling pathways that lead to a reconfiguration of tight junctions (TJs).[15,16]

Recent advances in our understanding of genetic mutations associated with IBD, starting with the discovery of the nucleotide-binding oligomerization domain-containing protein (NOD)-2 mutation through to multiple genome-wide array studies, have greatly illuminated the genetic component of IBD pathogenesis. A number of studies have identified single nucleotide polymorphisms (SNP) in the gene locus encoding *PTPN2* that are associated not just with CD and UC, but also with type I diabetes and celiac disease.[17–20] However, the functional role(s) for TCPTP and mutations in the *PTPN2* gene in the pathogenesis of these conditions have not been determined. Given the role of TCPTP as a negative regulator of IFN-γ signaling, we therefore set out to investigate if *PTPN2* (TCPTP) plays a protective role in preserving a critical function of intestinal epithelial cells, the ability to form an effective barrier, from the deleterious effects of IFN-γ.

We identified that TCPTP is upregulated both in intestinal epithelial cell (IECs) following IFN-γ treatment and in CD patient biopsies. We also determined that TCPTP acts as a protective factor to restrict IFN-γ–induced epithelial barrier dysfunction in part through prevention of expression of the pore-forming protein, claudin-2. These data suggest that *PTPN2* mutations, leading to a loss of function, could contribute to the pathogenesis of IBD.

Methods

Human colonic T_{84} and HT29.cl19A epithelial cell lines were cultured on 12 mm Millicell-HA semipermeable filter supports. siRNA transfection was performed using the Amaxa (Lonza; Allendale, NJ, USA) nucleofector system. Human studies were conducted with full Institutional Review Board approval. Western blotting, real-time polymerase chain reaction, immunofluorescence staining, and fluorescein isothiocyanate (FITC)-dextran (10 kD) permeability assays were performed as described previously.[21] Transepithelial electrical resistance (TER) was measured using an Evom voltohmeter (World Precision Instruments, Inc., Sarasota, FL, USA).

Results

TCPTP expression is increased in CD and by IFN-γ in IEC

To establish a clinical association between TCPTP expression and IBD, we determined TCPTP mRNA levels in human biopsies obtained from (1) macroscopically noninflamed areas of the terminal ileum, colon, or rectum of CD patients in clinical and macroscopic remission; (2) macroscopically inflamed areas of the terminal ileum and colon of CD patients with clinically and macroscopically active disease; or (3) healthy control subjects undergoing routine colon cancer screening. We found that TCPTP mRNA was significantly elevated in tissue samples of CD patients with acute inflammation ($P < 0.05$ vs. control), but not in CD patients in remission.[21] We next investigated whether a major inflammatory cytokine involved in CD, IFN-γ, directly affects TCPTP expression in intestinal epithelial cells. Confluent monolayers of T_{84} cells, a colonic crypt epithelial cell line, were treated with IFN-γ (10; 100; 1000; 3000 U/mL) for 24 hours. IFN-γ significantly increased TCPTP mRNA expression with a peak occurring at 1,000–3,000 U/mL.[21] Using 1,000 U/mL of IFN-γ, a concentration that has been detected in biopsy cultures from a patient with active CD,[22] we also observed that a significant increase in TCPTP mRNA occurred not just at 24 h but also at 48- and 72-h posttreatment versus control cells treated with medium alone. Protein expression was also assessed by Western blotting. Dose–response studies revealed that a concentration of 1,000 U/mL IFN-γ caused a maximal increase in TCPTP protein by 24-h IFN-γ treatment. A sustained increase in cytoplasmic TCPTP protein was observed from 24 to 72 hours, but a drop in nuclear TCPTP occurred after 24 h.[21] The decrease in nuclear TCPTP at later time points correlated with immunofluorescent staining, which indicated nuclear exit of TCPTP at 72 h (Fig. 1).

TCPTP activity negatively regulates IFN-γ signaling in IECs

In addition to demonstrating that IFN-γ increases transcription and expression of TCPTP, we also investigated whether IFN-γ alters the enzymatic activity of TCPTP. We observed that IFN-γ increases TCPTP activity maximally within 72 hours of treatment in TCPTP immunoprecipitates isolated from T_{84} whole cell lysates.[21] To study the

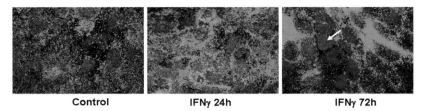

Control   **IFNγ 24h**   **IFNγ 72h**

Figure 1. IFN-γ affects TCPTP distribution over time in T84 IEC. Confocal microscopy shows TCPTP (green) distribution in control (24 hours) and IFN-γ (1000 U/mL) treated (24 and 72 hours) T84 cells. Nuclear staining is blue. Individual panels show an image for one experiment representative of three separate experiments for each condition. TCPTP is increased in both cytoplasmic and nuclear compartments after 24-hour IFN-γ but a loss of nuclear TCPTP occurs by 72 hours (arrow).

effects of TCPTP activity on IFN-γ signaling, we examined the tyrosine phosphorylation status of the IFN-γ downstream targets, and TCPTP substrates, STAT1 and STAT3. IFN-γ treatment for 24 h increased both cytoplasmic and nuclear STAT1 phosphorylation on tyrosine,[701] but also elevated STAT1 expression as determined by Western blotting.[21] Nuclear STAT1 phosphorylation declined markedly at 48-hour IFN-γ treatment and returned to control levels by 72 h, suggesting that TCPTP significantly dephosphorylates STAT1 in the nucleus as early as 24–48 h after IFN-γ treatment ($P < 0.001$; not shown). In contrast, cytoplasmic STAT1 phosphorylation was still significantly elevated even at 48-h treatment ($P < 0.05$). This difference in dephosphorylation in nuclear versus cytoplasmic fractions may partly be explained by the increased enzymatic activity of the nuclear 45 kD isoform of TCPTP.[8] Dephosphorylation of cytoplasmic p-STAT1 after 48 h may also be assisted by nuclear exit of the 45 kD form of TCPTP. By extension, the delayed increase in TCPTP activity, which conflicts with increased TCPTP expression at 24 h, may be due to whole cell lysates used for the TCPTP activity assay containing only a small portion of the nuclear protein fraction (not shown). Similar data were obtained regarding IFN-γ–induced phosphorylation of tyrosine[705] on cytoplasmic STAT3.[21] Overall, these data indicated that basal and IFN-γ–induced TCPTP activity correlates with decreased phosphorylation of STAT proteins.

We next investigated whether knockdown of TCPTP influenced IFN-γ–induced STAT1 and STAT3 activity. T_{84} cells were transfected with either *PTPN2*-specific siRNA or nonspecific control siRNA sequences, cultured on semipermeable supports for 48 h and then treated with IFN-γ (24 hours) or medium. Transfection of *PTPN2* siRNA

reduced TCPTP expression with a maximal reduction of $92 \pm 3\%$ (Fig. 2A), while levels of the nuclear envelope protein lamin A/C and total STAT1 and STAT3 were unaffected (Fig. 2A–C). As expected, IFN-γ increased TCPTP protein (Fig. 3A) as well as STAT1 and STAT3 phosphorylation, while nuclear and cytoplasmic STAT1 and 3 phosphorylation were significantly increased in TCPTP-deficient cells treated with IFN-γ (Fig. 2B and C). siRNA-induced knockdown of TCPTP in HT29cl.19a cells revealed the same regulatory influence of TCPTP on STAT1 phosphorylation (not shown).[21] These data suggest that TCPTP is a key negative regulator of IFN-γ–induced STAT signaling in IEC.

Loss of TCPTP elevates epithelial permeability

We next investigated the role of TCPTP in regulating IFN-γ–induced alterations in barrier function. We measured epithelial permeability by quantifying the flux of 10 kD FITC-Dextran across T_{84} and HT29cl.19a monolayers, transfected with either control or *PTPN2*-specific siRNA and subsequently treated with IFN-γ (72 hours). Although IFN-γ significantly increased permeability in cells transfected with nonspecific siRNA, TCPTP knockdown further exacerbated IFN-γ–induced barrier defects in both T_{84} ($P < 0.001$; $n = 3$; Fig. 3A) and HT29cl.19A ($P < 0.01$; $n = 4$) cell lines.[21] Moreover, TCPTP knockdown further exaggerated the IFN-γ–induced decrease in TER across T_{84} monolayers ($P < 0.05$; $n = 3$; Fig. 3B). These data indicate that TCPTP plays a key role in restricting intestinal epithelial barrier defects induced by inflammatory cytokines.

We subsequently explored molecular targets that may be affected more prominently in TCPTP-deficient cells versus TCPTP-competent cells in

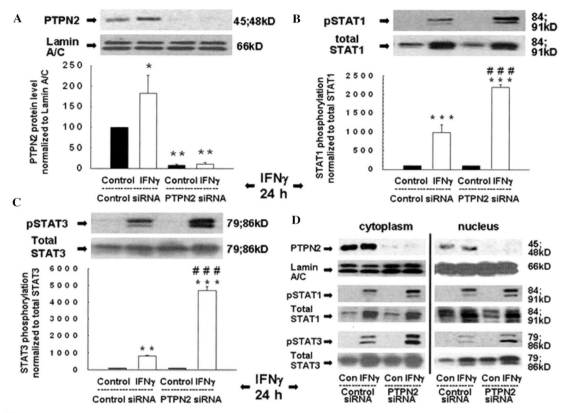

Figure 2. Loss of TCPTP promotes IFN-γ–stimulated STAT activation. T84 cells were transfected with either nonspecific siRNA or *PTPN2* siRNA and treated with IFN-γ (24 h) before generation of whole cell lysates. (A) Representative Western blots and densitometry for TCPTP (*PTPN2*) and lamin A/C ($n = 3$); (B, C) STAT1 and STAT3 phosphorylation and expression, respectively ($n = 3$). Asterisks indicate significant differences versus the respective control (*$P < 0.05$, **$P < 0.01$, ***$P < 0.001$). ###$P < 0.001$ versus 24-h IFN-γ treatment of control siRNA-transfected cells. (D) Representative Western blots showing TCPTP (*PTPN2*) and lamin A/C expression, STAT1 and STAT3 expression and phosphorylation, in cytoplasmic and nuclear lysates ($n = 2$). Reproduced from Ref. 21, with permission from Elsevier.

response to IFN-γ and may contribute to the greater defect in barrier function. Members of the claudin family can act either as sealing proteins, to enhance barrier function, or to form cation-selective pores that increase the electrolyte permeability of the TJ. Claudin-2 is a pore-forming protein that plays an important role in regulating epithelial permeability in CD and UC and whose expression is increased in IBD.[23–25] IFN-γ does not increase claudin-2 expression in IEC lines and we observed that in T$_{84}$ cells transfected with control siRNA, IFN-γ treatment did not affect claudin-2 expression.[25,26] However, in parallel with the rise in epithelial permeability, IFN-γ significantly increased expression of claudin-2 in TCPTP-deficient cells (Fig. 3C). The TJ protein, occludin, and the TJ-associated molecule zonula occludens-1 (ZO-1) are known to be affected by inflammatory cytokines, including IFN-γ.[27,28] Although the expression of ZO-1 and occludin was reduced by IFN-γ as expected, levels were not further decreased by TCPTP knockdown. We also examined expression of claudin-4, a prominent sealing claudin whose expression is decreased in IBD, and observed that claudin-4 expression was not affected by TCPTP knockdown with or without IFN-γ treatment (Fig. 3D).[25] Overall, these data suggest that loss of TCPTP uncovers IFN-γ–stimulated expression of claudin-2, and this may contribute to increased permeability in TCPTP-deficient cells treated with IFN-γ and possibly to the increased permeability associated with in IBD. These data thus identify a novel role for

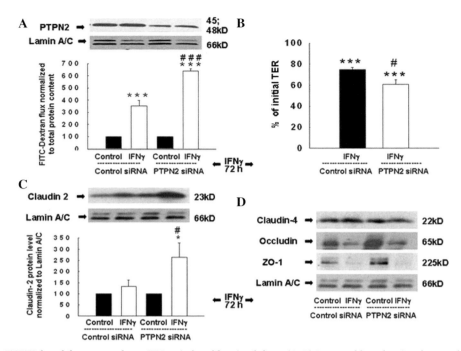

Figure 3. TCPTP knockdown exacerbates IFN-γ–induced barrier defects. (A, B) Western blots showing decreased TCPTP but not lamin A/C protein in TCPTP-deficient T84 cells after IFN-γ (72 h) treatment. (A) FITC-dextran flux ($n = 3$) and (B) TER ($n = 3$) across T84 monolayers transfected with control or *PTPN2* siRNA. (C) Western blot and densitometric analysis of claudin-2 expression in control or *PTPN2* siRNA-transfected cells in response to IFN-γ ($n = 3$). (D) Western blots of claudin-4, occludin, and ZO-1 expression in control or *PTPN2* siRNA-transfected T84 cells ($n = 3$). Asterisks indicate significant differences versus the respective control (*$P < 0.05$, ***$P < 0.001$). #$P < 0.05$, ###$P < 0.001$ versus 72-h IFN-γ treatment of control siRNA cells. Adapted from Ref. 21, with permission from Elsevier.

PTPN2 (TCPTP) in protecting epithelial barrier function.

Discussion

We observed that TCPTP plays an important role in negatively regulating IFN-γ–induced signaling as assessed by inhibition of STAT signaling, in intestinal epithelial cells, and is thus in agreement with previous studies demonstrating that TCPTP exerts a similar function in immune cells.[3,8,9] Moreover, *PTPN2*[−/−] mice die within three–five weeks from progressive systemic inflammation accompanied by elevated serum levels of IFN-γ and TNF-α.[29] These data were supported by *PTPN2* siRNA knockdown studies. Thus, it appears that TCPTP participates in a negative feedback mechanism activated by IFN-γ to limit cytokine signaling in IEC.

An observed functional consequence of elevated IFN signaling in TCPTP-deficient cells was a significant increase in epithelial permeability and a decrease TER. This was accompanied by a signif-

icant induction of claudin-2, which is not usually induced by IFN-γ in IEC.[25,26] The claudin family of TJ proteins are key regulators of paracellular permeability, with many claudins functioning as sealing proteins.[23,30] Interestingly, one such sealing claudin, claudin-4, was unaffected by IFN-γ in TCPTP-deficient cells, indicating a relatively specific effect on claudin-2, with the caveat that these were the only members of the claudin family investigated. Moreover, decreased expression of ZO-1 and occludin by IFN-γ was not further exacerbated by *PTPN2* knockdown. Changes in expression or localization of claudins can result in increased epithelial permeability, which is believed to contribute substantially to the pathophysiology of CD.[31] Although many claudins function to seal the TJ, claudin-2 is able to form a cation-selective membrane pore for substrates smaller than 4 Å and may therefore directly account for the greater decrease in TER observed in IFN-γ–treated TCPTP-deficient T$_{84}$ monolayers. Although the pore size of claudin-2 is not sufficiently large enough to permit passage of FITC-dextran (10 kD), and thus an increase in pore number is likely not responsible for the increase in IFN-γ–stimulated permeability to FITC-dextran observed in TCPTP-deficient cells, claudin-2 may indirectly contribute to increased macromolecule permeability.[23] Claudin-2 is upregulated in intestinal crypt epithelium in CD, while in active CD there is an increase in the number of TJ strand breaks in colonic epithelium.[24,25] Large macromolecules such as the 10 kD FITC-dextran used in our study can pass through these breaks, of ≥ 25 nm.[24] In addition, increased claudin-2 expression correlates with the appearance of discontinuous TJ strands.[32] Therefore, a possible explanation for the increased FITC-dextran flux observed in TCPTP-deficient cells is that this may be due to claudin-2 influenced changes in the structural composition of TJs, and a consequent increase in epithelial permeability.[23–26,33] These findings suggest that TCPTP normally prevents a rise in claudin-2 expression in response to IFN-γ. Alternatively, additional signaling events involved in "leak"-type barrier events, such as internalization of TJ proteins in response to myosin light chain kinase or Rho-associated kinase activity, could explain the increase in permeability to FITC-dextran in TCPTP-deficient cells.[34] This is currently under investigation in our laboratory.

It is possible that the IFN-γ–induced elevation of claudin-2 expression in TCPTP-deficient cells was mediated by STAT proteins. In support of this, we previously identified a putative STAT1/STAT3 binding sequence in the claudin-2 promoter region.[21] Whether TCPTP restricts the influence of STAT proteins in claudin-2 expression, perhaps by reducing their phosphorylation, remains to be determined. This will be a particularly important issue to address as the involvement of STAT proteins in IFN-γ –induced barrier defects is controversial.[16] Nevertheless, our findings indicate not only that TCPTP may protect epithelial barrier function during exposure to inflammatory cytokines, but also that loss of functional TCPTP may contribute to the pathogenesis of CD, a chronic inflammatory disease featuring elevated levels of IFN-γ. Given the association of mutations in the *PTPN2* gene locus with CD, it is possible that mutations could lead to a loss of functional TCPTP in IBD.[18,20,35] We have observed that (1) TCPTP expression is increased in biopsies from CD patients with active inflammation versus normal subjects; (2) IFN-γ increases TCPTP expression *in vitro*; and (3) loss of TCPTP exacerbates barrier defects. These findings are consistent with the hypothesis that increased TCPTP is normally a protective response to inflammation and that mutations that compromise TCPTP expression or activity are likely detrimental. This interpretation is supported by one study that identified that heterozygous *PTPN2*-deficient mice exhibit a more severe inflammation than wild-type mice following exposure to dextran sulfate sodium.[36] From a clinical perspective, identifying signaling events and pathophysiological outcomes that are amplified by reduced TCPTP expression, or loss-of-function *PTPN2* mutations, could lead to the isolation of molecular targets that may be particularly effective therapeutic targets in IBD patients harboring *PTPN2* SNPs. Therapeutic agents that "mimic" the effect of TCPTP by reducing the activity of these amplified signaling events could theoretically help to circumvent the biological consequences of *PTPN2* SNPs.

In summary, these observations indicate a crucial role for *PTPN2* (TCPTP) activity in the regulation of inflammatory responses and in the preservation of intestinal epithelial barrier properties in the setting of inflammation.

Acknowledgments

The studies in this paper were supported by a Senior Research Award from the Crohn's and Colitis Foundation of America and by the UCSD Digestive Diseases Research Development Center, NIH grant #DK080506.

Conflicts of interest

The author declares no conflict of interest.

References

1. Cool, D.E., N.K. Tonks, H. Charbonneau, *et al.* 1989. cDNA isolated from a human T-cell library encodes a member of the protein-tyrosine-phosphatase family. *Proc. Natl. Acad. Sci. USA* **86:** 5257–5261.
2. Pao, L.I., K. Badour, K.A. Siminovitch & B.G. Neel. 2007. Nonreceptor protein-tyrosine phosphatases in immune cell signaling. *Annu. Rev. Immunol.* **25:** 473–523.
3. ten Hoeve, J., M.J. Ibarra-Sanchez, Y. Fu, *et al.* 2002. Identification of a nuclear Stat1 protein tyrosine phosphatase. *Mol. Cell Biol.* **22:** 5662–5668.
4. Tiganis, T., A.M. Bennett, K.S. Ravichandran, N.K. Tonks. 1998. Epidermal growth factor receptor and the adaptor protein p52Shc are specific substrates of T-cell protein tyrosine phosphatase. *Mol. Cell Biol.* **18:** 1622–1634.

5. Ibarra-Sanchez, M.J., P.D. Simoncic, F.R. Nestel, P. Duplay, *et al.* 2000. The T-cell protein tyrosine phosphatase. *Semin. Immunol.* **12:** 379–386.

6. Mattila, E., T. Pellinen, J. Nevo, *et al.* 2005. Negative regulation of EGFR signalling through integrin-alpha1beta1-mediated activation of protein tyrosine phosphatase TCPTP. *Nat. Cell Biol.* **7:** 78–85.

7. Galic, S., M. Klingler-Hoffmann, M.T. Fodero-Tavoletti, *et al.* 2003. Regulation of insulin receptor signaling by the protein tyrosine phosphatase TCPTP. *Mol. Cell Biol.* **23:** 2096–2108.

8. Yamamoto, T., Y. Sekine, K. Kashima, *et al.* 2002. The nuclear isoform of protein-tyrosine phosphatase TC-PTP regulates interleukin-6-mediated signaling pathway through STAT3 dephosphorylation. *Biochem. Biophys. Res. Commun.* **297:** 811–817.

9. Zhu, W., T. Mustelin & M. David. 2002. Arginine methylation of STAT1 regulates its dephosphorylation by T cell protein tyrosine phosphatase. *J. Biol. Chem.* **277:** 35787–35790.

10. Scharl, M., P. Hruz & D.F. McCole. 2010. Protein tyrosine phosphatase non-receptor Type 2 regulates IFN-γ-induced cytokine signaling in THP-1 monocytes. *Inflamm. Bowel Dis.* **16:** 2055–64.

11. Kaser, A., S. Zeissig & R.S. Blumberg. 2010. Inflammatory bowel disease. *Annu. Rev. Immunol.* **28:** 573–621.

12. Podolsky, D.K. 2002. Inflammatory bowel disease. *N. Engl. J. Med.* **347:** 417–429.

13. Fuss, I.J., M. Neurath, M. Boirivant, *et al.* 1996. Disparate CD4+ lamina propria (LP) lymphokine secretion profiles in inflammatory bowel disease. Crohn's disease LP cells manifest increased secretion of IFN-gamma, whereas ulcerative colitis LP cells manifest increased secretion of IL-5. *J. Immunol.* **157:** 1261–1270.

14. Strober, W., F. Zhang, A. Kitani, *et al.* 2010. Proinflammatory cytokines underlying the inflammation of Crohn's disease. *Curr. Opin. Gastroenterol.* **26:** 310–317.

15. Madara, J.L. & J. Stafford. 1989. Interferon-gamma directly affects barrier function of cultured intestinal epithelial monolayers. *J. Clin. Invest.* **83:** 724–727.

16. Beaurepaire, C., D. Smyth & D.M. McKay. 2009. Interferon-gamma regulation of intestinal epithelial permeability. *J. Interferon Cytokine Res.* **29:** 133–144.

17. Franke, A., T. Balschun, T.H. Karlsen, *et al.* 2008. Replication of signals from recent studies of Crohn's disease identifies previously unknown disease loci for ulcerative colitis. *Nat. Genet.* **40:** 713–712.

18. Parkes, M., J.C. Barrett, N.J. Prescott, *et al.* 2007. Sequence variants in the autophagy gene IRGM and multiple other replicating loci contribute to Crohn's disease susceptibility. *Nat Genet.* **39:** 830–832.

19. Fisher, S.A., M. Tremelling, C.A. Anderson, *et al.* 2008. Genetic determinants of ulcerative colitis include the ECM1 locus and five loci implicated in Crohn's disease. *Nat Genet.* **40:** 710–712.

20. The Welcome Trust Case Control Consortium. 2007. Genome-wide association study of 14,000 cases of seven common diseases and 3,000 shared controls. *Nature* **447:** 661–678.

21. Scharl, M., G. Paul, A. Weber, *et al.* 2009. Protection of epithelial barrier function by the Crohn's disease associated gene protein tyrosine phosphatase n2. *Gastroenterology* **137:** 2030–2040.

22. Sasaki, T., N. Hiwatashi, H. Yamazaki, *et al.* 1992. The role of interferon gamma in the pathogenesis of Crohn's disease. *Gastroenterol. Jpn.* **27:** 29–36.

23. Van Itallie, C.M., J. Holmes, A. Bridges, *et al.* 2008. The density of small tight junction pores varies among cell types and is increased by expression of claudin-2. *J. Cell Sci.* **121:** 298–305.

24. Zeissig, S., N. Burgel, D. Gunzel, *et al.* 2007. Changes in expression and distribution of claudin 2, 5 and 8 lead to discontinuous tight junctions and barrier dysfunction in active Crohn's disease. *Gut* **56:** 61–72.

25. Prasad, S., R. Mingrino, K. Kaukinen, *et al.* 2005. Inflammatory processes have differential effects on claudins 2, 3 and 4 in colonic epithelial cells. *Lab. Invest.* **85:** 1139–1162.

26. Watson, C.J., C.J. Hoare, D.R. Garrod, *et al.* 2005. Interferon-gamma selectively increases epithelial permeability to large molecules by activating different populations of paracellular pores. *J. Cell Sci.* **118:** 5221–5230.

27. Bruewer, M., M. Utech, A.I. Ivanov, *et al.* 2005. Interferon-gamma induces internalization of epithelial tight junction proteins via a macropinocytosis-like process. *FASEB J.* **19:** 923–33.

28. Wang, F., W.V. Graham, Y. Wang, *et al.* 2005. Interferon-gamma and tumor necrosis factor-alpha synergize to induce intestinal epithelial barrier dysfunction by up-regulating myosin light chain kinase expression. *Am. J. Pathol.* **166:** 409–419.

29. Heinonen, K.M., F.P. Nestel, E.W. Newell, *et al.* 2004. T-cell protein tyrosine phosphatase deletion results in progressive systemic inflammatory disease. *Blood* **103:** 3457–3464.

30. Amasheh, S., N. Meiri, A.H. Gitter, *et al.* 2002. Claudin-2 expression induces cation-selective channels in tight junctions of epithelial cells. *J. Cell Sci.* **115:** 4969–4976.

31. Mankertz, J. & J.D. Schulzke. 2007. Altered permeability in inflammatory bowel disease: pathophysiology and clinical implications. *Curr. Opin. Gastroenterol.* **23:** 379–383.

32. Furuse, M., H. Sasaki & S. Tsukita. 1999. Manner of interaction of heterogeneous claudin species within and between tight junction strands. *J. Cell Biol.* **147:** 891–903.

33. Kinugasa, T., T. Sakaguchi, X. Gu & H.C. Reinecker. 2000. Claudins regulate the intestinal barrier in response to immune mediators. *Gastroenterology* **118:** 1001–1011.

34. Utech, M., A.I. Ivanov, S.N. Samarin, *et al.* 2005. Mechanism of IFN-gamma-induced endocytosis of tight junction proteins: myosin II-dependent vacuolarization of the apical plasma membrane. *Mol. Biol. Cell.* **16:** 5040–5052.

35. Weersma, R.K., P.C. Stokkers, I. Cleynen, *et al.* 2009. Confirmation of multiple Crohn's disease susceptibility loci in a large Dutch-Belgian cohort. *Am. J. Gastroenterol.* **104:** 630–638.

36. Hassan, S.W., K.M. Doody, S. Hardy, *et al.* 2010. Increased susceptibility to dextran sulfate sodium induced colitis in the T cell protein tyrosine phosphatase heterozygous mouse. *PLoS One* **5:** 1–7.

Ann. N.Y. Acad. Sci. ISSN 0077-8923

ANNALS OF THE NEW YORK ACADEMY OF SCIENCES
Issue: *Barriers and Channels Formed by Tight Junction Proteins*

Intracellular mediators of JAM-A–dependent epithelial barrier function

Ana C. Monteiro and Charles A. Parkos

Department of Pathology and Laboratory Medicine, Epithelial Pathobiology Research Unit, Emory University, Atlanta, Georgia

Address for correspondence: Charles A. Parkos, Department of Pathology and Laboratory Medicine, Epithelial Pathobiology Research Unit, Emory University, 615 Michael St., Atlanta, GA 30322. cparkos@emory.edu

Junctional adhesion molecule-A (JAM-A) is a critical signaling component of the apical junctional complex, a structure composed of several transmembrane and scaffold molecules that controls the passage of nutrients and solutes across epithelial surfaces. Observations from JAM-A–deficient epithelial cells and JAM-A knockout animals indicate that JAM-A is an important regulator of epithelial paracellular permeability; however, the mechanism(s) linking JAM-A to barrier function are not understood. This review highlights recent findings relevant to JAM-A–mediated regulation of epithelial permeability, focusing on the role of upstream and downstream signaling candidates. We draw on what is known about proteins reported to associate with JAM-A in other pathways and on known modulators of barrier function to propose candidate effectors that may mediate JAM-A regulation of epithelial paracellular permeability. Further investigation of pathways highlighted in this review may provide ideas for novel therapeutics that target debilitating conditions associated with barrier dysfunction, such as inflammatory bowel disease.

Keywords: JAM-A; barrier function; scaffold proteins; apical junctional complex (AJC); permeability

Introduction

The epithelial barrier is a critical component of tissue homeostasis as it selectively controls the passage of nutrients and solutes across mucosal surfaces while deterring the passage of pathogens and toxins. Functional regulation of the epithelial barrier is determined by a collection of tight junction (TJ)– and adherens junction (AJ)–associated transmembrane and scaffold proteins, termed the *apical junctional complex* (AJC). One transmembrane component of the TJ is the junctional adhesion molecule-A (JAM-A). Evidence suggests that JAM-A does not directly regulate barrier by forming a "seal" between cells but rather functions as a signaling molecule with divergent downstream target proteins.[1,2] Indeed, recent reports have implicated JAM-A–mediated signaling events in regulating a diverse array of epithelial functions, including epithelial proliferation, migration, and barrier function.[2–4] Although the evidence linking expression of JAM-A to TJ regulation and barrier maintenance

is accepted, insights into downstream mechanisms linking JAM-A to regulation of barrier are limited. In this review, we summarize current findings that are relevant to how JAM-A might control barrier function, focusing on the role of upstream and downstream signaling components linking JAM-A to paracellular permeability. We draw on what is known about JAM-A effectors from other pathways to speculate on attractive candidate molecules that regulate the epithelial barrier.

JAM-A expression affects epithelial permeability

The importance of JAM-A in regulating barrier function is best illustrated by reported observations in JAM-A–deficient intestinal cell lines and knockout (KO) animals.[2,5] In cell lines, it has been clearly shown that siRNA mediated loss of JAM-A expression results in enhanced permeability as determined by transepithelial resistance (TER) and paracellular flux of labeled dextran. Such observations have been confirmed in multiple epithelial

doi: 10.1111/j.1749-6632.2012.06521.x

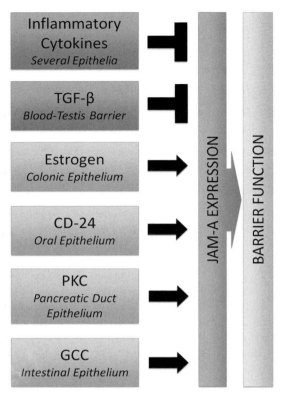

Figure 1. Examples of cues that affect epithelial barrier function and alter JAM-A expression.

and endothelial cell types,[1,5–8] including primary rat alveolar epithelial cells, which exhibit decreased TER after treatment with JAM-A shRNA (Mitchell and Koval, personal communication). *In vivo*, JAM-A–deficient mice have a leaky colonic epithelium.[2] In addition, JAM-A–deficient mice are more susceptible to dextran sulfate sodium (DSS)–induced colitis compared to control mice, presenting with a higher disease activity index and more severe weight loss. Intriguingly, JAM-A KO mice also present with increased mucosal infiltration of leukocytes in the colonic mucosa and altered levels of proinflammatory cytokines.[2] This finding is consistent with current views that link impaired barrier function with increased susceptibility to mucosal inflammation. Future studies should determine whether enhanced permeability and leukocyte infiltration are also observed in other epithelial compartments of JAM-A–deficient mice.

Epithelial barrier function is established by a complex series of poorly understood signaling events that culminate in the formation of mature TJs. Evidence suggests that JAM-A is important in early events required for TJ assembly. Such early events begin with emergence of nascent puncta containing the AJ proteins E-cadherin and nectin, providing initial points of cell–cell contact that recruit other junctional proteins necessary to establish a mature AJC. JAM-A seems to be an early mediator of this process because it is recruited along with occludin immediately after puncta are established.[9] These observations are consistent with our findings demonstrating that antibody blockade of JAM-A in cultures of subconfluent epithelial cells delays the development of a tight barrier, as determined by TER measurements (unpublished observations).

Dimerization of JAM-A is necessary for regulation of barrier

JAM-A is composed of a short cytoplasmic tail, a single-pass transmembrane region, and two extracellular Ig-like loops. A number of reports, including crystallography data,[10] indicate that JAM-A forms functionally significant homodimers through ionic interactions within a conserved motif in its distalmost Ig (D1) loop.[5,11] These reports suggest that such ionic interactions within the D1 loop of JAM-A mediates homodimerization in a cis configuration (on the surface of the same cell); however, structural data from murine JAM-A[12] suggests that JAM-A also homodimerizes in *trans*, or across cells, at distinct but yet undefined site(s) in the same D1 loop. We observed that mutagenesis of residues promoting *cis*-dimerization of JAM-A or treatment with a monoclonal antibody that binds to the dimerization domain results in attenuated JAM-A–dependent regulation of epithelial cell migration.[11] Interestingly, the same dimerization disrupting antibodies delay barrier development in monolayers of epithelial cells with disassembled AJC after transient calcium depletion (calcium switch). These observations suggest that JAM-A homodimerization is necessary for assembly of functional TJs.[1,5] Studies on JAM-A–mediated effects on cell migration eventuated in a model of JAM-A function involving dimerization-mediated activation of a signaling module that leads to cell migration. Although mechanistic insights detailing how JAM-A dimerization leads to regulation of paracellular permeability are not understood, some clues are provided by the above model and from reports implicating known

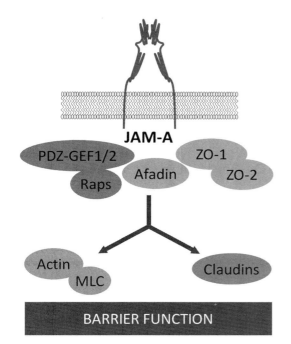

Figure 2. Possible downstream mechanisms linking JAM-A dimerization to barrier function. Although PDZ-GEF2, Afadin, and Rap1 have been reported to associate with JAM-A, their potential roles in regulation of epithelial barrier require further elucidation. Other key components reported to affect epithelial barrier, including ZO-1, claudins, and the epithelial cytoskeleton, may also act downstream of JAM-A; however, association(s) with JAM-A have not yet been established.

JAM-A–interacting molecules with the regulation of barrier function. From these observations, we can begin to assemble a potential model by which JAM-A dimerization controls epithelial barrier function.

Several stimuli that alter epithelial barrier also affect JAM-A expression

Although there are limited studies on the signaling pathways that link JAM-A to regulation of barrier, there are several recent reports describing paracrine and autocrine cues that affect junctional integrity and alter JAM-A expression and localization (Fig. 1). Such studies underscore the fluidity of the AJC, which is constantly restructured to accommodate physiological events, such as leukocyte transmigration across epithelia and varying demands of fluid and nutrient absorption in the gastrointestinal tract.

Cytokine-mediated internalization of epithelial TJ proteins exacerbates inflammatory conditions by maintaining an open entryway for leukocytes to the inflamed region and enhancing leukocyte exposure to luminal antigens. Inflammatory cytokines such as INF-γ and TNF-α enhance permeability of endothelial and epithelial barriers by inducing internalization of JAM-A and other AJC proteins,[13,14] while local administration of TGF-β and TNF-α to the blood–testis barrier also induces clathrin-dependent internalization of AJC proteins that include JAM-A, occluding, and N-cadherin.[15] Conversely, cues that enhance barriers may be protective against chronic inflammation, again by altering the AJC. For example, estrogen, thought to have antiinflammatory effects in the gut,[16,17] was reported to reduce the permeability of the intestinal epithelium *in vivo* and *in vitro* by upregulating TJ protein levels of JAM-A and occludin.[18] CD-24, a ligand for p-selectin implicated in epithelial restitution in mouse models of inflammatory bowel disease (IBD), was also reported to enhance barrier function of the oral epithelium by upregulating JAM-A and claudins 4 and 15 in a src-kinase–dependent manner.[19] These examples indicate a potential reciprocal influence of inflammatory signals on mucosal permeability, which may act to perpetuate a pathological inflammatory response.

Other studies implicating a role of paracrine signaling in JAM-A expression provide mechanistic insights into JAM-A recruitment to TJs, which may be important for JAM-A stability and function. Studies using immortalized primary pancreatic duct cells[20] revealed that inclusion of fetal bovine serum (FBS) after serum starvation enhanced the expression and TJ localization of several TJ proteins, including JAM-A, occludin, zonula occludens-1 (ZO-1), and several claudins in a protein kinase C (PKC)-dependent manner. Cells formed no functional barrier during serum starvation, but develop a functionally tight barrier after the addition of serum. Interestingly, inhibition of PKC reduced JAM-A expression and TER to that of serum-free levels. In serum-free media, addition of TPA, a DAG pharmacomimetic that activates typical PKCs, enhanced levels and TJ localization of ZO-1, ZO-2, and occludin; however, JAM-A expression and TER remained unchanged. The study does not further explore the pathway regulating JAM-A expression, but it is tempting to speculate that JAM-A recruitment to TJs may be dependent on an atypical PKC, one not activated by TPA/DAG. This is consistent with observations of JAM-A association with aPKC[21] in the context of cell polarity. An understudied aspect of JAM-A is related to the multiple potential phosphorylation sites on the relatively short cytoplasmic tail that may be important for JAM-A recruitment and function, five of which are likely targets for PKC, as determined by netPhosK analysis. Notably, JAM-A has been shown to be phosphorylated by PKC in platelets.[22] Furthermore, during the final editorial review of this manuscript, Ebnet et al. published a study demonstrating that the cytoplasmic tail of JAM-A is indeed phosphorylated by aPKCζ at serine 285 to affect TJ assembly and epithelial barrier function.[23] Further investigation of JAM-A phosphorylation by aPKC may provide additional insights on mechanisms controlling the stability and localization of JAM-A to the TJ, which may be an important event in the transition from nascent to mature TJ formation leading to a stable epithelial barrier.

Studies of other barrier-forming pathways have clearly demonstrated that cytoskeletal dynamics play an important role in barrier function. One study has provided a potential link between JAM-A, cytoskeletal dynamics, and barrier function. Mice lacking guanylyl cyclase C (GCC), a transmembrane receptor to endogenous ligands that modulates epithelial chloride conductance, demonstrate an intriguingly similar phenotype to JAM-A KO mice and have altered phosphorylation of actin-associated proteins. Compared to wild-type mice, GCC-null animals have a more permeable gut mucosa, increased levels of pro-inflammatory cytokines, and large amounts of lymphocytes in the intestinal epithelial compartment, suggesting a concomitant inflammatory phenotype.[24] Importantly, GCC-null mice and GCC-deficient colonic epithelial cells have decreased levels of JAM-A and claudin-2 with increased phosphorylation of myosin light chain (pMLC), suggesting that the barrier deficiency observed in the GCC-null mice may be related to the loss of JAM-A and claudin-2 as well as the phosphorylation of MLC. Notably, MLC phosphorylation has been implicated in increased epithelial permeability by inducing contraction of the epithelial actomyosin ring and the enlargement of the intercellular space, thereby enhancing epithelial leak.[25] However, Han et al. propose instead that pMLC is important for TJ assembly by recruiting JAM-A and other proteins to the AJC. Future studies are required to clarify the relationship between JAM-A signaling and actomyosin contraction to better understand the role of cytoskeletal dynamics in JAM-A–dependent regulation of epithelial permeability.

Putative signaling effectors downstream of JAM-A that regulate barrier function

Although the importance of JAM-A in endothelial and epithelial barrier function is appreciated,[1,5,7,8] downstream pathways linking JAM-A to paracellular permeability are unknown. As mentioned above, signaling pathways regulating JAM-A–dependent cell migration have been described.[4] From these studies, it is reasonable to postulate that similar pathway(s) may regulate barrier function. In migration studies, it was found that JAM-A associates with the scaffold protein afadin and the guanine nucleotide exchange factor PDZ–GEF2, resulting in the activation of the small GTPase Rap1a, stabilization of β1 integrins and enhanced cell migration. Although these effector molecules have not been reported to directly affect barrier in epithelia, afadin, PDZ-GEFs, and Rap1 have been widely implicated in regulation of endothelial barrier, as will be discussed below.

Afadin, a large PDZ-containing scaffold protein shown to associate with JAM-A,[4] has been strongly implicated in the regulation of barrier function. Mice with intestinal epithelial-targeted loss of afadin have increased intestinal permeability[26] and a phenotype similar to that observed with JAM-A KO mice. JAM-A KO mice have normal intestinal mucosal architecture but a leaky colonic epithelium, increased mucosal lymphoid follicles, and enhanced susceptibility to acute injury–induced colitis.[2] Although complete genomic deletion of afadin is lethal, mice with intestinal epithelial-targeted loss of afadin (cKO) are viable, have a similar increase in gut permeability, and exhibit seemingly normal intestinal morphology. Similarly, afadin cKO mice show enhanced susceptibility to acute injury–induced colitis. Although the above parallels between JAM-A KO and afadin cKO animals are consistent with afadin regulation of barrier function downstream of JAM-A, afadin has also been regarded as a cadherin-associated scaffold that mediates outside-in signaling after nectin-driven nascent junctions are initiated.[27] Although the latter observation might implicate afadin in controlling barrier downstream of nectin, mice lacking nectins-2 and -3 in the intestinal epithelium have no increase in intestinal permeability compared to wildtype counterparts.[26] Furthermore, the intestinal epithelium of such nectin-deficient mice has normal localization of afadin. These findings suggest that intestinal barrier function and junctional localization of afadin can occur by nectin-independent mechanism(s).

In addition to afadin, the guanine nucleotide exchange factor PDZ–GEF2 has been reported to associate with JAM-A and mediate β1 integrin–dependent epithelial cell migration,[4] presumably through activation of the small GTPase Rap1a. Despite this observation, the role for PDZ–GEF2 or the closely related PDZ–GEF1 in regulating epithelial barrier is not understood. Loss of PDZ–GEF1/2 in epithelial and endothelial cells has been shown to affect the composition and architecture of the AJ so that its morphology resembles nascent puncta,[28,29] suggesting that PDZ-GEF1/2 may be important in the maturation of initial puncta into functionally developed AJCs. On the other hand, the differences in AJ architecture described in these reports were only detectable when cells were recultured at low confluence, suggesting that the role of PDZ–GEFs in barrier function may be limited to early events in junction formation. In addition, loss of PDZ–GEF2 in epithelial cells did not affect the localization of the TJ proteins occludin or ZO-1,[28] supporting the idea that altered AJ morphology does not necessarily translate to defects in TJ composition. In endothelial cells, Pannekoek et al. reported that PDZ–GEF1/2 depletion resulted in decreased transendothelial impedance, a proxy measure of barrier function,[29] but no analogous functional observations were reported for epithelial cells. Such observations support the concept of distinct regulatory mechanisms governing endothelial and epithelial barrier, which may be explained in part by the fact that the AJC in endothelial cells differs from that of epithelial cells by having a less defined separation between AJs and TJs.[30] It is therefore important to confirm the functional importance of PDZ–GEFs in epithelial barrier function so as to determine whether they may act downstream of JAM-A to affect epithelial permeability.

As is the case for PDZ–GEFs, the small GTPase Rap1 has been implicated in downstream signaling events from JAM-A–mediated regulation of cell migration. Interestingly, Rap1a/b have been widely implicated in regulation of endothelial barrier function,[29,31,32] but much less is known regarding the role of Rap1 as an effector of barrier in epithelial cells. In epithelial cells, Rap1a has been implicated in mediating trans-dimerization of E-cadherin[32] and in organization of E-cadherin along cell–cell contacts.[28,33] However, inactivation of Rap1 by the guanine nucleotide activating protein RapGAP does not affect the localization of ZO-1 to cell–cell contacts,[33] suggesting that TJ formation does not require Rap1 in epithelial cells. Furthermore an in vivo study on the role of the oxidized phospholipid OXPAPC in lung epithelial permeability reported no changes in TER after downregulation of Rap1.[34] In addition, Rap1 null Caenorhabditis elegans display normal epithelial architecture of the epidermis and gut.[35] Notably, most studies claiming a role for Rap1 in regulating endothelial and epithelial barrier are in fact describing functions of proteins known to alter the activation of Rap1, such as EPAC, RapGAP, and PDZ–GEF1/2.[29,33] Because these mediators lack specificity for Rap1,[36–39] the possibility of involvement of other small GTPases has not been excluded. The paucity of data directly relating Rap1 to functional measures of epithelial permeability leaves more questions than answers regarding a link between Rap1– and JAM-A–dependent regulation of barrier function.

ZO-1 is an important TJ-associated scaffold protein and one of three zonula occludens proteins. ZO-1 has three PDZ domains (PDZ1-3), an Src homology 3 domain (SH3) and a guanylate kinase homology domain (GUK), and has been reported to associate with JAM-A.[40] A recent crystallography study identified PDZ3 as the putative binding pocket in ZO-1 responsible for JAM-A association and that this association required the presence of the SH3 domain.[41] Interestingly, mice lacking ZO-1 or ZO-2 do not survive, suggesting that both proteins are required for embryonic development.[42,43] ZO-3 null mice, however, have little or very subtle phenotypic differences compared to their wildtype counterparts, suggesting a redundant role of ZO-3 in epithelial function.[42] Cell culture studies have further defined barrier-inducing roles of ZO proteins through simultaneous silencing of ZO-1, -2, and -3 followed by their replacement one at a time.[44,45] Epithelial cells lacking all three ZO proteins have no TJs, highlighting the importance of these scaffold proteins to barrier formation. Interestingly, addition of either ZO-1 or ZO-2 to ZO-null cells is sufficient for establishing TJs. In addition, it was found that protein levels of JAM-A, claudins, and occludin are unchanged in ZO-depleted cells, suggesting that ZO-1 or -2 are likely necessary and sufficient for recruitment of TJ components to the AJC during barrier formation, but that synthesis of TJ proteins occurs independently of ZO-1 or -2 expression.[46] ZO-1 has also been shown to associate with claudins via PDZ1, occludin through its GUK domain, afadin via its SH3 domain, and with other ZO proteins via PDZ2, allowing it to cluster several scaffold proteins, potentially leading to TJ maturation. It is possible that ZO-1, afadin, PDZ-GEFs, and other PDZ-containing scaffold proteins associate with transmembrane PDZ motif-containing proteins, such as JAM-A, to direct maturation and maintenance of barrier function. Ebnet and others have reported coassociation between JAM-A and ZO-1 from cell lysates,[40,47] and Nomme et al. have shown in vitro direct association of these proteins via crystallography studies.[41] However, there are no reports showing a direct association of JAM-A and ZO-1 in cells. In addition, there is limited data on whether ZO-2 can associate with JAM-A or participate in the same signaling module that ZO-1 and JAM-A are reported to share. Although the role of ZO-1 and -2 proteins in barrier formation are well appreciated, the mechanisms linking ZO-1 and -2 to JAM-A–mediated regulation of barrier function require further elucidation.

The GEFS and scaffold proteins discussed above are attractive candidate mediators regulating JAM-A–dependent barrier function. However, potential distal signaling elements downstream of JAM-A, which are intimately associated with regulation of epithelial permeability, merit careful consideration. In particular, cytoskeletal dynamics and claudin composition of TJs directly affect paracellular permeability (Fig. 2).[48–50] The tetraspan TJ-forming claudins are classified as either leaky or tight and dimerize across the apical intercellular space to form channels that control the permeability of monolayers.[48,50,51] It was previously reported that JAM-A null mice and JAM-A–depleted intestinal epithelial cells demonstrate enhanced levels of the leaky claudins 10 and 15,[2] but not of claudin-2 or occludin, suggesting that JAM-A affects the claudin composition of TJs. It is not known how JAM-A does this; however, previous studies indicate that JAM-A affects levels of β1 integrin by maintaining its stability at the cell surface.[52] Given the likely over-

lap in function of some of the signaling elements discussed above, it is reasonable to hypothesize that JAM-A may regulate barrier function through effects on the stability of certain claudin family members at the TJ. Clearly, further studies should help to answer this important question.

JAM-A–associated AJC scaffold proteins, such as afadin and ZO molecules, have actin-binding domains that allow for communication between the AJC and the apically positioned actomyosin ring, a critical component of barrier integrity.[25] The association of the AJC with cytoskeletal components is necessary for maintaining AJC structure.[53,54] An attractive potential mechanism for JAM-A regulation of barrier would involve signaling through effectors such as afadin to induce cytoskeletal changes that control paracellular flux and epithelial permeability. JAM-A modulates epithelial cell migration, a process dependent on dynamic restructuring of actin, and induces activation of Rap GTPases, which have important cytoskeletal regulatory properties.[11,55] Analogous JAM-A–dependent pathways

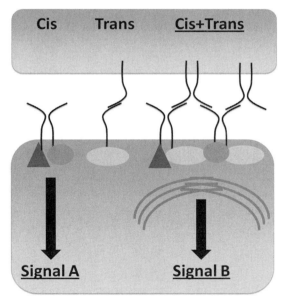

Figure 3. Model of how JAM-A may differentially induce divergent signaling modules based on JAM-A dimerization. When JAM-A homodimerizes exclusively in *cis*, as may be the case in subconfluent epithelial sheets, a proliferative or migratory pathway may be initiated. When JAM-A homodimerizes in *cis* and *trans*, as may be the case in confluent, fully polarized epithelial monolayers, JAM-A may associate with other scaffold proteins that promote barrier function and possibly senescence.

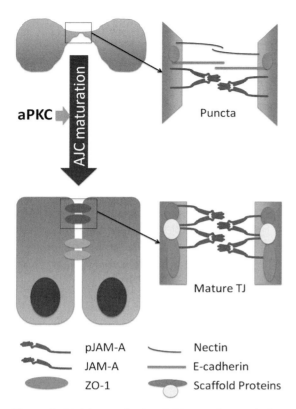

Figure 4. Model proposing how JAM-A may be recruited to differential subcellular compartments based on JAM-A phosphorylation. In the upper part of the figure, unphosphorylated JAM-A preferentially localizes to the basolateral membrane, where it can participate in events such as initiation of nascent puncta. JAM-A phosphorylation by an atypical PKC may then lead to its colocalization with TJ-associated proteins and maturation of the AJC (lower panels).

that regulate barriers are therefore easily envisioned and require further exploration.

Possible mechanisms that may differentiate divergent JAM-A signaling modules

A major challenge has been to understand how JAM-A mediates diverse cellular functions, such as altered permeability,[55] cell migration, and enhanced proliferation.[3] It is assumed that functional specificity of JAM-A signaling is determined by the interaction of JAM-A with specific scaffold proteins, which may in turn be differentially distributed to subcompartments of epithelial cells. Because previous observations have demonstrated that JAM-A concentrates not only at the AJC but also

along the basolateral membrane,[56] it is possible that JAM-A dimerization or phosphorylation may determine the localization of JAM-A in either compartment, allowing for specific activation of a particular JAM-A–dependent signaling cascade. For example, dimerization of JAM-A in a *cis* but not *trans* configuration, as would be expected in spreading or subconfluent cells, might favor proliferation and/or migration. However, *trans* interactions between cells, as would be expected in confluent epithelia, would favor close apposition of PDZ-bound molecules that promote barrier forming and perhaps even senescence-inducing cues (Fig. 3).

Likewise, JAM-A phosphorylation events may be important for determining subcellular localization. JAM-A localizes to nascent puncta with E-cadherin, nectin, and ZO-1 to initiate AJC formation; however, it also colocalizes with ZO-1 in mature TJs away from the AJ proteins E-cadherin and nectin. It is possible that JAM-A phosphorylation event(s) could facilitate movement of JAM-A toward TJ maturation after puncta establishment (Fig. 4). Strong support for this latter proposed mechanism was very recently shown in a report published while this manuscript was in editorial review. In particular, aPKCζ -mediated phosphorylation of ser285 in the JAM-A cytoplasmic tail was shown to regulate JAM-A localization to mature TJs.[23] It will be interesting to investigate whether such phosphorylation events alter binding affinities between JAM-A and different scaffold proteins, thereby determining activation of a specific JAM-A–dependent cascade in epithelial cells. These two proposed mechanisms may be key elements determining specificity of JAM-A–dependent cues and are attractive areas to be investigated as novel therapeutic targets may emerge that can target JAM-A–dependent function.

Conclusions

JAM-A is a transmembrane, TJ-associated Ig superfamily member that regulates epithelial barrier function, cell migration, and proliferation. Current evidence indicates that JAM-A dimerization is necessary for functional regulation of barrier; however, the mechanisms linking JAM-A to barrier function have not been elucidated. Although it has been shown that JAM-A regulates cell migration through dimerization-dependent clustering of the cytoplasmic scaffold molecules afadin and PDZ–GEF2, resulting in activation of Rap1a to stabilize cell surface $\beta 1$ integrin, the involvement of these pathways in regulation of barrier is unclear. Moreover, association of JAM-A with cellular components known to affect epithelial permeability, such as claudins, ZO-1/2, and the epithelial cytoskeleton remains to be established. Understanding mechanisms that regulate epithelial permeability downstream of JAM-A may be useful in identifying therapeutic targets for diseases associated with barrier dysfunction and may lead to novel approaches toward *trans*-epithelial drug delivery. Finally, future studies should aim to elucidate mechanisms that determine JAM-A expression and its subcellular localization, because such cues may be key to deciphering how divergent JAM-A–modulated signaling cascades are differentially induced. Identification of effectors that specifically modulate epithelial permeability downstream of JAM-A without affecting proliferation, migration, or other pathways could lead to the development of specific pharmaceutical interventions with fewer off-target effects.

Acknowledgments

We thank Robert Rankin, Stefanie Ritter, Christopher Capaldo, and Asma Nusrat for useful comments. This study was supported by National Institutes of Health Grants R01-DK61379, R01-DK72564, and R24-DK064399.

Conflicts of interest

The authors declare no conflicts of interest.

References

1. Liu, Y., A. Nusrat, F.J. Schnell, *et al.* 2000. Human junction adhesion molecule regulates tight junction resealing in epithelia. *J. Cell Sci.* **113:** 2363–2374.

2. Laukoetter, M.G., P. Nava, W.Y. Lee, *et al.* 2007. JAM-A regulates permeability and inflammation in the intestine *in vivo*. *J. Exp. Med.* **204:** 3067–3076.

3. Nava, P., C.T. Capaldo, S. Koch, *et al.* 2011. JAM-A regulates epithelial proliferation through Akt/β-catenin signalling. *Nat. Publ. Group.* **12:** 314–320.

4. Severson, E., W. Lee, C. Capaldo, *et al.* 2009. Junctional adhesion molecule A interacts with Afadin and PDZ-GEF2 to activate Rap1A, regulate {beta} 1 integrin levels, and enhance cell migration. *Mol. Biol. Cell.* **20:** 1916–1925.

5. Mandell, K.J., I.C. McCall & C.A. Parkos. 2004. Involvement of the junctional adhesion molecule-1 (JAM1) homodimer interface in regulation of epithelial barrier function. *J. Biol. Chem.* **279:** 16254–16262.

6. Haarmann, A., A. Deiß, J. Prochaska, *et al.* 2010. Evaluation of soluble junctional adhesion molecule-A as a biomarker of

human brain endothelial barrier breakdown. *PLoS ONE* **5:** e13568.

7. Mandell, K.J., G.P. Holley, C.A. Parkos & H.F. Edelhauser. 2006. Antibody blockade of junctional adhesion molecule-A in rabbit corneal endothelial tight junctions produces corneal swelling. *Invest. Ophth. Vis. Sci.* **47:** 2408–2416.

8. Mandell, K.J., L. Berglin, E.A. Severson, *et al.* 2007. Expression of JAM-A in the human corneal endothelium and retinal pigment epithelium: localization and evidence for role in barrier function. *Invest. Ophth. Vis. Sci* **48:** 3928–3936.

9. Cereijido, M., R.G. Contreras, L. Shoshani, *et al.* 2008. Tight junction and polarity interaction in the transporting epithelial phenotype. *Biochimica et biophysica acta* **1778:** 770–793.

10. Prota, A.E., J.A. Campbell, P. Schelling, *et al.* 2003. Crystal structure of human junctional adhesion molecule 1: implications for reovirus binding. *PNAS* **100:** 5366–5371.

11. Severson, E.A., L. Jiang, A.I. Ivanov, *et al.* 2008. Cis-dimerization mediates function of junctional adhesion molecule A. *Mol. Biol. Cell* **19:** 1862–1872.

12. Kostrewa, D., M. Brockhaus, A. D'Arcy & G. Dale. 2001. X-ray structure of junctional adhesion molecule: structural basis for homophilic adhesion via a novel dimerization motif. *EMBO J* **20:** 4391–4398.

13. Bruewer, M., A. Luegering, T. Kucharzik, *et al.* 2003. Proinflammatory cytokines disrupt epithelial barrier function by apoptosis-independent mechanisms. *J. Immunol.* **171:** 6164–6172.

14. Capaldo, C.T. & A. Nusrat. 2009. Cytokine regulation of tight junctions. *Biochimica et Biophysica Acta (BBA)— Biomembranes* **1788:** 864–871.

15. Xia, W., E.W.P. Wong, D.D. Mruk & C.Y. Cheng. 2009. TGF-β3 and TNFα perturb blood–testis barrier (BTB) dynamics by accelerating the clathrin-mediated endocytosis of integral membrane proteins: a new concept of BTB regulation during spermatogenesis. *Dev. Biol.* **327:** 48–61.

16. Houdeau, E., R. Moriez, M. Leveque, *et al.* 2007. Sex steroid regulation of macrophage migration inhibitory factor in normal and inflamed colon in the female rat. *Gastroenterology* **132:** 982–993.

17. Harnish, D.C., L.M. Albert, Y. Leathurby, *et al.* 2004. Beneficial effects of estrogen treatment in the HLA-B27 transgenic rat model of inflammatory bowel disease. *Am. J. Physiol. Gastrointest. Liver Physiol.* **286:** G118–G125.

18. Braniste, V., M. Leveque, C. Buisson-Brenac, *et al.* 2009. Oestradiol decreases colonic permeability through oestrogen receptor-mediated up-regulation of occludin and junctional adhesion molecule-A in epithelial cells. *J. Physiol.* **587:** 3317–3328.

19. Ye, P., H. Yu, M. Simonian & N. Hunter. 2011. Ligation of CD24 expressed by oral epithelial cells induces kinase dependent decrease in paracellular permeability mediated by tight junction proteins. *Biochem. Biophys. Res. Comm.* **412:** 165–169.

20. Yamaguchi, H., T. Kojima, T. Ito, *et al.* 2010. Transcriptional control of tight Junction proteins via a protein kinase C signal pathway in human telomerase reverse transcriptase-transfected human pancreatic duct epithelial cells. *Am. J. Pathol.* **177:** 698–712.

21. Ebnet, K., A. Suzuki, Y. Horikoshi, *et al.* 2001. The cell polarity protein ASIP/PAR-3 directly associates with junctional adhesion molecule (JAM). *The EMBO J.* **20:** 3738–3748.

22. Ozaki, H., K. Ishii, H. Arai, *et al.* 2000. Junctional adhesion molecule (JAM) is phosphorylated by protein kinase C upon platelet activation. *Biochem. Biophys. Res. Comm.* **276:** 873–878.

23. Iden, S., S. Misselwitz, S.S.D. Peddibhotla, *et al.* 2012. aPKC phosphorylates JAM-A at Ser285 to promote cell contact maturation and tight junction formation. *J. Cell Biol.*

24. Han, X., E. Mann, S. Gilbert, Y. Guan, K.A. Steinbrecher, M.H. Montrose & M.B. Cohen. 2011. Loss of guanylyl cyclase C (GCC) signaling leads to dysfunctional intestinal barrier. *PLoS ONE* **6:** e16139.

25. Nusrat, A., J.R. Turner & J.L. Madara. 2000. Molecular physiology and pathophysiology of tight junctions. IV. Regulation of tight junctions by extracellular stimuli: nutrients, cytokines, and immune cells. *Am. J. Physiol. Gastrointest. Liver Physiol.* **279:** G851–G857.

26. Tanaka-Okamoto, M., K. Hori, H. Ishizaki, *et al.* 2011. Involvement of afadin in barrier function and homeostasis of mouse intestinal epithelia. *J. Cell Sci.* **124:** 2231–2240.

27. Takai, Y. & H. Nakanishi. 2003. Nectin and afadin: novel organizers of intercellular junctions. *J. Cell Sci.* **116:** 17–27.

28. Dube, N., M. Kooistra, W. Pannekoek, *et al.* 2008. The RapGEF PDZ-GEF2 is required for maturation of cell-cell junctions. *Cell. Signal.* **20:** 1608–1615.

29. Pannekoek, W.-J., J.J.G. van Dijk, O.Y.A. Chan, *et al.* 2011. Epac1 and PDZ-GEF cooperate in Rap1 mediated endothelial junction control. *Cell. Signal.* **23:** 2056–2064.

30. Schulze, C. & J.A. Firth. 1993. Immunohistochemical localization of adherens junction components in blood-brain barrier microvessels of the rat. *J. Cell Sci.* **104:** 773–782.

31. Bos, J.L. 2005. Linking Rap to cell adhesion. *Curr. Opin. Cell Biol.* **17:** 123–128.

32. Price, L., A. Hajdo-Milasinovic, J. Zhao, *et al.* 2004. Rap1 regulates E-cadherin-mediated cell-cell adhesion. *J. Biol. Chem.* **279:** 35127–35132.

33. Hogan, C., N. Serpente, P. Cogram, *et al.* 2004. Rap1 regulates the formation of E-cadherin-based cell-cell contacts. *Mol. Cell. Biol.* **24:** 6690–6700.

34. Birukova, A.A., N. Zebda, P. Fu, *et al.* 2011. Association between adherens junctions and tight junctions via rap1 promotes barrier protective effects of oxidized phospholipids. *J. Cell. Physiol.* **226:** 2052–2062.

35. Frische, E.W., W. Pellis-van Berkel, G. van Haaften, *et al.* 2007. RAP-1 and the RAL-1/exocyst pathway coordinate hypodermal cell organization in Caenorhabditis elegans. *EMBO J.* **26:** 5083–5092.

36. Roscioni, S.S., C.R.S. Elzinga & M. Schmidt. 2008. Epac: effectors and biological functions. *Naunyn-Schmiedeberg's Arch. Pharm.* **377:** 345–357.

37. Janoueix-Lerosey, I., P. Polakis, A. Tavitian & J. de Gunzburg. 1992. Regulation of the GTPase activity of the ras-related rap2 protein. *Biochem. Biophys. Res. Comm.* **189:** 455–464.

38. Kuiperij, H., J. De Rooij, H. Rehmann, *et al.* 2003. Characterisation of PDZ-GEFs, a family of guanine nucleotide exchange factors specific for Rap1 and Rap2. *Biochimica et biophysica acta* **1593:** 141–149.

39. De Rooij, J., N.M. Boenink, M. van Triest, *et al.* 1999. PDZ-GEF1, a guanine nucleotide exchange factor specific for Rap1 and Rap2. *J. Biol. Chem.* **274:** 38125–38130.

40. Bazzoni, G., O. Martínez-Estrada, F. Orsenigo, *et al.* 2000. Interaction of junctional adhesion molecule with the tight junction components ZO-1, Cingulin, and Occludin. *J. Biol. Chem.* **275:** 20520–20526.

41. Nomme, J., A.S. Fanning, M. Caffrey, *et al.* 2011. The Src homology 3 domain is required for junctional adhesion molecule binding to the third PDZ domain of the scaffolding protein ZO-1. *J. Biol. Chem.* **286:** 43352–43360.

42. Xu, J., P. Kausalya, D. Phua, *et al.* 2008. Early embryonic lethality of mice lacking ZO-2, but not ZO-3, reveals critical and nonredundant roles for individual zonula occludens proteins in mammalian development. *Mol. Cell. Biol.* **28:** 1669–1678.

43. Katsuno, T., K. Umeda, T. Matsui, *et al.* 2008. Deficiency of zonula occludens-1 causes embryonic lethal phenotype associated with defected yolk sac angiogenesis and apoptosis of embryonic cells. *Mol. Biol. Cell* **19:** 2465–2475.

44. Umeda, K., J. Ikenouchi, S. Katahira-Tayama, *et al.* 2006. ZO-1 and ZO-2 independently determine where claudins are polymerized in tight-junction strand formation. *Cell* **126:** 741–754.

45. Yamazaki, Y., K. Umeda, M. Wada, *et al.* 2008. ZO-1- and ZO-2-dependent integration of myosin-2 to epithelial zonula adherens. *Mol. Biol. Cell* **19:** 3801–3811.

46. Tsukita, S., T. Katsuno, Y. Yamazaki, *et al.* 2009. Roles of ZO-1 and ZO-2 in establishment of the belt-like adherens and tight junctions with paracellular permselective barrier function. *Ann. NY. Acad. Sci.* **1165:** 44–52.

47. Ebnet, K., C. Schulz, M.Z. Brickwedde, *et al.* 2000. Junctional adhesion molecule interacts with the PDZ domain-containing proteins AF-6 and ZO-1. *J. Biol. Chem.* **275:** 27979–27988.

48. Overgaard, C.E., B.L. Daugherty, L.A. Mitchell & M. Koval. 2011. Claudins: control of barrier function and regulation in response to oxidant stress. *Antioxid. Redox. Sign.* **15:** 1179–1193.

49. Shen, L., C.R. Weber, D.R. Raleigh, D. Yu & J.R. Turner. 2011. Tight junction pore and leak pathways: a dynamic duo. *Ann. Rev. Physiol.* **73:** 283–309.

50. Furuse, M. & S. Tsukita. 2006. Claudins in occluding junctions of humans and flies. *Trends Cell Biol.* **16:** 181–188.

51. Amasheh, S., M. Fromm & D. Günzel. 2011. Claudins of intestine and nephron—a correlation of molecular tight junction structure and barrier function. *Acta physiologica (Oxford, England)* **201:** 133–140.

52. Severson, E.A., W.Y. Lee, C.T. Capaldo, *et al.* 2009. Junctional adhesion molecule A interacts with afadin and PDZ-GEF2 to activate Rap1A, regulate b1 integrin levels, and enhance cell migration. *Mol. Biol. Cell* **20:** 1916–1925.

53. Fanning, A.S., B.J. Jameson, L.A. Jesaitis & J.M. Anderson. 1998. The Tight junction protein ZO-1 establishes a link between the transmembrane protein occludin and the actin cytoskeleton. *J. Biol. Chem.* **273:** 29745–29753.

54. Mandai, K., H. Nakanishi, A. Satoh, *et al.* 1997. Afadin: A novel actin filament-binding protein with one PDZ domain localized at cadherin-based cell-to-cell adherens junction. *J. Cell Biol.* **139:** 517–528.

55. Mandell, K.J., B.A. Babbin, A. Nusrat & C.A. Parkos. 2005. Junctional adhesion molecule 1 regulates epithelial cell mor-phology through effects on beta1 integrins and Rap1 activity. *J. Biol. Chem.* **280:** 11665–11674.

56. Liang, T.W., R.A. DeMarco, R.J. Mrsny, *et al.* 2000. Char-acterization of huJAM: evidence for involvement in cell-cell contact and tight junction regulation. *Am. J. Physiol. Cell Physiol.* **279:** C1733–43.

Ann. N.Y. Acad. Sci. ISSN 0077-8923

ANNALS OF THE NEW YORK ACADEMY OF SCIENCES

Issue: *Barriers and Channels Formed by Tight Junction Proteins*

Cingulin, paracingulin, and PLEKHA7: signaling and cytoskeletal adaptors at the apical junctional complex

Sandra Citi, Pamela Pulimeno, and Serge Paschoud

Department of Molecular Biology, University of Geneva, Geneva, Switzerland

Address for correspondence: Sandra Citi, Department of Molecular Biology, 4 Boulevard d'Yvoy, 1211–4 Geneva, Switzerland. sandra.citi@unige.ch

Cingulin, paracingulin, and PLEKHA7 are proteins localized in the cytoplasmic region of the apical junctional complex of vertebrate epithelial cells. Cingulin has been detected at tight junctions (TJs), whereas paracingulin has been detected at both TJs and adherens junctions (AJs) and PLEKHA7 has been detected at AJs. One function of cingulin and paracingulin is to regulate the activity of Rho family GTPases at junctions through their direct interaction with guanidine exchange factors of RhoA and Rac1. Cingulin also contributes to the regulation of transcription of several genes in different types of cultured cells, in part through its ability to modulate RhoA activity. PLEKHA7, together with paracingulin, is part of a protein complex that links E-cadherin to the microtubule cytoskeleton at AJs. In this paper, we review the current knowledge about these proteins, including their discovery, the characterization of their expression, localization, structure, molecular interactions, and their roles in different developmental and disease model systems.

Keywords: cingulin; paracingulin; PLEKHA7; ZO-1; p120ctn; junctions

Cingulin and paracingulin

Cingulin was discovered as a M_r 140 kDa protein that copurified with myosin II from intestinal epithelial cells and was localized in an apical circumferential "belt" in these cells[1]—hence the name cingulin, from the Latin *cingere* (to form a belt around). Immunoelectron microscopy demonstrated that cingulin is localized at the cytoplasmic surface of tight junctions (TJs) in intestinal epithelial cells.[1,2] Furthermore, the subcellular localization and tissue distribution of cingulin indicate a close correlation with the continuous TJ belt of differentiated epithelia[2–4] and endothelia,[2,5] as well as the continuous and discontinuous junctions of stratified epithelia.[2,6,7] Interestingly, expression of cingulin can be induced in fibroblasts by differentiating agents, resulting in its targeting to spot-like adherens junctions (AJs).[8] Moreover, cingulin is detectable at sites, such as the kidney glomerular slit diaphragms[9] and the outer limiting membrane of the retina,[2] which do not have the characteristic morphological features of TJs, confirming that TJ proteins, together with AJ proteins, can be part of atypical junctional structures.

The sequence of cingulin, its biochemical behavior, and rotary shadowing electron microscopy show that the molecule exists as a parallel homodimer of two subunits, each comprising a globular head domain, a coiled-coil "rod," and a small globular tail (Fig. 1).[10,11] Although such domain organization is similar to that of myosin II, the sequence of cingulin does not indicate either the presence of an actin-activated MgATPase activity or a propensity of the coiled-coil rod to form higher order multimolecular assemblies, such as myosin.[11] Cingulin binds to and bundles actin filaments *in vitro* and also interacts *in vitro* with myosin, suggesting that it could link TJ proteins to the actin cytoskeleton.[10,12] However, since no major changes in the organization of the actin cytoskeleton have been detected in cingulin-depleted cells, it seems that cingulin does not play a significant role in controlling the architecture of actin filaments.

doi: 10.1111/j.1749-6632.2012.06506.x

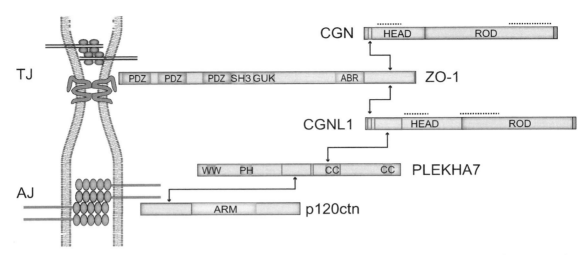

Figure 1. A molecular machinery linking TJs to AJs. Simplified diagram representing the molecular interactions between cingulin (CGN), paracingulin (CGNL1), PLEKHA7, the TJ protein ZO-1, and the AJ protein p120ctn. The TJ and AJ domains of the plasma membrane are represented on the left, with schematics of JAM-A (purple) and claudins/occludin (blue) at TJs and E-cadherin (blue) at AJs. The major structural domains of the proteins are illustrated by colored boxes, whereas the regions involved in mutual interactions are highlighted by red rectangles, linked by lines with arrows. The head and rod domains are shown for CGN and CGNL1, with the ZIM (ZO-1–interaction motif) regions in the N-terminal part of the head,[16,30] and the PLEKHA7-interacting region C-terminal to the ZIM in CGNL1.[16] Dotted lines above the schemes of CGN and CGNL1 indicate the regions that interact with GEFs *in vitro*.[21,22,27] ZO-1 contains PDZ domains that mediate the interaction with transmembrane proteins, Src homology 3 (SH3), guanylate kinase (GUK), and actin-binding-region (ABR) domains.[33] The CGN- and CGNL1-binding region of ZO-1, as identified by yeast two-hybrid screens, is composed within the region C-terminal to the ABR. The CGNL1- and p120ctn-interacting regions of PLEKHA7 are distinct and located in the central region of the molecule. p120ctn contains a central region with armadillo (ARM) repeats. Afadin is another key component of the apical AJ belt but has been omitted from this scheme for the sake of clarity.

Paracingulin (M_r 150–160 kDa) is likely to be a paralogue of cingulin—e.g., the two proteins probably arose from a gene duplication event and hence the name "para"-cingulin.[13,14] Cingulin and paracingulin show a good degree of sequence homology, especially in the coiled-coil rod domain (39% identity) and in the N-terminal ZO-1–interacting motif (ZIM; 73% identity; Fig. 1).[13,15,16] Thus, we speculate that paracingulin also exists as a parallel dimer. The head region of paracingulin is larger than the head domain of cingulin (approximately 600 versus approximately 350 residues, depending on species), and the coiled-coil rod is smaller (approximately 660 residues versus approximately 800), accounting for the difference in their apparent molecular size. Paracingulin was independently characterized as a junction-associated coiled-coil protein (JACOP).[15] Immunoelectron microscopy analysis showed that unlike cingulin, the localization of paracingulin is more promiscuous. In kidney tissue, where paracingulin mRNA is detected at high levels, paracingulin is localized both at TJs and AJs.[15] In liver tissue, immunofluo-

rescence shows junctional and apical localizations, whereas immunoelectron microscopy shows an exclusive TJ localization. Furthermore, in intestinal tissue, paracingulin is associated with nonjunctional actin filaments in the basal region of the cells.[15] A nonjunctional localization of paracingulin along actin stress fibers has also been observed in fibroblasts expressing exogenous paracingulin.[15,16] Therefore, although the actin- and myosin-binding properties of paracingulin are not known, paracingulin may have a more direct association with the actin cytoskeleton than cingulin.

Unlike cingulin, paracingulin is structurally linked to the microtubule cytoskeleton. Perturbing the organization of microtubules results in loss of junctional paracingulin, whereas cingulin is unaffected.[14] This is probably due to the interaction of paracingulin with PLEKHA7,[16] which is indirectly bound to microtubules[17] (see later). In contrast, both cingulin and paracingulin localizations are dramatically affected by actin filament-disrupting drugs,[14] confirming the association of both proteins with the actin cytoskeleton. Indeed, the dynamics of junction exchange of cingulin and paracingulin are similar to actin and more rapid than those of ZO-1.[14] Cingulin and paracingulin can be detected in a complex together, and with other TJ proteins.[14,16] However, since the junctional recruitment and dynamics of cingulin are independent of paracingulin and vice-versa, the two proteins likely do not function as a unit, but rather as independent proteins, with partially redundant functions.[14]

In summary, cingulin and paracingulin are structurally homologous proteins, with an asymmetric shape, similar dynamics, partially overlapping subcellular localizations, and distinct interactions with the actin and microtubule cytoskeletons.

Cingulin and paracingulin as adaptors for guanidine exchange factors

The first direct approach to study the function of cingulin was to generate embryonic stem (ES) cells lacking cingulin through homologous recombination. The results showed that embryoid bodies (EB) obtained by differentiation of cingulin knockout (KO) ES cells had apparently normal TJs, based on immunofluorescence, electron microscopy, and permeability assays, but had a remarkably altered pattern of gene expression: over 800 genes showed greater than twofold change in expression.[18,19] Since these genes include several transcription factors, downstream target genes, and proteins such as claudin-2—whose expression is correlated to cell differentiation—their alteration indicates that cingulin is a regulator of signaling pathways that ultimately affects transcription and differentiation. Interestingly, the cingulin promoter has been identified as a target of HNF-α, a transcription factor regarded as a key regulator of intestinal differentiation.[20]

To address the molecular mechanisms through which cingulin regulates gene expression and cell differentiation, we isolated cingulin-depleted kidney epithelial cell clones and discovered that another effect of cingulin depletion is an increase in RhoA activation in confluent monolayers, due to the junctional sequestration of the RhoA activator GEF-H1.[21,22] Cingulin binds directly to GEF-H1 *in vitro*[19,22] and has also been implicated in the junctional recruitment of a second Rho GEF, p114RhoGEF, which is important for junction assembly in corneal cells.[23] However, depletion or KO of cingulin does not have any detectable effect on junction assembly and development of the TJ barrier in kidney cells, and the junctional targeting and activity of p114RhoGEF in these cells depends on the FERM domain–containing protein Lulu2, rather than cingulin.[24]

Significantly, the increase in cell proliferation, cell density at confluence, and claudin-2 expression in cingulin-depleted kidney epithelial cells are rescued by the inhibition of RhoA, indicating that RhoA mediates, at least in part, the effects of cingulin on transcriptional and cell cycle regulation.[22] It is important to note that RhoA is modulated by several other junctional molecules,[25] and that transcription factors such as ZONAB and YAP are putative downstream effectors of RhoA.[26] Therefore, the activity of cingulin as a RhoA regulator must be viewed in the context of a wider signaling network that is likely to exist in different cell-specific configurations.

Paracingulin also interacts with the RhoA activator GEF-H1; consequently, depletion of paracingulin in confluent cells results in increased RhoA activity, an effect that phenocopies cingulin depletion. However, unlike cingulin, paracingulin also interacts with the Rac1 activator Tiam1. This results in a delay in the establishment of the TJ barrier, reduced peak of transepithelial resistance, and decreased Rac1 activation during the process of junction assembly/barrier formation in MDCK cells.[27] Indeed, Tiam1 recruitment at junctions of MDCK cells is impaired in paracingulin-depleted cells, indicating an important role of paracingulin in Rac1 activation via the junctional recruitment of Tiam1.[27] Tiam1 has been shown to play a key role in Rac1 regulation in different cell model systems, to interact with the Par polarity complex, and to regulate Ras-induced oncogenesis.[28,29]

In summary, although it is not clear to what extent paracingulin is implicated in the regulation of gene expression and cell proliferation, through RhoA or other mechanisms, the available data indicate that cingulin and paracingulin contribute to fine-tuning the activity of Rho family GTPases both during junction assembly (paracingulin) and at confluence (cingulin and paracingulin). Furthermore, modulation of RhoA activity by cingulin is involved in its ability to control cell proliferation and gene expression.

Identification of PLEKHA7 as a paracingulin-interacting protein

Cingulin and paracingulin are both localized at TJs, but paracingulin is also detected at AJs and in association with extra-junctional actin filaments. What is the molecular basis for these different localizations? Recent studies have provided some answers to these questions. ZO-1 contributes to TJ recruitment of both cingulin and paracingulin, as indicated in experiments showing that in cells depleted of ZO-1, but not ZO-2, the junctional recruitment of cingulin is decreased and the junctional recruitment of paracingulin is delayed.[16] Both proteins bind to ZO-1 *in vitro* through their conserved ZIM in the N-terminal region of the globular head domain (Fig. 1), and deletion of the same region abolishes cingulin junctional targeting in transfected cells.[16,30] Thus, it is reasonable to conclude that cingulin and paracingulin directly bind to ZO-1 at TJs. However, despite the observation that in an epithelial breast cancer line ZO-1 KO completely abolishes cingulin junctional localization,[31] a normal junctional localization of cingulin was observed in tissues from ZO-1 KO mouse embryos.[32] Therefore, the mechanisms of cingulin junctional recruitment appear to be cell-context dependent, and redundant interactions can target cingulin, and possibly also paracingulin, to TJs. Indeed, *in vitro* cingulin binds not only to ZO-1 and ZO-2 but also to ZO-3 and other TJ proteins.[10] Interestingly, the region of ZO-1 sufficient for interaction with either cingulin or paracingulin (by the yeast two-hybrid technique) is the C-terminal region adjacent to the actin-binding region (Fig. 1). Thus, the C-terminal region of ZO-1 appears to be a structural module dedicated to interaction with actin, actin-binding proteins, and other cytoskeletal proteins.[33]

A yeast two-hybrid screen revealed PLEKHA7 as a very high-confidence interactor of paracingulin; and we showed that the AJ targeting of paracingulin in MDCK cells is mediated by its interaction with PLEKHA7.[16] PLEKHA7 is an AJ protein[34] comprising WW, proline-rich, PH, and coiled-coil domains (Fig. 1). An N-terminal region of the head domain of paracingulin, distinct from the ZO-1–interacting region, interacts with a region in the C-terminal domain of PLEKHA7, the first coiled-coil domain[16] (Fig. 1). Depletion of PLEKHA7 in kidney epithelial cells results in a loss of junctional staining and de-creased expression of paracingulin, indicating that PLEKHA7 plays a major role in the recruitment and stabilization of paracingulin to AJ. Indeed, the association of paracingulin with PLEKHA7 at AJs can explain why ZO-1 depletion only delays, but does not significantly decrease, paracingulin junctional recruitment.[16]

The mechanisms controlling subcellular localization of paracingulin should be investigated in additional cell types, considering the heterogeneous localization of paracingulin in tissues. For example, it is not clear which molecular interactions target paracingulin to actin stress fibers in epithelial and nonepithelial cells.[15,16] This targeting occurs independently of the interaction with ZO-1– and PLEKHA7, since it is observed in Rat-1 fibroblasts using exogenous mutant constructs lacking the ZO-1– and PLEKHA7-interacting regions.[16] The functional role of paracingulin at stress fibers remains unknown, but considering its ability to interact with GEFs that regulate Rho family GTPases, a putative role in regulating RhoA and Rac1 activities at this site could be hypothesized.

PLEKHA7: a new component of the apical AJ belt

PLEKHA7 was discovered independently by Takeichi's and our laboratory as a novel protein component of epithelial AJs.[16,17,34] The distinguishing feature of PLEKHA7, with respect to most other AJ proteins, except afadin, is that it is confined to a continuous apical belt and is absent from the lateral membrane.[34] In the pancreas, for example, p120ctn and PLEKHA7 labeling are only partially overlapping, with PLEKHA7 strongly labeling ducts, whereas p120ctn is localized along both junctional and lateral surfaces of exocrine acinar cells (Fig. 2). The localization of PLEKHA7 in several tissues is highly reminiscent of the localization of TJ proteins and similar to that of afadin,[34] another AJ protein that has a belt-like distribution and has alternatively been described as a TJ protein.[35,36] Double labeling of PLEKHA7 with ZO-1 shows a partial co-localization, which may depend on tissue type and section orientation (Fig. 2). However, the overall localization and tissue distribution of ZO-1 and PLEKHA7 are not identical, and immunoelectron microscopy of intestinal cells shows a concentration of PLEKHA7 labeling at AJs.[34]

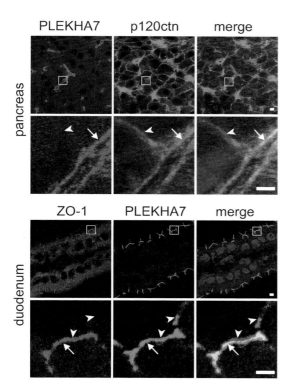

PLEKHA7   **p120ctn**   **merge**

pancreas

ZO-1   **PLEKHA7**   **merge**

duodenum

Figure 2. The localization of PLEKHA7 at the apical AJ belt. Double immunofluorescent localization of PLEKHA7 (red) and p120ctn (green) in frozen sections of pancreas (top), and of ZO-1 (red) and PLEKHA7 (green) in frozen sections of duodenum. Samples were obtained and processed as described.[34] The boxed areas in the low magnification images are shown enlarged below. Arrowheads indicate lack of co-localization, and arrows indicate co-localization. Nuclei are labeled in blue by DAPI in merged images. Bar = 2 μm. Note that PLEKHA7 is not co-localized with p120ctn along the lateral surface of acinar epithelial cells (see also Ref. 34) but co-localizes with p120ctn along intralobular and interlobular ducts. ZO-1 is not highly expressed in the duodenum, but in some sections of cells expressing sufficient amounts of ZO-1, there is a partial co-localization of ZO-1 with PLEKHA7, indicating that the apical, PLEKHA7-containing AJ belt is immediately adjacent to the ZO-1–containing TJ belt.

Besides recruiting paracingulin to kidney epithelial AJs, PLEKHA7 is part of a molecular complex linking E-cadherin to microtubules, through p120ctn and nezha, and stabilizing AJs.[17] PLEKHA7 was indeed identified as a protein binding to the N-terminal domain of p120ctn, and which is required to maintain the junctional localization of E-cadherin, nezha, and KIFC3, a minus-ended kinesin motor that binds to nezha.[17] In cadherin-deficient neuroblastoma cells, the expression of a mutant form of N-cadherin, lacking the juxta-

membrane domain, results in the redistribution of both p120ctn and PLEKHA7, indicating that PLEKHA7 may also be associated with N-cadherin based junctions, through p120ctn.[17] However, the observation that PLEKHA7 is only associated with p120ctn along the AJ belt, but not along the lateral surfaces of polarized epithelial cells (Fig. 2),[34] suggests that additional interactions are essential to target PLEKHA7 to the junctional belt.

Cingulin and PLEKHA7 in development and disease

Insights about the functions of junctional proteins can be obtained from studies about their roles in development and disease. Cingulin is maternally expressed and cortically localized both in mouse and in *Xenopus laevis* embryos.[37–40] In the mouse early embryo, cytocortical maternal cingulin is degraded by endocytic turnover, and zygotically expressed cingulin accumulates at TJs after the 16-cell stage, just after ZO-1.[37,38] Moreover, cingulin is upregulated in the mouse blastocyst trophectoderm, and its stability requires cell–cell contact.[38] In the frog embryo, maternal cingulin is recruited from the apical cortex into the newly forming TJs starting from the first cleavage, and defines the border between apical and lateral membranes.[39,40] Depletion or KO of cingulin from either *Xenopus* embryos or mice, using either morpholino oligos or gene targeting techniques, respectively, does not impair embryo viability (our unpublished results). However, it was recently reported that morpholino-mediated depletion of cingulin from the developing neural crest of the chick midbrain expands the size of the migratory neural crest cell domain.[41] Moreover, in the chicken embryo, overexpression of cingulin is correlated with changes in RhoA protein distribution in ventrolateral neuroepithelial cells, although it is not clear whether RhoA activity is altered in this model.[41] Interestingly, overexpression of either full-length cingulin or its domains does not result in changes to RhoA activity in cultured cells.[42] Nevertheless, the observations on chick embryos[41] suggest a possible role of cingulin in regulating neuroepithelial differentiation and neural crest development.

Little is known about the role of cingulin in human disease. The expression of cingulin in human cancers indicates that its expression is maintained and sometimes increased in benign and malignant adenocarcinomas, but is typically lost in squamous carcinomas, whereas it is undetectable in nonepithelial tumors.[43,44] In undifferentiated adenocarcinomas, the expression of cingulin and other TJ proteins can be detected in areas bordering the stroma, in the absence of a lumen.[43,45] Therefore, cingulin, together with other TJ proteins, can be used as an additional marker to identify epithelial tumors, and distinguish adenocarcinomas from squamous carcinomas.[44]

PLEKHA7, on the other hand, has been implicated in heart development and hypertension. Single nucleotide polymorphisms of the PLEKHA7 locus are associated with diastolic high blood pressure in genome-wide studies on caucasian and asian ethnic groups.[46–48] However, the cellular mechanisms whereby mutations in the PLEKHA7 locus lead to increased blood pressure are unknown. Knockdown studies show that the zebrafish homolog of PLEKHA7, Hadp1, is required for cardiac contractility and morphogenesis, through a mechanism involving the regulation of intracellular Ca^{2+} handling, and possibly phosphatidylinositols, in the absence of morphological defects.[49] However, the subcellular localization of PLEKHA7/Hadp1 in heart tissue has not been clarified.[34,49] Therefore, PLEKHA7 functions as a cytoskeletal adaptor in epithelial cells, and controls cardiac function through mechanisms that remain to be investigated.

Conclusion

Cingulin and paracingulin are essential components of a regulatory network controlling the activity of Rho family GTPases at the apical junctional complex, and possibly at other cellular sites. This activity may be important to control gene expression, cell proliferation, and hence differentiation and morphogenesis, although additional work is required to confirm these functions at the whole organism level. Furthermore, both proteins are part of multimolecular complexes linking cell–cell junctions to the actin and microtubule cytoskeletons. PLEKHA7, a recently characterized protein, may have multiple roles in epithelial and cardiac cells, both as a cytoskeletal linker and a regulator of intracellular Ca^{2+} signaling.

We propose that cingulin, ZO-1, paracingulin, and PLEKHA7, together with afadin and other regulatory and cytoskeletal proteins,[50–52] are part of a molecular framework that links TJ to AJ (Fig. 1), and thus the apical to the lateral domain of polarized epithelial cells, by orchestrating the formation of a continuous circumferential belt at the apical junctional complex.

It is noteworthy that homologues of cingulin, paracingulin, and PLEKHA7 have not been described in invertebrate organisms. On the one hand, this limits the availability of genetic experimental tools, but on the other hand, this underlines the potential importance of these proteins for the evolution of highly complex signaling networks, typical of vertebrates.

Acknowledgments

We thank Rocio Tapia and Domenica Spadaro for comments on the manuscript, and the Swiss National Foundation, the State of Geneva, and the Section of Biology of the Faculty of Sciences of the University for financial support.

Conflicts of interest

The authors declare no conflicts of interest.

References

1. Citi, S., H. Sabanay, R. Jakes, B. Geiger & J. Kendrick-Jones. 1988. Cingulin, a new peripheral component of tight junctions. *Nature* **333:** 272–276.
2. Citi, S., H. Sabanay, J. Kendrick-Jones & B. Geiger. 1989. Cingulin: characterization and localization. *J. Cell Sci.* **93:** 107–122.
3. Halbleib, J.M., A.M. Saaf, P.O. Brown & W.J. Nelson. 2007. Transcriptional modulation of genes encoding structural characteristics of differentiating enterocytes during development of a polarized epithelium in vitro. *Mol. Biol. Cell.* **18:** 4261–4278.
4. Erickson, D.R., S.R. Schwarze, J.K. Dixon, C.J. Clark & M.A. Hersh. 2008. Differentiation associated changes in gene expression profiles of interstitial cystitis and control urothelial cells. *J. Urol.* **180:** 2681–2687.
5. Corada, M. *et al.* 1999. Vascular endothelial-cadherin is an important determinant of microvascular integrity in vivo. *Proc. Natl. Acad. Sci. USA* **96:** 9815–9820.
6. Schluter, H., I. Moll, H. Wolburg & W.W. Franke. 2007. The different structures containing tight junction proteins in epidermal and other stratified epithelial cells, including squamous cell metaplasia. *Eur. J. Cell Biol.* **86:** 645–655.
7. Langbein, L. *et al.* 2002. Tight junctions and compositionally related junctional structures in mammalian stratified epithelia and cell cultures derived therefrom. *Eur. J. Cell Biol.* **81:** 419–435.
8. Bordin, M., F. D' Atri, L. Guillemot & S. Citi. 2004. Histone deacetylase inhibitors up-regulate the expression of tight junction proteins. *Mol. Cancer Res.* **2:** 692–701.
9. Fukasawa, H., S. Bornheimer, K. Kudlicka & M.G. Farquhar. 2009. Slit diaphragms contain tight junction proteins. *J Am Soc Nephrol.* **20:** 1491–1503.

10. Cordenonsi, M. *et al.* 1999. Cingulin contains globular and coiled-coil domains and interacts with ZO-1, ZO-2, ZO-3, and myosin. *J. Cell Biol.* **147:** 1569–1582.

11. Citi, S., F. D'Atri & D.A.D. Parry. 2000. Human and Xenopus cingulin share a modular organization of the coiled-coil rod domain: predictions for intra- and intermolecular assembly. *J. Struct. Biol.* **131:** 135–145.

12. D'Atri, F. & S. Citi. 2001. Cingulin interacts with F-actin in vitro. *FEBS Lett.* **507:** 21–24.

13. Guillemot, L. & S. Citi. 2006. Cingulin, a cytoskeleton-associated protein of the tight junction. In *Tight Junctions.* L. Gonzalez-Mariscal, Ed.: 54–63. Landes Bioscience-Springer Science. New York.

14. Paschoud, S. *et al.* 2010. Cingulin and paracingulin show similar dynamic behaviour, but are recruited independently to junctions. *Mol. Membr. Biol.* **28:** 123–135.

15. Ohnishi, H. *et al.* 2004. JACOP, a novel plaque protein localizing at the apical junctional complex with sequence similarity to cingulin. *J. Biol. Chem.* **279:** 46014–46022.

16. Pulimeno, P., S. Paschoud & S. Citi. 2011. A role of ZO-1 and PLEKHA7 in recruiting paracingulin to epithelial junctions. *J. Biol. Chem.* **286:** 16743–16750.

17. Meng, W., Y. Mushika, T. Ichii & M. Takeichi. 2008. Anchorage of microtubule minus ends to adherens junctions regulates epithelial cell-cell contacts. *Cell.* **135:** 948–959.

18. Guillemot, L. *et al.* 2004. Disruption of the cingulin gene does not prevent tight junction formation but alters gene expression. *J. Cell Sci.* **117:** 5245–5256.

19. Citi, S. *et al.* 2009. The tight junction protein cingulin regulates gene expression and RhoA signalling. *Ann. N.Y. Acad. Sci.* **1165:** 88–98.

20. Boyd, M., S. Bressendorff, J. Moller, J. Olsen & J.T. Troelsen. 2009. Mapping of HNF4alpha target genes in intestinal epithelial cells. *BMC Gastroenterol.* **9:** 68.

21. Aijaz, S., F. D'Atri, S. Citi, M.S. Balda & K. Matter. 2005. Binding of GEF-H1 to the tight junction-associated adaptor cingulin results in inhibition of Rho signaling and G1/S phase transition. *Dev. Cell.* **8:** 777–786.

22. Guillemot, L. & S. Citi. 2006. Cingulin regulates claudin-2 expression and cell proliferation through the small GTPase RhoA. *Mol. Biol. Cell.* **17:** 3569–3577.

23. Terry, S.J. *et al.* 2011. Spatially restricted activation of RhoA signalling at epithelial junctions by p114RhoGEF drives junction formation and morphogenesis. *Nat. Cell Biol.* **13:** 159–166.

24. Nakajima, H. & T. Tanoue. 2011. Lulu2 regulates the circumferential actomyosin tensile system in epithelial cells through p114RhoGEF. *J. Cell Biol.* **195:** 245–261.

25. Citi, S., D. Spadaro, Y. Schneider, J. Stutz & P. Pulimeno. 2011. Regulation of small GTPases at epithelial cell-cell junctions. *Mol. Membr. Biol.* **28:** 427–444.

26. Spadaro, D., R. Tapia, P. Pulimeno & S. Citi. 2012. The control of gene expression and cell proliferation by the apical junctional complex. *Essays Biochem.* In press.

27. Guillemot, L., S. Paschoud, L. Jond, A. Foglia & S. Citi. 2008. Paracingulin regulates the activity of Rac1 and RhoA GTPases by recruiting Tiam1 and GEF-H1 to epithelial junctions. *Mol. Biol. Cell.* **19:** 4442–4453.

28. Mertens, A.E., T.P. Rygiel, C. Olivo, R. van der Kammen & J.G. Collard. 2005. The Rac activator Tiam1 controls tight junction biogenesis in keratinocytes through binding to and activation of the Par polarity complex. *J. Cell Biol.* **170:** 1029–1037.

29. Strumane, K., T.P. Rygiel & J.G. Collard. 2006. The Rac activator Tiam1 and Ras-induced oncogenesis. *Methods Enzymol.* **407:** 269–281.

30. D'Atri, F., F. Nadalutti & S. Citi. 2002. Evidence for a functional interaction between cingulin and ZO-1 in cultured cells. *J. Biol. Chem.* **277:** 27757–27764.

31. Umeda, K. *et al.* 2004. Establishment and characterization of cultured epithelial cells lacking expression of ZO-1. *J. Biol. Chem.* **279:** 44785–44794.

32. Katsuno, T. *et al.* 2008. Deficiency of zonula occludens-1 causes embryonic lethal phenotype associated with defected yolk sac angiogenesis and apoptosis of embryonic cells. *Mol. Biol. Cell.* **19:** 2465–2475.

33. Fanning, A.S. & J.M. Anderson. 2009. Zonula occludens-1 and -2 are cytosolic scaffolds that regulate the assembly of cellular junctions. *Ann. N. Y. Acad. Sci.* **1165:** 113–120.

34. Pulimeno, P., C. Bauer, J. Stutz & S. Citi. 2010. PLEKHA7 is an adherens junction protein with a tissue distribution and subcellular localization distinct from ZO-1 and E-Cadherin. *PLoS One.* **5:** doi:10.1371/journal.pone.0012207.

35. Mandai, K. *et al.* 1997. Afadin: a novel actin filament-binding protein with one PDZ domain localized at cadherin-based cell-to-cell adherens junction. *J. Cell Biol.* **139:** 517–528.

36. Yamamoto, T. *et al.* 1997. The Ras target AF-6 interacts with ZO-1 and serves as a peripheral component of tight junctions in epithelial cells. *J. Cell Biol.* **139:** 785–795.

37. Fleming, T. P., M. Hay, Q. Javed & S. Citi. 1993. Localisation of tight junction protein cingulin is temporally and spatially regulated during early mouse development. *Development* **117:** 1135–1144.

38. Javed, Q., T.P. Fleming, M.J. Hay & S. Citi. 1993. Tight junction protein cingulin is expressed by maternal and embryonic genomes during early mouse development. *Development* **117:** 1145–1151.

39. Cardellini, P., G. Davanzo & S. Citi. 1996. Tight junctions in early amphibian development: detection of junctional cingulin from the 2-cell stage and its localization at the boundary of distinct membrane domains in dividing blastomeres in low calcium. *Dev. Dyn.* **207:** 104–113.

40. Fesenko, I. *et al.* 2000. Tight junction biogenesis in the early Xenopus embryo. *Mech. Dev.* **96:** 51–65.

41. Wu, C.Y., S. Jhingory & L.A. Taneyhill. 2011. The tight junction scaffolding protein cingulin regulates neural crest cell migration. *Dev. Dyn.* **240:** 2309–2323.

42. Paschoud, S. & S. Citi. 2008. Inducible overexpression of cingulin in stably transfected MDCK cells does not affect tight junction organization and gene expression. *Mol. Membr. Biol.* **25:** 1–13.

43. Citi, S., A. Amorosi, F. Franconi, A. Giotti & G. Zampi. 1991. Cingulin, a specific protein component of tight junctions, is expressed in normal and neoplastic human epithelial tissues. *Am. J. Pathol.* **138:** 781–789.

44. Paschoud, S., M. Bongiovanni, J.C. Pache & S. Citi. 2007. Claudin-1 and claudin-5 expression patterns differentiate lung squamous cell carcinomas from adenocarcinomas. *Mod. Pathol.* **20:** 947–954.

45. Langbein, L. *et al.* 2003. Tight junction-related structures in the absence of a lumen: occludin, claudins and tight junction plaque proteins in densely packed cell formations of stratified epithelia and squamous cell carcinomas. *Eur. J. Cell Biol.* **82:** 385–400.

46. Levy, D. *et al.* 2009. Genome-wide association study of blood pressure and hypertension. *Nat. Genet.* **41:** 677–687.

47. Hong, K.W. *et al.* 2010. Recapitulation of two genomewide association studies on blood pressure and essential hypertension in the Korean population. *J. Hum. Genet.* **55:** 336–341.

48. Lin, Y. *et al.* 2011. Genetic variations in CYP17A1, CACNB2 and PLEKHA7 are associated with blood pressure and/or hypertension in she ethnic minority of China. *Atherosclerosis* **219:** 709–714.

49. Wythe, J. D. *et al.* 2011. Hadp1, a newly identified pleckstrin homology domain protein, is required for cardiac contractility in zebrafish. *Dis. Model Mech.* **4:** 607–621.

50. Ooshio, T. *et al.* 2010. Involvement of the interaction of afadin with ZO-1 in the formation of tight junctions in Madin-Darby canine kidney cells. *J. Biol. Chem.* **285:** 5003–5012.

51. Suzuki, A. *et al.* 2002. aPKC kinase activity is required for the asymmetric differentiation of the premature junctional complex during epithelial cell polarization. *J. Cell Sci.* **115:** 3565–3573.

52. Choi, W. *et al.* 2011. The single Drosophila ZO-1 protein Polychaetoid regulates embryonic morphogenesis in coordination with Canoe/afadin and Enabled. *Mol. Biol. Cell.* **22:** 2010–2030.

Ann. N.Y. Acad. Sci. ISSN 0077-8923

ANNALS OF THE NEW YORK ACADEMY OF SCIENCES
Issue: *Barriers and Channels Formed by Tight Junction Proteins*

ZO-2, a tight junction scaffold protein involved in the regulation of cell proliferation and apoptosis

Lorenza Gonzalez-Mariscal, Pablo Bautista, Susana Lechuga, and Miguel Quiros

Center of Research and Advanced Studies (Cinvestav), Department of Physiology, Biophysics and Neuroscience, Mexico D.F., Mexico

Address for correspondence: Lorenza Gonzalez-Mariscal, Ph.D., Center of Research and Advanced Studies (Cinvestav), Department of Physiology, Biophysics, and Neuroscience, Ave. IPN 2508, Mexico, D.F. 07360, Mexico. lorenza@fisio.cinvestav.mx

ZO-2 is a membrane-associated guanylate kinase homologue (MAGUK) tight protein associated with the cytoplasmic surface of tight junctions. Here, we describe how ZO-2 is a multidomain molecule that binds to a variety of cell signaling proteins, to the actin cytoskeleton, and to gap, tight, and adherens junction proteins. In sparse cultures, ZO-2 is present at the nucleus and associates with molecules active in gene transcription and pre-mRNA processing. ZO-2 inhibits the Wnt signaling pathway, reduces cell proliferation, and promotes apoptosis; its absence, mutation, or overexpression is present in various human diseases, including deafness and cancer.

Keywords: tight junctions; ZO-2; scaffold; cell proliferation; apoptosis

Introduction

ZO is an acronym of zonula occludens, the Latin name for tight junctions. ZO-2 is a peripheral protein associated with the cytoplasmic surface of tight junctions, originally identified in 1991 in a ZO-1 immunoprecipitate.[1] Here, we describe how ZO-2 is a multidomain scaffold that at the cell membrane binds to gap, tight, and adherens proteins and at the nucleus associates with molecules active in gene transcription and pre-mRNA processing. Then, we will detail how ZO-2 inhibits the Wnt signaling pathway, reduces cell proliferation, and promotes apoptosis, and will relate these observations to human disease.

Molecular structure and protein–protein interactions

ZO-2 is a 160 kDa scaffold molecule with several different protein-binding domains (Fig. 1). ZO-2 belongs to the MAGUK (membrane-associated guanylate kinase homologue) family as it has three PDZ domains, one SH3 module, and a GuK domain. In addition, the carboxyl segment of ZO-2 has acidic and proline-rich domains. Between these domains, ZO-2 exhibits linker regions also named unique (U). At the carboxyl terminal, the last three amino acids, TEL, constitute a PDZ-binding motif.

The first PDZ domain of ZO-2 associates with the scaffold attachment factor SAF-B,[2] a DNA-binding protein that serves as a molecular base to assemble a transcriptosome complex in the vicinity of actively transcribed genes, and to the PDZ-binding motifs of: the tight junction integral proteins claudins 1 to 8;[3,4] the adenovirus type 9 oncogenic E4-ORF1 protein;[5] and the transcriptional coactivator TAZ implicated in cell proliferation and epithelial to mesenchymal transition[6] and of its paralog YAP that promotes cell detachment and apoptosis.[7]

The second PDZ domain of ZO proteins is responsible for the homo and hetero dimerization of these molecules.[8,9] The crystal structure of the second PDZ domain of human ZO-2 reveals that homo dimerization takes place by domain swapping of $\beta 1$ and $\beta 2$ strands.[10] In addition, NMR spectroscopy has demonstrated the formation of ZO-1-PDZ-2/ZO-2-PDZ-2 heterodimers by domain swapping, owing to the 68% identity present between the PDZ-2 domains of these proteins.[11] ZO-2-PDZ-2 homodimers have a similar binding pattern to that exhibited in the ZO-1-PDZ-2/connexin 43 complex,[10] and interestingly, ZO-2-PDZ-2 additionally

doi: 10.1111/j.1749-6632.2012.06537.x

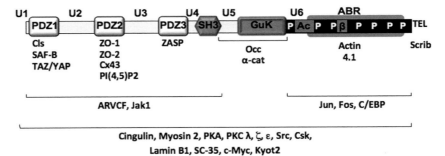

Figure 1. ZO-2 is a MAGuK protein that associates to several molecules. ZO-2 contains domains (PDZ, SH3, and GuK), regions (U, unique; Ac, acidic; ABR, actin binding; and P, proline rich), and a carboxyl PDZ-binding motif (TEL). Molecules known to interact with different segments of ZO-2 are indicated.

associates with the gap junction protein connexin 43.[12] ZO-2-PDZ-2 also interacts with phosphatidylinositol 4.5-biphosphate.[13]

The third PDZ domain of ZO-2 is presumed to be the site of interaction with the JAM family of tight junction integral proteins. Although this association has been demonstrated for ZO-1,[14,15] it remains to be tested for ZO-2. We recently demonstrated that the third PDZ domain of ZO-2, interacts with the internal and non-C-terminal, PDZ-binding motif S-X-V of a novel protein present in nuclear speckles, named ZASP.[16]

In addition, we know that the segment of ZO-2 containing the three PDZ domains interacts with the non-receptor tyrosine kinase Jak 1;[17] and ARVCF, an armadillo repeat protein of the p120 catenin family that associates with the membrane proximal region cytosolic tail of classical cadherins at adherens junctions.[18]

In ZO-1, the SH3 domain, homologous to a non-catalytic region of Src tyrosine kinase, forms with the GuK domain, and the U5 and U6 linkers, respectively located before and after the GuK domain, an integrated functional module with inter and intramolecular interactions.[19] Thus, SH3 and GuK domains do not work as independently folded domains but as a unit assembled by subdomains separated in sequence, where, for example, the U6 domain binds to a basic surface on the GuK domain. ZO-2 binds directly to occludin and α-catenin.[20] Due to the high-sequence homology, the C-terminal coiled-coil domain of occludin exhibits a common structure with the four α-helical elements present in the crystal structure of α-catenin. Therefore, it was not surprising to find that occludin and

α-catenin interact with the same epitopes of ZO-1.[21] Due to the similarity of the SH3-U5-GuK-U6 segment among ZO-1 and ZO-2, these interactions are presumed to be mediated by the U5-GuK segment,[21,22] and to be negatively regulated by U6, as has been demonstrated for ZO-1.[22]

At the carboxyl segment of ZO-2, residues present at the proline-rich region, mediate the interaction with the actin-binding protein 4.1R, a member of the 4.1 superfamily that includes proteins like ezrin, radixin, moesin, merlin, and myosin.[23] In mammalian cells, protein 4.1 colocalizes in nuclear speckles with the splicing essential factor SC35.[24] ZO-2 interacts directly with F-actin but does not act as a cross-linking protein nor binds to actin-filament ends.[20] In the case of ZO-1, a 220 amino acid region present at the C-terminus of the protein, has been identified as the actin-binding region (ABR) for being both necessary and sufficient for interaction with F-actin both *in vitro* and *in vivo*. This region is also present at the proline-rich domain of ZO-2. We also know that the AP segment of ZO-2 containing the acidic and proline-rich domains of the molecule associates with transcription factors Jun, Fos, and C/EBP.[25]

The carboxyl terminal PDZ-binding motif of ZO-2, TEL, associates with PDZ-2 and -3 domains of scribble, a tumor suppressor protein involved in cell migration and in the definition of the lateral versus the apical surface of epithelial cells.[26]

For several proteins that associate with ZO-2, the precise binding regions in ZO-2 remain to be established. Such is the case of cingulin, a peripheral tight junction protein;[27] non-muscle myosin II;[28] the serine/threonine kinases PKA and

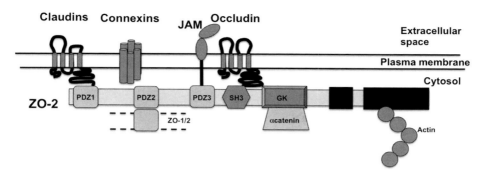

Figure 2. At the plasma membrane, ZO-2 forms a scaffold that interacts with both the actin cytoskeleton and the integral proteins of the tight and gap junctions.

PKC isoforms ε, λ, and ζ;[29] the tyrosine kinase c-Src and its negative-regulator Csk;[30] the nuclear matrix protein lamin B1;[31] and the transcription factors c-myc[32] and KyoT2.[33]

Localization and intracellular trafficking of ZO-2

The observation that ZO-1 localizes to the nucleus before the maturation of cell–cell contacts in epithelial cell,[34] prompted us to explore the subcellular localization of ZO-2. In sparse cultures of epithelial cells, ZO-2 is a dual localization protein, present both at the nucleus and the tight junction (Fig. 2). Instead, in confluent cultures and tissues ZO-2 concentrates at the tight junction.[35]

ZO-2 enters the nucleus at the late G1 phase of the cell cycle, whereas loss of nuclear ZO-2 happens when the nuclear envelope breaks during mitosis,[36] hence explaining why in confluent quiescent cells or in serum-starved cultures arrested at G0, ZO-2 is absent from the nucleus.

Intracellular trafficking of ZO-2 is regulated by several nuclear localization and exportation signals (NLS and NES).[37] The reason for such redundancy is unknown, and speculatively it was proposed to guarantee the movement of the protein even when binding to associated molecules could block the exposure of a given signal. However, it has been observed that mutation of any of the four NES present in canine ZO-2 (cZO-2), two at PDZ-2, and the rest at the GuK domain, is sufficient for inducing nuclear accumulation of ZO-2.[38] The intracellular movement of ZO-2 is a target of posttranslational modifications, thus cZO-2 NES-1 contains a serine (S369) whose phosphorylation by novel PKCε is necessary for the exportation of the protein from the nucleus.[39]

cZO-2 has two bipartite and one monopartite NLS localized at the end of PDZ-1, and the beginning of linker U1. When these signals are eliminated from the amino segment of cZO-2, the nuclear entry of the protein is impaired.[31] Employing an assay in which the antibodies against ZO-2 were microinjected into the nucleus, it was demonstrated that in sparse cultures, newly synthesized ZO-2 first travels to the nucleus and then to the plasma membrane.[39]

ZO-2 functions as a nucleo-cytoplasmic shuttle that allows the delivery of diverse factors into the nucleus (Fig. 3). For example, the overexpression of ZO-2 enhances the nuclear accumulation of the catenin ARVCF and of the transcriptional coactivator YAP, whereas the presence of these proteins at the nucleus is greatly diminished in cells expressing a mutant ZO-2 lacking the NLSs.[7,18] Interestingly, ZO-2 forms a tripartite complex with YAP and AmotL1, a member of the angiomontin family of proteins that through a PPXY motif binds the WW domain of YAP. AmotL1 inhibits YAP nuclear translocation and proapoptotic function, whereas ZO-2 enhances it.[40]

ZO-2 localizes at the nuclei of hamster and mouse blastomeres in all stages of embryonic development.[41] In mice, ZO-2 first appears at the cell border of the blastomeres at the morula stage (16-cell), while in hamster ZO-2 localizes at the cell membrane from the two-cell stage onward. In mice, ZO-2 co-localizes with E-cadherin at the 16-cell stage and later at the early blastocyst stage segregates to tight junctions colocalizing with ZO-1α+ and occludin.[42]

In nonepithelial cells, such as fibroblasts and cardiac muscle cells lacking tight junctions, ZO-2 concentrates at adherens junction.[8] In vascular smooth muscle cells (VSMC), ZO-2 mediates homotypic

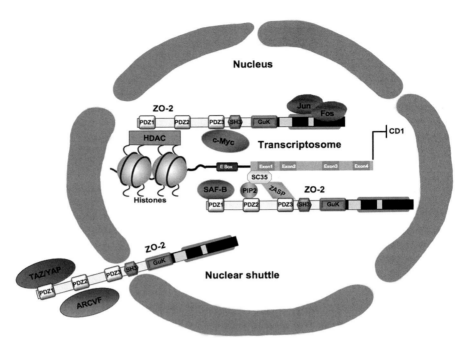

Figure 3. Functions of ZO-2 at the nucleus. ZO-2 serves as a shuttle for the transport of TAZ/YAP and ARVCF to the nucleus. At the nucleus, ZO-2 is present in transcriptosomes inhibiting gene transcription and associating with molecules involved in pre-mRNA splicing.

cell contacts and the non-receptor tyrosine kinase Jak1 induces ZO-2 phosphorylation and membrane localization through a signaling pathway mediated by the urokinase receptor.[17] The transcript for ZO-2 exhibits a strong and sustained upregulation after vascular injury at both the intimal and medial layers.[43] One of the functions of ZO-2 in VSMC is the regulation of cell proliferation via the transcription factor Stat1. Thus, upon ZO-2 silencing, STAT-1 becomes upregulated, leading to the inhibition of cell proliferation, hence implicating ZO-2 as a key player in vascular remodeling in cardiovascular disease.[44]

ZO-2 function

ZO proteins have a high degree of identity (hZO-1 vs. hZO-2 = 44%; hZO-1 vs. hZO-3 = 40%; hZO-2 vs. hZO-3 = 40%) among themselves.[45] Therefore, it is not surprising to find that the three ZOs are functionally redundant and at the same time involved in certain specific functions.

Lessons learned from ZO-2 silencing and knockdown

ZO-2 silencing (KD, knockdown) using small interfering RNA (siRNA) has no effect in Eph4 cells,[46] while in MDCK cells, it compromises the barrier function, delays *de novo* formation of tight junctions in an assay modulated by calcium, widens the intercellular spaces, and generates the apical mislocalization of E-cadherin.[47] It is important to mention that while ZO-1 silencing in MDCK cells with short hairpin RNAs (shRNAs) results in an increased permeability for solutes larger than ∼4° and reorganization of actin and myosin, ZO-2 KD with shRNAs does not replicate the changes observed, indicating that in these functions ZO-1 and ZO-2 are not redundant.[48] In contrast, when Eph4 cells are deficient in both ZO-1 and ZO-2 (ZO-1 KO/ZO-1 KD) by homologous recombination (KO, knockout) and siRNA, cell polarity is maintained, but no tight junction strands are formed and the paracellular barrier function is lost. The introduction of exogenous ZO-1 or ZO-2 to ZO-1 KO/ZO-1 KD cells restores tight junction strands formation at the uppermost region of the lateral membrane.[46] These results hence indicate that ZO-1 and ZO-2 have a redundant role determining the polymerization of claudins at the tight junction region.

During mouse embryonic development, ZO-2 siRNA microinjection into zygotes or two-cell embryos, results in a delayed blastocoel cavity formation and increased assembly of ZO-1 as a compensatory mechanism. ZO-1 KD or a combined ZO-1 and ZO-2 KD generates a more severe inhibition of blastocoel formation,[42] hence indicating that both ZO-1 and ZO-2 participate in blastocoel development, albeit the role of ZO-2 appears to be more modest in comparison to that of ZO-1.

Additional insight on the function of ZO-2 has been obtained generating KO mice. These embryos show decreased proliferation at embryonic day 6.5 and increased apoptosis at E7.5 in comparison to wild-type embryos, and die shortly after implantation.[49] Interestingly, embryonic lethality is rescued when ZO KO embryonic stem cells are injected into wild-type blastocysts, suggesting that while ZO-2 is crucial for the proper function of the extra embryonic tissue it is not for the development of the embryo. Although ZO-2 chimeras are viable, the males show a reduced fertility and a compromised blood testis barrier, hence demonstrating the unique and non-redundant role of ZO-2 at tight junctions between Sertoli cells.[50]

Lessons learned from embryonic development

The differential expression of ZO-1, -2, and -3, in diverse tissues is indicative of special and unique roles for each of these proteins. To date the best studied case is that of zebra fish during embryonic development, where the highest degree of expression heterogeneity of all ZOs transcripts is observed during early development. Thus, axial mesoderm and the hind-brain region express ZO-1, somatic mesoderm exhibits ZO-2, and sensory placodes express ZO-3. At later stages the expression pattern converges.[51]

Lessons learned from ZO-2 presence at the nucleus

The conspicuous presence of ZO-2 at nuclear speckles[2] prompted us to investigate the physiological role of ZO-2 in gene transcription. The sequence of ZO-2 contains no DNA-binding sites but is rich in protein–protein binding domains and motifs including a leucine zipper (Leu823, 830, 837, and 844 in cZO-2) located within the GuK domain,[25] which in other proteins is necessary and sufficient for the heterodimerization of DNA-binding proteins. ZO-2 associates with the transcription factors Jun and Fos and inhibits in a dose-dependent manner the transcription of reporter genes under the control of a promoter with AP-1 sites.[25] ZO-2 overexpression inhibits cell proliferation[32] and blocks cell-cycle progression from GI to S.[36] This inhibition is due to a decreased transcription[32] and increased degradation at the proteosome of cyclin D1 (CD1). ZO-2 bound to the transcription factor Myc and the histone deacetylase 1 associates to an E box present in the promoter of human CD1, inhibiting gene transcription.[32] The latter is reversed when epithelial cells are cotransfected with ZO-2 and ZASP, a recently identified nuclear factor that concentrates in speckles and associates to the E box, and together with SC-35, to the intron/exon region of CD1 gene (Fig. 3).[16]

ZO-2 also plays an inhibitory role on TAZ induced reporter gene expression and ZO-2 overexpression results in a near depletion of nuclear TAZ.[6] The latter is a transcriptional coactivator that through its WW domain recognizes L/PPxY motifs in transcription factors, promoting cell proliferation and epithelial to mesenchymal transition (EMT).

ZO-2 acts as an inhibitor of the Wnt pathway, a route that promotes EMT. In the canonical Wnt signaling pathway the key determinant is the presence of a cytosolic pool of β-catenin that associates with a degradation complex comprising the core proteins axin, APC, and the glycogen synthase kinase-3 (GSK3β). We have demonstrated, that ZO-2 overexpression decreases the inhibitory phosphorylation of GSK3β, presumably increasing in consequence the activity of the β-catenin degradation complex; decreases the mRNA content of axin2 whose gene is a negative feedback target of the Wnt pathway,[52] and represses in a dose-dependent manner, the transcription of reporter genes driven either by artificial promoters with LEF/TCF-binding sites or by the Siamois promoter,[36] whose homeobox gene is another target of the Wnt pathway. More recently, we have demonstrated that ZO-2 overexpression prevents podocyte injury mediated by the Wnt pathway in a mouse model of nephrosis induced by the anthracycline antibiotic Adriamycin (unpublished results).

The addition of three repetitive SV40 NLS to the sequence of ZO-2, overrides the endogenous NES of ZO-2 and triggers the nuclear accumulation of the protein. Surprisingly, this condition leads to an increased proliferative activity and a reduced intercellular junction stability.[53,54] The reason remains poorly defined, but may be related to the increased expression of M2 type pyruvate kinase, an enzyme overexpressed in carcinoma and embryonic stem cells.

At the nucleus, ZO-2 associates, in addition to transcription factors, with a complex set of molecules, including nuclear matrix proteins lamin B1 and actin,[31] the essential nuclear splicing factor SC-35,[35] the DNA-binding scaffold protein SAFB,[2,4] and the lipid PtdIns4,5P2. The latter that gives a dispersed nuclear staining pattern when ZO-2 expression is diminished,[13] is present in nuclear domains containing RNA polymerase II and the splicing factor SC-35.[55] Protein 4.1 is also expected to be present in this nuclear complex, as it is known to associate with ZO-2 and SC-35.[24] Due to the complex array of molecules that participate in gene transcription and pre-mRNA processing that associate with nuclear ZO-2, it is tempting to suggest that ZO-2 at the nucleus, serves as a platform for transcriptosome assembly (Fig. 3).

Lessons learned from human diseases

Familial hypercholanemia. Mutation V48A of human ZO-2 induces the loss of the alpha-helical structure of PDZ-1, reducing in consequence binding to claudins 1, 2, 3, 5, and 7-carboxyl terminal amino acids. V48A mutation is associated in humans to familial hypercholanemia, a disease characterized by an elevated concentration of bile acid in serum, itching, and malabsorption of fats. Why the mutation only apparently affects the tightness of biliary ducts and not other epithelial tight junctions remains to be determined; however, it is interesting to observe that the clinical appearance of the disease requires a concomitant mutation in the liver coenzyme A amino acid *N*-acetyltransferase.[56]

Nonsyndromic progressive hearing loss. An inverted genomic duplication of ZO-2 gene in humans, leads to an overexpression of the protein that results in nonsyndromic, autosomal dominant, progressive hearing loss that starts at high frequency but eventually alters the whole auditory spectrum. *In vitro* overexpression of ZO-2 leads to a reduced phosphorylation of GSK3β, which results in activation of this kinase. In patients, the overexpression of ZO-2 gives rise to the expression of various apoptosis related genes like BCL2L11, IL-6, Rel, and TSPO. In the human ear, ZO-2 is mainly expressed at the organ of Corti, at the membrane between hairy and

supporting cells. Therefore, it is presumed that the altered activity of GSK3β tips the balance in hairy cells to apoptosis causing hearing loss.[57] Why ZO-2 overexpression only causes dysfunction in the auditory epithelium may be related to the fact that hairy cells are very few and terminally differentiated, and to the observation that outer hairy cells are very susceptible to apoptotic stimuli, such as excessive noise exposure, aminoglycoside antibiotics, and the chemotherapeutic agent cisplatinum.[58]

Cancer. The relationship of ZO proteins to cancer was first highlighted after observing the high homology existing between ZO-1 and the *Drosophila*, tumor suppressor protein Dlg.[59] Although it has long been recognized that tumor cells lose cell–cell contacts, it was not known if the loss of cell–cell adhesion proteins was a consequence or a trigger of the disease. The inhibitory action of ZO-2 on the Wnt pathway, and on cell proliferation as well as its role in the promotion of apoptosis, favor the idea of ZO-2 as a tumor suppressor protein. In further support of this proposal is the discovery of the E4-ORF1 oncogenic protein of human adenovirus type 9, which elicits mammary tumors in animals and sequesters through its PDZ-binding motif the tight junction proteins PATJ and ZO-2 in the cytoplasm of epithelial and fibroblast cells.[5] In epithelia, this provokes the disruption of the tight junction barrier and causes a loss of apicobasal polarity.[60] This result hence implies that by nullifying a putative tumor suppressor protein like ZO-2, E4-ORF1 enhances its oncogenic potential. In a similar fashion, it has been observed that the overexpression of ZO-2 suppresses the transformation induced by adenovirus 9 E4-ORF1, even when the experiment is done with a ZO-2 mutant-lacking PDZ-1, the binding target of the oncogenic protein, indicating that by elevating the amount of the free tumor suppressor protein ZO-2, cell transformation is suppressed. In support of this idea comes the observation that ZO-2 overexpression similarly diminishes transformed focus formation by the activated RasV12 and polyomavirus middle T-proteins that lack PDZ-binding motifs.[5]

ZO-2 is silenced in certain types of cancer. For example, in lung squamous cell carcinomas and in adenocarcinomas, there is a significant decrease in the mRNA level of several tight junction proteins, including ZO-2.[61] In contrast, in tumor tissue of human breast, the expression of ZO-2 protein by Western blot shows no difference compared to normal tissue, whereas the expression of ZO-2 mRNA augments as the prognosis indicator of the disease becomes more unfavorable.[62] This trend is opposite of that found for ZO-1 and may be a consequence of a compensatory mechanism set off to cope with ZO-1 reduction, which leads to massive redistribution of ZO-2 within the cell and a reduced immunohistochemical staining of the protein at the cell borders. In a somewhat similar manner, in testicular carcinoma *in situ*, ZO-2 immunoreactivity diminishes at the blood–testis barrier and spreads to the Sertoli cell cytoplasm.[63]

ZO-2 isoforms A and C are present in normal human pancreatic duct cells, whereas in pancreatic duct adenocarcinoma, only the latter is expressed. The difference between both isoforms is the absence of 23 amino acid residues at the amino terminus of ZO-2C compared with ZO-2A, suggesting that the 23 amino acid motif, plays a role-limiting tumor development in pancreatic cells.[64] In the earlier stages of pancreatic tumor development, promoter inactivation is not due to mutation, lack of transcription factors, or methylation and is proposed to be the result of alterations in regulatory elements outside of the promoter region.[65] Instead, at the late stages of pancreatic tumor development, inactivation of promoter A is due to methylation. Accordingly, in prostate cancer cells, ZO-2 promoter is methylated.[66] Loss of ZO-2 isoform A is observed in breast ductal carcinomas but not in breast benign ducts or in colon or prostate carcinoma, thus suggesting that the loss of isoform A is associated with cancer of the ductal type.[67]

Conflict of interest

The authors declare no conflicts of interest.

References

1. Gumbiner, B., T. Lowenkopf & D. Apatira. 1991. Identification of a 160-kDa polypeptide that binds to the tight junction protein ZO-1. *Proc. Natl. Acad. Sci. USA* **88:** 3460–3464.

2. Traweger, A., R. Fuchs, I.A. Krizbai, *et al.* 2003. The tight junction protein ZO-2 localizes to the nucleus and interacts with the heterogeneous nuclear ribonucleoprotein scaffold attachment factor-B. *J. Biol. Chem.* **278:** 2692–2700.

3. Itoh, M., M. Furuse, K. Morita, *et al.* 1999. Direct binding of three tight junction-associated MAGUKs, ZO-1, ZO-2, and ZO-3, with the COOH termini of claudins. *J. Cell Biol.* **147:** 1351–1363.

4. Nayler, O., W. Stratling, J.P. Bourquin, *et al.* 1998. SAF-B protein couples transcription and pre-mRNA splicing to SAR/MAR elements. *Nucleic Acids Res.* **26:** 3542–3549.

5. Glaunsinger, B.A., R.S. Weiss, S.S. Lee & R. Javier. 2001. Link of the unique oncogenic properties of adenovirus type 9 E4-ORF1 to a select interaction with the candidate tumor suppressor protein ZO-2. *EMBO J.* **20:** 5578–5586.

6. Remue, E., K. Meerschaert, T. Oka, *et al.* 2010. TAZ interacts with zonula occludens-1 and -2 proteins in a PDZ-1 dependent manner. *FEBS Lett.* **584:** 4175–4180.

7. Oka, T., E. Remue, K. Meerschaert, *et al.* 2010. Functional complexes between YAP2 and ZO-2 are PDZ domain-dependent, and regulate YAP2 nuclear localization and signalling. *Biochem. J.* **432:** 461–472.

8. Itoh, M., K. Morita & S. Tsukita. 1999. Characterization of ZO-2 as a MAGUK family member associated with tight as well as adherens junctions with a binding affinity to occludin and alpha catenin. *J. Biol. Chem.* **274:** 5981–5986.

9. Utepbergenov, D.I., A.S. Fanning & J.M. Anderson. 2006. Dimerization of the scaffolding protein ZO-1 through the second PDZ domain. *J. Biol. Chem.* **281:** 24671–24677.

10. Chen, H., S. Tong, X. Li, *et al.* 2009. Structure of the second PDZ domain from human zonula occludens 2. *Acta Crystallogr. Sect. F. Struct. Biol. Cryst. Commun.* **65:** 327–330.

11. Wu, J., Y. Yang, J. Zhang, *et al.* 2007. Domain-swapped dimerization of the second PDZ domain of ZO2 may provide a structural basis for the polymerization of claudins. *J. Biol. Chem.* **282:** 35988–35999.

12. Singh, D., J.L. Solan, S.M. Taffet, *et al.* 2005. Connexin 43 interacts with zona occludens-1 and -2 proteins in a cell cycle stage-specific manner. *J. Biol. Chem.* **280:** 30416–30421.

13. Meerschaert, K., M.P. Tun, E. Remue, *et al.* 2009. The PDZ2 domain of zonula occludens-1 and -2 is a phosphoinositide-binding domain. *Cell Mol. Life Sci.* **66:** 3951–3966.

14. Ebnet, K., C.U. Schulz, M.K. Meyer Zu Brickwedde, *et al.* 2000. Junctional adhesion molecule interacts with the PDZ domain-containing proteins AF-6 and ZO-1. *J. Biol. Chem.* **275:** 27979–27988.

15. Itoh, M., H. Sasaki, M. Furuse, *et al.* 2001. Junctional adhesion molecule (JAM) binds to PAR-3: a possible mechanism for the recruitment of PAR-3 to tight junctions. *J. Cell Biol.* **154:** 491–497.

16. Lechuga, S., L. Alarcon, J. Solano, *et al.* 2010. Identification of ZASP, a novel protein associated to Zona occludens-2. *Exp. Cell Res.* **316:** 3124–3139.

17. Tkachuk, N., S. Tkachuk, M. Patecki, *et al.* 2011. The tight junction protein ZO-2 and Janus kinase 1 mediate intercellular communications in vascular smooth muscle cells. *Biochem. Biophys. Res. Commun.* **410:** 531–536.

18. Kausalya, P.J., D.C. Phua & W. Hunziker. 2004. Association of ARVCF with zonula occludens (ZO)-1 and ZO-2: binding to PDZ-domain proteins and cell-cell adhesion regulate plasma membrane and nuclear localization of ARVCF. *Mol. Biol. Cell* **15:** 5503–5515.

19. Lye, M.F., A.S. Fanning, Y. Su, *et al.* 2010. Insights into regulated ligand-binding sites from the structure of ZO-1 Src homology 3-guanylate kinase module. *J. Biol. Chem.* **285:** 13907–13917.

20. Wittchen, E.S., J. Haskins & B.R. Stevenson. 1999. Protein interactions at the tight junction. Actin has multiple binding

partners, and ZO-1 forms independent complexes with ZO-2 and ZO-3. *J. Biol. Chem.* **274:** 35179–35185.

21. Muller, S.L., M. Portwich, A. Schmidt, *et al.* 2005. The tight junction protein occludin and the adherens junction protein alpha-catenin share a common interaction mechanism with ZO-1. *J. Biol. Chem.* **280:** 3747–3756.

22. Schmidt, A., D.I. Utepbergenov, S.L. Mueller, *et al.* 2004. Occludin binds to the SH3-hinge-GuK unit of zonula occludens protein 1: potential mechanism of tight junction regulation. *Cell Mol. Life Sci.* **61:** 1354–1365.

23. Mattagajasingh, S.N., S.C. Huang, J.S. Hartenstein & E.J. Benz, Jr. 2000. Characterization of the interaction between protein 4.1R and ZO-2. A possible link between the tight junction and the actin cytoskeleton. *J. Biol. Chem.* **275:** 30573–30585.

24. Lallena, M.J. & I. Correas. 1997. Transcription-dependent redistribution of nuclear protein 4.1 to SC35-enriched nuclear domains. *J. Cell Sci.* **110**(Pt 2): 239–247.

25. Betanzos, A., M. Huerta, E. Lopez-Bayghen, *et al.* 2004. The tight junction protein ZO-2 associates with Jun, Fos, and C/EBP transcription factors in epithelial cells. *Exp. Cell Res.* **292:** 51–66.

26. Metais, J.Y., C. Navarro, M.J. Santoni, *et al.* 2005. hScrib interacts with ZO-2 at the cell–cell junctions of epithelial cells. *FEBS Lett.* **579:** 3725–3730.

27. Cordenonsi, M., F. D'Atri, E. Hammar, *et al.* 1999. Cingulin contains globular and coiled-coil domains and interacts with ZO-1, ZO-2, ZO-3, and myosin. *J. Cell Biol.* **147:** 1569–1582.

28. Yamazaki, Y., K. Umeda, M. Wada, *et al.* 2008. ZO-1- and ZO-2-dependent integration of myosin-2 to epithelial zonula adherens. *Mol. Biol. Cell* **19:** 3801–3811.

29. Avila-Flores, A., E. Rendon-Huerta, J. Moreno, *et al.* 2001. Tight-junction protein zonula occludens 2 is a target of phosphorylation by protein kinase C. *Biochem. J.* **360:** 295–304.

30. Saito, K., K. Enya, C. Oneyama, *et al.* 2008. Proteomic identification of ZO-1/2 as a novel scaffold for Src/Csk regulatory circuit. *Biochem. Biophys. Res. Commun.* **366:** 969–975.

31. Jaramillo, B.E., A. Ponce, J. Moreno, *et al.* 2004. Characterization of the tight junction protein ZO-2 localized at the nucleus of epithelial cells. *Exp. Cell Res.* **297:** 247–258.

32. Huerta, M., R. Munoz, R. Tapia, *et al.* 2007. Cyclin D1 is transcriptionally down-regulated by ZO-2 via an E box and the transcription factor c-Myc. *Mol. Biol. Cell* **18:** 4826–4836.

33. Huang, H.Y., R. Li, Q. Sun, *et al.* 2002. [LIM protein KyoT2 interacts with human tight junction protein ZO-2-i3]. *Yi. Chuan Xue. Bao.* **29:** 953–958.

34. Gottardi, C.J., M. Arpin, A.S. Fanning & D. Louvard. 1996. The junction-associated protein, zonula occludens-1, localizes to the nucleus before the maturation and during the remodeling of cell–cell contacts. *Proc. Natl. Acad. Sci. USA* **93:** 10779–10784.

35. Islas, S., J. Vega, L. Ponce & L. Gonzalez-Mariscal. 2002. Nuclear localization of the tight junction protein ZO-2 in epithelial cells. *Exp. Cell Res.* **274:** 138–148.

36. Tapia, R., M. Huerta, S. Islas, *et al.* 2009. Zona occludens-2 inhibits cyclin D1 expression and cell proliferation and exhibits changes in localization along the cell cycle. *Mol. Biol. Cell* **20:** 1102–1117.

37. Lopez-Bayghen, E., B. Jaramillo, M. Huerta, *et al.* 2006. TJ proteins that make round trips to the nucleus. In *Tight Junctions*. L. Gonzalez-Mariscal, Ed.: 76–100. Springer Science and Landes Bioscience. New York and Georgetown.

38. Gonzalez-Mariscal, L., A. Ponce, L. Alarcon & B.E. Jaramillo. 2006. The tight junction protein ZO-2 has several functional nuclear export signals. *Exp. Cell Res.* **312:** 3323–3335.

39. Chamorro, D., L. Alarcon, A. Ponce, *et al.* 2009. Phosphorylation of zona occludens-2 by protein kinase C epsilon regulates its nuclear exportation. *Mol. Biol. Cell* **20:** 4120–4129.

40. Oka, T., A.P. Schmitt & M. Sudol. 2012. Opposing roles of angiomotin-like-1 and zona occludens-2 on pro-apoptotic function of YAP. *Oncogene* **31:** 128–134.

41. Wang, H., L. Luan, T. Ding, *et al.* 2011. Dynamics of zonula occludens-2 expression during pre-implantation embryonic development in the hamster. *Theriogenology* **76:** 678–686.

42. Sheth, B., R.L. Nowak, R. Anderson, *et al.* 2008. Tight junction protein ZO-2 expression and relative function of ZO-1 and ZO-2 during mouse blastocyst formation. *Exp. Cell Res.* **314:** 3356–3368.

43. Adams, L.D., J.M. Lemire & S.M. Schwartz. 1999. A systematic analysis of 40 random genes in cultured vascular smooth muscle subtypes reveals a heterogeneity of gene expression and identifies the tight junction gene zonula occludens 2 as a marker of epithelioid "pup" smooth muscle cells and a participant in carotid neointimal formation. *Arterioscler. Thromb. Vasc. Biol.* **19:** 2600–2608.

44. Kusch, A., S. Tkachuk, N. Tkachuk, *et al.* 2009. The tight junction protein ZO-2 mediates proliferation of vascular smooth muscle cells via regulation of Stat1. *Cardiovasc. Res.* **83:** 115–122.

45. Adachi, M., A. Inoko, M. Hata, *et al.* 2006. Normal establishment of epithelial tight junctions in mice and cultured cells lacking expression of ZO-3, a tight-junction MAGUK protein. *Mol. Cell Biol.* **26:** 9003–9015.

46. Umeda, K., J. Ikenouchi, S. Katahira-Tayama, *et al.* 2006. ZO-1 and ZO-2 independently determine where claudins are polymerized in tight junction strand formation. *Cell* **126:** 741–754.

47. Hernandez, S., M.B. Chavez & L. Gonzalez-Mariscal. 2007. ZO-2 silencing in epithelial cells perturbs the gate and fence function of tight junctions and leads to an atypical monolayer architecture. *Exp. Cell Res.* **313:** 1533–1547.

48. Van Itallie, C.M., A.S. Fanning, A. Bridges & J.M. Anderson. 2009. ZO-1 stabilizes the tight junction solute barrier through coupling to the perijunctional cytoskeleton. *Mol. Biol. Cell* **20:** 3930–3940.

49. Xu, J., P.J. Kausalya, D.C. Phua, *et al.* 2008. Early embryonic lethality of mice lacking ZO-2, but Not ZO-3, reveals critical and non-redundant roles for individual zonula occludens proteins in mammalian development. *Mol. Cell Biol.* **28:** 1669–1678.

50. Xu, J., F. Anuar, S.M. Ali, *et al.* 2009. Zona occludens-2 is critical for blood-testis barrier integrity and male fertility. *Mol. Biol. Cell* **20:** 4268–4277.

51. Kiener, T.K., I. Sleptsova-Friedrich & W. Hunziker. 2007. Identification, tissue distribution, and developmental expression of tjp1/zo-1, tjp2/zo-2 and tjp3/zo-3 in the zebrafish, Danio rerio. *Gene Expr. Patterns.* **7:** 767–776.

52. Jho, E.H., T. Zhang, C. Domon, *et al.* 2002. Wnt/beta-catenin/Tcf signaling induces the transcription of Axin2, a negative regulator of the signaling pathway. *Mol. Cell Biol.* **22:** 1172–1183.

53. Traweger, A., C. Lehner, A. Farkas, *et al.* 2008. Nuclear Zonula occludens-2 alters gene expression and junctional stability in epithelial and endothelial cells. *Differentiation* **76:** 99–106.

54. Bauer, H., J. Zweimueller-Mayer, P. Steinbacher, *et al.* 2010. The dual role of zonula occludens (ZO) proteins. *J. Biomed. Biotechnol.* **2010:** 402593.

55. Osborne, S.L., C.L. Thomas, S. Gschmeissner & G. Schiavo. 2001. Nuclear PtdIns(4,5)P2 assembles in a mitotically regulated particle involved in pre-mRNA splicing. *J. Cell Sci.* **114:** 2501–2511.

56. Carlton, V.E., B. Z.Harris, E.G. Puffenberger, *et al.* 2003. Complex inheritance of familial hypercholanemia with associated mutations in TJP2 and BAAT. *Nat. Genet.* **34:** 91–96.

57. Walsh, T., S.B. Pierce, D.R. Lenz, *et al.* 2010. Genomic duplication and overexpression of TJP2/ZO-2 leads to altered expression of apoptosis genes in progressive non-syndromic hearing loss DFNA51. *Am. J. Hum. Genet.* **87:** 101–109.

58. Op de Beeck, K., J. Schacht & G. Van Camp. 2011. Apoptosis in acquired and genetic hearing impairment: the programmed death of the hair cell. *Hearing Res.* **281:** 18–27.

59. Willott, E., M.S. Balda, A.S. Fanning, *et al.* 1993. The tight junction protein ZO-1 is homologous to the Drosophila discs-large tumor suppressor protein of septate junctions. *Proc. Natl. Acad. Sci. USA* **90:** 7834–7838.

60. Latorre, I.J., M.H. Roh, K.K. Frese, *et al.* 2005. Viral oncoprotein-induced mislocalization of select PDZ proteins disrupts tight junctions and causes polarity defects in epithelial cells. *J. Cell Sci.* **118:** 4283–4293.

61. Paschoud, S., M. Bongiovanni, J.C. Pache & S. Citi. 2007. Claudin-1 and claudin-5 expression patterns differentiate lung squamous cell carcinomas from adenocarcinomas. *Mod. Pathol.* **20:** 947–954.

62. Martin, T.A., G. Watkins, R.E. Mansel & W.G. Jiang. 2004. Loss of tight junction plaque molecules in breast cancer tissues is associated with a poor prognosis in patients with breast cancer. *Eur. J. Cancer* **40:** 2717–2725.

63. Fink, C., R. Weigel, T. Hembes, *et al.* 2006. Altered expression of ZO-1 and ZO-2 in Sertoli cells and loss of blood-testis barrier integrity in testicular carcinoma in situ. *Neoplasia* **8:** 1019–1027.

64. Chlenski, A., K.V. Ketels, M.S. Tsao, *et al.* 1999. Tight junction protein ZO-2 is differentially expressed in normal pancreatic ducts compared to human pancreatic adenocarcinoma. *Int. J. Cancer* **82:** 137–144.

65. Chlenski, A., K.V. Ketels, J.L. Engeriser, *et al.* 1999. zo-2 gene alternative promoters in normal and neoplastic human pancreatic duct cells. *Int. J. Cancer* **83:** 349–358.

66. Wang, Y., Q. Yu, A.H. Cho, *et al.* 2005. Survey of differentially methylated promoters in prostate cancer cell lines. *Neoplasia* **7:** 748–760.

67. Chlenski, A., K.V. Ketels, G.I. Korovaitseva, *et al.* 2000. Organization and expression of the human zo-2 gene (tjp-2) in normal and neoplastic tissues. *Biochim. Biophys. Acta* **1493:** 319–324.

Ann. N.Y. Acad. Sci. ISSN 0077-8923

ANNALS OF THE NEW YORK ACADEMY OF SCIENCES

Issue: *Barriers and Channels Formed by Tight Junction Proteins*

From TER to trans- and paracellular resistance: lessons from impedance spectroscopy

Dorothee Günzel,[1] Silke S. Zakrzewski,[1] Thomas Schmid,[1,2] Maria Pangalos,[1,3] John Wiedenhoeft,[1] Corinna Blasse,[1] Christopher Ozboda,[1] and Susanne M. Krug[1]

[1]Institute of Clinical Physiology, Charité—Universitätsmedizin Berlin, Berlin, Germany. [2]Department of Computer Engineering, Faculty of Mathematics and Informatics, University of Leipzig, Leipzig, Germany

Address for correspondence: Dr. Dorothee Günzel, Institute of Clinical Physiology, Charité – Universitätsmedizin Berlin, Campus Benjamin Franklin, Hindenburgdamm 30, 12203 Berlin, Germany. dorothee.guenzel@charite.de. [3]Present address: Neuroscience Research Center, Charité–Universitätsmedizin Berlin, Berlin, Germany

In epithelia and endothelia, overall resistance (TER) is determined by all ion-conductive structures, such as membrane channels, tight junctions, and the intercellular space, whereas the epithelial capacitance is due to the hydrophobic phase of the plasma membrane. Impedance means alternating current resistance and, in contrast to ohmic resistance, takes into account that, e.g., capacitors become increasingly conductive with increasing frequency. Impedance spectroscopy uses the association of the capacitance with the transcellular pathway to distinguish between this capacitive pathway and purely conductive components (tight junctions, subepithelium). In detail, one-path impedance spectroscopy distinguishes the resistance of the epithelium from the resistance of subepithelial tissues. Beyond that, two-path impedance spectroscopy allows for the separation of paracellular resistance (governed by tight junctional properties) from transcellular resistance (determined by conductive structures residing in the cell membranes). The present paper reviews the basic principles of these techniques, some historic milestones, as well as recent developments in epithelial physiology.

Keywords: epithelium; endothelium; tight junction; paracellular transport; two-path impedance spectroscopy

Introduction

Epithelia differ from other tissues in their ability to regulate transport of ions and uncharged solutes in the paracellular space between neighboring cells. This is the prerequisite for creating separate compartments with differing ionic concentrations and/or osmolarities. Rigorous sealing of the paracellular space allows the maintenance of large gradients and is typical for tight epithelia like distal nephron or distal colon. In contrast, moderate sealing is characteristic for leaky epithelia exhibiting high transepithelial transport rates like proximal nephron or ileum. The degree of tightening is determined by the molecular composition of the tight junction (TJ). Since the discovery of channel-forming TJ proteins, such as the cation channel–formers claudin-2, -10b, and -15 or the anion channel–formers claudin-10a and -17 (for review, see Refs. 1 and 2), it is clear that

moderate sealing properties are not due to "imperfect" TJs, but rather are built-in properties that may not only be size selective but also be charge selective.

Among the solutes transported across epithelia, inorganic ions and small organic compounds (e.g., monosaccharides) possess the smallest diameters. Ions in aqueous solution are surrounded by hydration shells, and therefore diameters of hydrated inorganic ions and various uncharged organic compounds with molecular weights < 200 Da do not differ greatly (Table 1). Hence, in epithelia without pronounced tight junctional charge preference, paracellular permeabilities for these substances are often found to be in the same range.

Due to this similarity of permeabilities for small charged and uncharged solutes and based on the assumption that, at fixed ionic concentrations, permeabilities, and conductances should be proportional electrophysiological techniques can be

doi: 10.1111/j.1749-6632.2012.06540.x

 Ann. N.Y. Acad. Sci. 1257 (2012) 142–151 © 2012 New York Academy of Sciences.

Table 1. Molar mass, effective radius, and apparent permeability

	Cl⁻	Na⁺	K⁺	Mg²⁺	Glucose	Mannitol	Fluorescein
Molar mass (g/mol)	35.453	22.989	39.098	24.305	180.16	182.17	332.31
Radius (Å)							
Crystal	1.81	0.95	1.33	0.65	–	–	–
Stokes	1.21	1.84	1.25	3.47	3.5	3.6	4.5(5.5)
Hydrated	3.32	3.58	3.31	4.28	4.2	4.2	4.5(5.5)
Apparent permeability (10^{-6} cm/sec)							
MDCK II	13.8	43.3	43.3	15.9		4.7	4.8
MDCK C7	0.9	1.1	1.2	1.1		0.9	0.6
HT-29/B6	2.8	2.0				2.3	0.4

Radius values for cations from Ref. 47; for glucose and mannitol, see Ref. 48; for fluorescein, see Ref. 49 (Ref. 50 for values in brackets). Permeability values from Refs. 36, 51–53 and unpublished observations. MDCK C7 and HT-29/B6: cell lines forming layers with little charge preference (predominant transport of hydrated ions) and thus similar permeabilities for inorganic ions and mannitol. MDCK II: cell layers with strong cation selectivity (transport of at least partially dehydrated ions) and hence much larger permeabilities for inorganic ions than for mannitol.

applied to determine fundamental epithelial properties.

Today, electrophysiology on epithelia is often reduced to measuring transepithelial resistance (TER). TER measurements are overall resistance measurements of all components lying between two electrodes: epithelial cells, subepithelial tissues or—in the case of cultured cells—filter supports, and bath solution/culture medium. Strictly, TER measurements are DC measurements, although they are often carried out under "near DC conditions," that is, using low-frequency AC conditions ($\sim$3–20 Hz) to avoid electrode polarization. For many applications during which TER remains low, such as simple monitoring of cell growth and cell layer formation during culture, these measurements provide sufficient information. However, they do not provide conclusive information on tight junctional properties, as TER is always a combined measure of the transcellular and paracellular pathway and, as illustrated below, both may be affected independently.

In contrast to TER measurements, impedance measurements determine the opposition to currents under AC conditions and impedance spectroscopy establishes its frequency dependence. For epithelia, typically a range of frequencies between $\sim$1 Hz and $\sim$100 kHz is employed. The fundamental difference between resistance and impedance is that some electrical components, such as capacitors, cause phase shifts between current and voltage. As cell membranes act as capacitors, impedance spectroscopy

reveals considerably more information about epithelial properties than TER measurements.

The present review deals with theoretical and experimental aspects of this technique and will illustrate conclusions that can be drawn from such measurements.

Early history

Impedance spectroscopy on biological materials has been established more than a century ago. Among the first publications are those by Höber and McClendon.[3,4] Höber followed suggestions by Nernst to develop a technique replacing the "usual procedure by Kohlrausch with bridge and telephone" (a Wheatstone bridge adapted for AC measurements and a telephone receiver to adjust the bridge to zero; for a detailed description of this technique, see Ref. 5) to demonstrate that red blood cells are surrounded by a high-resistance cell membrane and that the interior of these cells has a conductance similar to a 0.1 M KCl solution.[3] Höber concluded that the cell membrane must have properties to actively transport solutes that cannot freely diffuse across the membrane, and that the poisonous effects of some salts may be due to changes in these transport properties.

McClendon worked on a great variety of preparations, including sea urchin eggs.[4] Using Kohlrausch's method, he was able to demonstrate changes in membrane conductance upon fertilization of the eggs. His "observations indicate that the plasma membrane of the unfertilized egg is less permeable to anions than to cations" and "seem to indicate that the permeability of the egg is increased suddenly (in less than five minutes) on fertilization."[4] In other words, he was able to demonstrate that the resistance of cell membranes can be physiologically modulated. Using this technique, he was further able to demonstrate changes in membrane impedance in relaxed, stimulated, and fatigued muscle.[6]

A further milestone is the work by Fricke who used impedance measurements to determine the capacitance of erythrocyte plasma membranes and came up with the value of $\sim 1\ \mu F/cm^2$, which has since become a textbook value.[7] Fricke further used this value to estimate the thickness of the plasma membrane for which he obtained "the value 3.3×10^{-7} cm. This value corresponds to from 20 to 30 carbon atoms, if we assume that the distance between two neighboring carbon atoms of an organic molecule is $1–1.5 \times 10^{-8}$ cm."[7] His estimate of about 3 nm thickness of the hydrophobic phase was proved to be astoundingly accurate about a quarter of a century later, when the first electron micrographs of plasma membranes were published.[8–10]

From the late 1920s to the 1940s, the field of impedance measurements in biological preparations was dominated by the physiologist Kenneth S. Cole; his brother Robert H. Cole, a physicist; and his coworker Howard J. Curtis. They developed much of the theoretical background and carried out measurements in almost any preparation they could get hold of, ranging from algae to potatoes, from eggs to muscles to epithelia.[11–18] Most impedance measurements from that period, however, were conducted on nonepithelial tissues and are thus rather predecessors to today's bioimpedance measurements for determination of body fat mass or intra/extracellular volume imbalances.[19]

With their 1941 paper, Cole, and Cole made the visualization of impedance values as complex numbers in an Argand diagram (real part of the complex number on the *x*-axis, imaginary part on the *y*-axis) popular in physiology.[18] In their honor, even today this method of plotting impedance values is still often referred to as Cole–Cole plot. Alternatively, it is called Nyquist diagram, after a Swedish physicist who worked for the Bell Telephone Laboratories (even though it was probably invented by his colleague Carter[20]).

Basic principle

In a simplified model, a cell membrane is viewed as a resistor R and a capacitor C in parallel. Membrane capacitance is due to the insulating, hydrophobic part of the lipid bilayer, whereas the (ohmic) resistance corresponds to the total resistance of open ion channels within this lipid membrane of near infinite resistance (Fig. 1A). Under DC conditions, the capacitor, once fully charged, has an indefinite resistance; hence, all current will flow along the ohmic pathway (i.e., through open ion channels). Under AC conditions, charging/discharging of the capacitor may be so fast that the capacitor is never fully charged. Thus, at very high frequencies, f, the impedance of the capacitor drops to zero, short-circuiting the membrane resistor. In contrast to an ohmic resistor, where voltage and current are always in phase (phase angle $\theta = 0°$), a capacitor causes a phase shift, θ, of $-90°$ between voltage and current (voltage lagging behind the current). The impedance, Z, of a cell membrane can therefore be treated as a vector with the absolute value $|Z|$ and an angle, θ, the value of which lies between 0 and $-90°$ and depends on the circular frequency ω ($\omega = 2 \cdot \pi \cdot f$).

In the Bode plot,[21] $\log |Z|$ and θ, respectively, are plotted against $\log f$ (or ω; see Fig. 1E, F). Alternatively, impedance vectors can be plotted in a Cartesian coordination system, depicting θ as the angle relative to the *x*-axis (Fig. 1G). The locus of the impedance value on the *x–y* plane can be calculated most elegantly by treating impedances as complex numbers. The phase shift by $-90°$ is annotated by multiplication with the square root of -1, j. Hence, the impedance of the membrane resistor is $R + j \cdot 0$, as there is no phase shift, the impedance of the capacitor is $0 + 1/(j \cdot \omega \cdot C)$. As both are in parallel, total membrane impedance amounts to

$$Z = 1/(1/R + j \cdot \omega \cdot C)$$
$$= \frac{R \cdot (1 - j \cdot \omega \cdot R \cdot C)}{(1 + j \cdot \omega \cdot R \cdot C) \cdot (1 - j \cdot \omega \cdot R \cdot C)}$$
$$= \frac{R \cdot (1 - j \cdot \omega \cdot R \cdot C)}{1 + (\omega \cdot R \cdot C)^2}. \tag{1}$$

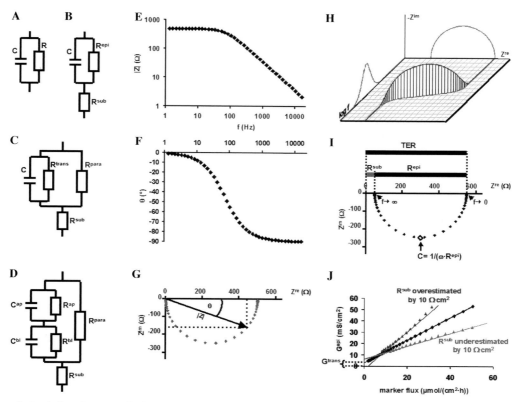

Figure 1. Equivalent circuits, Bode plot, and Cole–Cole plot of epithelial impedance spectra. (A to D) Equivalent circuits describing (A) a single membrane as a capacitor C and a resistor R in parallel; (B) an epithelium, distinguishing between epithelial capacitance C, epithelial resistance R^{epi}, and subepithelial resistance R^{sub}; (C) an epithelium as in B but additionally distinguishing between a transcellular and paracellular resistance (R^{trans}, R^{para}); (D) an epithelium as in B and C but additionally distinguishing between the apical (C^{ap}, R^{ap}) and basolateral (C^{bl}, R^{bl}) membrane. (E, F) Bode plot of an impedance spectrum. (E) log ($|Z|$) versus log (f); (F) θ versus log (f) with $|Z|$ being the absolute value and θ the phase angle of the impedance vector, and f the AC frequency. (G) Cartesian plot of the impedance vector, resulting in the Cole–Cole plot. (H) Blue line: three-dimensional Cole–Cole plot (redrawn after Cole[22]) plotting Z^{re} versus Z^{im} versus log (f). Red lines: projections onto the three planes, i.e., Z^{re} versus log (f), Z^{im} versus log (f), and Z^{re} versus Z^{im}. The latter corresponds to the two-dimensional Cole–Cole plot shown in G and I. (I) Cole–Cole representation of TER $= R^{sub} + R^{epi}$. C can be calculated as $1/(\omega^0 \cdot R^{epi})$, with ω^0 being the circular frequency at which Z^{im} reaches its minimum (◇). (J) Two-path impedance spectroscopy: Graphical determination of the transcellular conductance G^{trans} ($= 1/R^{trans}$) by plotting the epithelial conductance G^{epi} ($= 1/R^{epi}$) versus the flux of a paracellular marker substance. Even minor over- or underestimates of R^{sub} (▲) will cause considerable deviations of the data points from the ideal linear relationship.(●) Adapted from Ref. 32.

Regarding Z as a complex number of the form $Z^{re} + j \cdot Z^{im}$ (Z^{re}, real part of the impedance or resistance; Z^{im}, imaginary part of the impedance or reactance), it follows that

$$Z^{re} = \frac{R}{1 + (\omega \cdot R \cdot C)^2} \qquad (2)$$

$$Z^{im} = \frac{-\omega \cdot R^2 \cdot C}{1 + (\omega \cdot R \cdot C)^2}. \qquad (3)$$

Equations (2) and (3) result in $Z^{re} \rightarrow R$ and $Z^{im} \rightarrow 0$ for $\omega \rightarrow 0$ and in $Z^{re} \rightarrow 0$ and $Z^{im} \rightarrow 0$ for $\omega \rightarrow \infty$.

For $1/\omega = R \cdot C = \tau$ (the time constant of the cell membrane), it follows that

$$Z^{re} = Z^{im} = R/2. \qquad (4)$$

Plotting Z^{re} versus Z^{im} is the aforementioned Cole–Cole plot. As can be seen from Fig. 1G, both values can be obtained from the experimental readouts, $|Z|$ and θ, as

$$Z^{re} = |Z| \cdot \cos\theta \qquad (5)$$

$$Z^{im} = |Z| \cdot \sin\theta. \qquad (6)$$

By rearranging Eq. (1), it can be shown that for a resistor and a capacitor in parallel, the Cole–Cole plot will result in a semicircle, with a radius of $R/2$. C can be calculated as $1/(\omega_0 \cdot R)$ with ω_0 being the frequency at which Z^{im} reaches its minimum.

Compared to the Bode plot, a disadvantage of the Cole–Cole plot is that frequency information is not explicitly represented. Therefore, Cole suggested that impedances should be viewed as three dimensional, with the frequency f on the z-axis (see Fig. 1H).[22]

To obtain a nearly complete semicircle during an experiment, frequencies ω should at least range from $1/(10 \cdot \tau)$ to $10/\tau$. Because τ is initially unknown and may vary widely during an experiment, especially due to changes in R (activation/inhibition of membrane channels), a much wider range of ω is usually applied. The frequency dependence of $|Z|$ can also be used to illustrate why TER measurements may cause underestimations under "near DC" conditions, e.g., at 20 Hz, which correspond to an ω of 125/second. Thus, for $|Z|_{(20\ Hz)} \approx$ TER, τ would have to be at least 0.8 ms and consequently, at a capacitance of 4 μF/cm^2, only resistances < 200 $\Omega \cdot$ cm^2 will be measured accurately.

Impedance of epithelia

To explain basic features of impedance curves obtained from epithelia, at least one further resistor has to be introduced into the equivalent circuit: the subepithelial resistance, R^{sub}, which lies in series to the epithelial resistance R^{epi} and capacitance C^{epi} (Fig. 1B). Thus, Eqs. (1)–(3) still apply and only R^{sub} has to be added:

$$Z = R^{sub} + \frac{R^{epi} \cdot (1 - j \cdot \omega \cdot R^{epi} \cdot C^{epi})}{1 + (\omega \cdot R^{epi} \cdot C^{epi})^2} \qquad (1a)$$

$$Z^{re} = R^{sub} + \frac{R^{epi}}{1 + (\omega \cdot R^{epi} \cdot C^{epi})^2} \qquad (2a)$$

$$Z^{im} = \frac{-\omega \cdot R^{epi^2} \cdot C^{epi}}{1 + (\omega \cdot R^{epi} \cdot C^{epi})^2}. \qquad (3a)$$

In the Cole–Cole plot, this results in a shift of the semicircle along the x-axis to the right by the amount of R^{sub}. Thus, all three parameters—R^{sub}, R^{epi}, and C^{epi}—can be directly determined from these plots (Fig. 1I).

One-path impedance spectroscopy (1PI): R^{sub}, R^{epi}, C

In our laboratory, this type of measurement has been dubbed "one-path" impedance spectroscopy (1PI), because no distinction is made between the two major transport pathways (paracellular, transcellular) across the epithelium. Most cultured epithelial cell lines under resting conditions yield impedance spectra that follow the predicted semicircle almost ideally. Spectra from epithelial tissues, in contrast, usually deviate from this shape. According to Cole and Cole, such data can be fitted by a semicircle, the midpoint of which is moved above the x-axis (Cole–Cole fit).[18] A possible explanation for this deviation is that neighboring cells may differ in their apical and basolateral time constants.[23]

Despite these deviations, 1PI has proven to be a powerful tool, as it allows reliable determination of R^{sub} and R^{epi} from the intercepts of the impedance curve with the x-axis.[23,24] By applying this technique to biopsies from patients with inflammatory intestinal diseases, it could be shown that due to the inflammation R^{sub} increased whereas R^{epi} dramatically decreased.[25–27] The increase in R^{sub} was explained by an increased thickness of the subepithelial tissue. The thickness, however, does not contribute toward transepithelial tightness under *in situ* conditions, as blood capillaries make their way through this subepithelial tissue and lie in direct vicinity to the epithelium, thereby maintaining short diffusion distances. In contrast, the decrease in R^{epi} can explain the type of diarrhea (leak-flux diarrhea) these patients are suffering from.[25,26] This was demonstrated to be due to an upregulation of the channel-forming TJ protein claudin-2 together with a downregulation of the tightening claudin-3, -4, -5, and -8.[25–27] Simple TER measurements would have missed these effects because the sum, R^{sub} + R^{epi}, remained almost constant.[25–27]

Two-path impedance spectroscopy (2PI): R^{para}, R^{trans}

The results from gut epithelia described above already indicate that it would be desirable to measure paracellular and transcellular resistances separately and thus to quantify the anticipated changes in paracellular resistance. To reflect these two separate transport pathways within the previously described equivalent circuit, R^{epi} has to be replaced by two resistors in parallel, the paracellular resistance, R^{para}, and the transcellular resistance, R^{trans} (Fig. 1C). Consequently, the following equation holds true:

$$R^{epi} = R^{para} \cdot R^{trans}/(R^{para} + R^{trans}). \quad (7)$$

Equation (7) implies that R^{epi} will always be of smaller value than R^{para} as well as of smaller value than R^{trans}.

For the following considerations, it is more convenient to transform Eq. (7) by using the reciprocal values of the resistances, the conductances G^{para}, G^{trans}, and G^{epi}, respectively:

$$G^{epi} = G^{para} + G^{trans}. \quad (8)$$

Thus, there are two variables but only one equation and, therefore, further experiments are required to distinguish between G^{trans} and G^{para}. Since the 1970s, impedance spectroscopy has been performed extensively on epithelia, and various methods have been developed to separately measure G^{trans} and G^{para}, e.g., by impaling microelectrodes into epithelial cells[28,29] or by using "conductance scanning," a technique for measuring local conductances above an epithelial layer.[30,31]

Recently, our laboratory has published another, more convenient technique to measure G^{trans} and G^{para}, dubbed "two-path" impedance spectroscopy (2PI).[32] 2PI is based on the reasoning that G^{para} depends directly on the flux of the predominant ions that move along the paracellular space, that is, Na^+ and Cl^-. Therefore, G^{para} should be proportional to the flux of these ions.[32] In principle, these fluxes can be measured by employing radioactive isotopes. Considering handling inconvenience of these isotopes and the fact that both, Na^+ and Cl^-, may also be actively transported through the cells, an alternative paracellular marker substance should possess the following properties:

(1) it should be easy to detect,
(2) it should not be transported through the cells, and
(3) it should be small enough so that its movement along the paracellular pathway is proportional to Na^+ and Cl^-, and hence, to G^{para}.

One substance fulfilling these criteria in many cultured epithelial cells was found to be the ionic form of fluorescein.

By assuming that G^{para} is proportional to J_{fluo} (fluorescein flux), Eq. (5) becomes a linear equation

$$G^{epi} = G^{trans} + s \cdot J^{fluo}, \quad (8a)$$

with s being a proportionality constant.

If the TJ (and hence G^{para}) is experimentally modulated during determination of J_{fluo}, and the resulting G^{epi} values are plotted against the corresponding J_{fluo} values, then all data points should fall onto a straight line and the (extrapolated) y-intercept should equate G^{trans}. EGTA was used to chelate extracellular Ca^{2+} and thereby to modulate the paracellular pathway and to increase G^{para}, a paradigm that worked well in several cell lines (e.g., MDCK, HT-29/B6, T84).[33,32] Care has to be taken, however, that G^{epi} remains $< G^{sub}$, otherwise the subepithelium will restrict the flux of the marker substance and distort the relationship between G^{epi} and J_{fluo}.

All the above assumptions have to be carefully validated for each cell line or tissue, as cells may, e.g., possess paracellular channels that allow the passage of Na^+ and/or Cl^- but not of fluorescein, express membrane transporters that accept fluorescein and transport it along the transcellular route, or may activate transport processes upon the withdrawal of extracellular Ca^{2+}.[34]

In principle, this experimental approach does not require full impedance scans, as long as the pure epithelial resistance can be determined. However, it has to be emphasized that even small deviations, e.g., through inadequate determination of the subepithelial resistance, may cause considerable distortions of the G^{epi}–J_{fluo} relationship and thus make a determination of G^{trans} impossible (see Fig. 1J).

2PI applications: variations in R^{para}

Initial 2PI experiments were carried out on HT-29/B6, MDCK C7, and MDCK C11 cell layers, and results are summarized in Table 2.

According to the definition by Schultz,[35] HT-29/B6 cells as well as MDCK C7 cells form tight epithelia, as $R^{para} > R^{trans}$. As a consequence, such epithelial layers are able to maintain large transepithelial gradients. In contrast, MDCK C11 cells are leaky ($R^{para} < R^{trans}$). Interestingly and unexpectedly, the two MDCK clones C7 and C11 do not

Table 2. Membrane parameters of various cell lines

Cell type	R^{epi} ($\Omega \cdot cm^2$)	R^{para} ($\Omega \cdot cm^2$)	R^{trans} ($\Omega \cdot cm^2$)	C ($\mu F/cm^2$)
IPEC-J2 average	4 100			1
example 1	1 840	2 200	11 100	0.9
example 2	8 500	12 900	24 700	0.9
MDCK C7	1 600	4 500	2 650	1.3
+ cldn10b overexpression	750	1 100	2 500	1.1
MDCK C11	18	24	68	1.9
MDCK II control	25	44	69	3.4
+ tricellulin overexpression in tri- and bicellular TJ	72	610	83	3.7
HT-29/B6	750	1 800	1 300	3.5
HT-29/B6 control	390	1 800	550	4.5
+ 500 U/ml TNFα	120	200	400	3.9
Caco-2	250	900	340	7.5

Values from Refs. 32, 36, 54, as well as unpublished results.

only differ in their R^{para} but also in R^{trans}. However, they did not differ significantly in membrane capacitance. If the membrane capacitance is taken as a measure of membrane area, these results indicate that MDCK C11 cells differ from MDCK C7 cells in the number of (open) ion channels within their cell membranes.

Transfection of MDCK C7 cells with claudin-10b converted these cells into a leaky epithelium by specifically reducing R^{para}, whereas strong tricellulin overexpression in MDCK II cells converted the resulting layers into tight epithelia by specifically increasing R^{para}.[32,36]

Caco-2 cells were initially found to transport fluorescein along the transcellular route and could therefore not be used for 2PI measurements.[32,37] However, altering culture conditions appeared to change their transport properties and therefore, preliminary data are included in Table 2. Despite their relatively low epithelial resistances, Caco-2 cells form tight epithelia, i.e., $R^{para} > R^{trans}$. This is in contrast to the porcine jejunal cell line IPEC-J2 that is able to build up transepithelial resistances in the order of 10,000 Ω cm^2. Yet, preliminary data indicate that in these cell layers $R^{para} < R^{trans}$ and hence, these cell layers have to be regarded as leaky (unpublished results). This paradox may at least in part be explained by comparing epithelial capacitances. IPEC-J2 cells exhibit extremely low capacitances in the order of 1 μF/cm^2. This is in the order of the

value reported by Fricke for unit cell membranes.[7] Consequently, IPEC-J2 cells cannot be much more than two flat pieces of membrane joined at the TJ, even if it is taken into consideration that in an epithelium there are two capacitances in series (apical and basolateral cell membrane), which means that the minimum possible total capacitance is ~0.5 μF/cm^2. In contrast, Caco-2 cells exhibit large capacitances, indicating excessive folding of the membranes. It thus becomes obvious that simple TER values do not give reliable indications whether an epithelium is leaky or tight.

2PI applications: variations in R^{trans}, evaluation using artificial neural networks

Affecting transcellular transport processes alters R^{trans} and therefore TER. Using simple TER measurements, these changes cannot be distinguished from changes in R^{para} (compare Fig. 2A and B). In many cases, changes in transcellular transport affect the apical and basolateral membrane resistance by different amounts. This causes differences in the apical and basolateral time constants that become very obvious in Cole–Cole plots of the impedance spectra (compare Fig. 2C and D). Instead of one single semicircle, curves are now composed of two overlapping semicircles. These can only be explained, if the postulated equivalent circuit is again extended and apical and basolateral resistances (R^{ap},

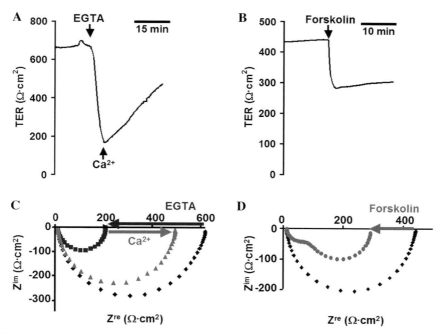

Figure 2. Effects of changes in R^{para} and R^{trans} on TER and impedance spectra. (A,B) TER recordings from HT-29/B6 cells. (A) Application of the Ca^{2+} chelator EGTA causes a drop in TER that is reversible upon application of Ca^{2+}. (B) Similarly, application of forskolin, an activator of the adenylate cyclase, causes a decrease in TER. (C,D) Impedance spectra recorded during the experiments shown in A and B, respectively. (C) Application of EGTA causes a decrease in R^{epi} (decrease in diameter of the semicircular impedance spectrum is indicated by the black arrow). Readdition of Ca^{2+} causes a decrease in R^{epi} (gray arrow). Under both conditions, the semicircular shapes of the spectra remain unaffected, indicating a decrease in R^{para} without concomitant change in R^{trans}. ◆, ■,▲, values before and after the application of EGTA and after re-addition of Ca^{2+}, respectively. (D) Application of forskolin induced a decrease in R^{epi} (gray arrow) that was accompanied by a clear deviation of the impedance spectrum (●) from the semi-circular shape observed before the application of forskolin (◆). This indicated a decrease in R^{trans}, which is probably mainly due to a decrease in R^{ap}.[32]

R^{bl}) and capacitances (C^{ap}, C^{bl}) are introduced (see Fig. 1D).

Similar impedance spectra are obtained if one of the membranes is permeabilized, for example, by the application of ionophores such as nystatin. Several authors used this approach to derive information about the apical and basolateral membrane properties.[38–40]

Application of the 2PI technique in the presence of substances that alter apical or basolateral membrane properties poses the problem that non-semicircular Cole–Cole fits cannot be used to determine R^{epi} from the x-intercepts of the impedance spectra with sufficient accuracy. We have therefore developed a computational approach based on the equivalent circuit presented in Fig. 1D. It could be shown that combining 2PI measurements with an application of nystatin to either the apical or basolateral mem-

brane and employing artificial neural networks for analysis is a promising concept for estimating all six parameters: R^{ap}, R^{bl}, C^{ap}, C^{bl}, R^{para}, and R^{sub}.[41]

Alternative techniques: variations in *C*

Changes in membrane capacitance occur if the membrane area is altered, e.g., through extensive endo- or exocytosis. These processes are usually too rapid to be measured by conventional impedance spectroscopy, as obtaining full spectra typically takes more than one minute. Therefore, alternative methods have been developed, such as application of square pulses and analysis of the resulting output transients,[42,43] using very brief impulses or simultaneously imposing different sine waves and using Fourier analysis.[44,45] These techniques allowed, for example, to analyze capacitance changes during the exocytotic insertion of ion channels induced by cAMP, purinergic, or cholinergic stimulation.[45,46]

Necessary equipment

Impedance measurements do not require much specialized equipment other than a frequency response analyzer that works in a range of about 1 Hz to 1 MHz, together with an impedance interface (see, e.g., Refs. 23, 32, 55, and 56). Both are readily commercially available from different companies. Any standard computer is sufficient for data acquisition, either using commercial software or, as in our lab, a somewhat extended Excel macro. A very straightforward method employing four silver electrodes integrated into the cover of a 12-well plate has recently been described,[56] although we usually prefer to mount preparation in a Using chamber to avoid unstirred layers and to facilitate access to the apical and basolateral solution for flux measurements.

Conclusions

Over the past century, impedance spectroscopy techniques for application on epithelial cell layers and tissues have been developed. These techniques allow detailed analysis and quantification of epithelial electrical parameters as well as discrimination between paracellular and transcellular processes.

Acknowledgments

We gratefully acknowledge financial support by the Deutsche Forschungsgemeinschaft FOR721 and SFB 852, FAZIT-Stiftung Gemeinnützige Verlagsgesellschaft mbH (Frankfurt am Main), and Sonnenfeld-Stiftung (Berlin), and cordially thank Prof. M. Fromm for critically reading the manuscript.

Conflicts of interest

The authors declare no conflicts of interest.

References

1. Günzel, D. & M. Fromm. 2012. Claudins and other tight junction proteins. *Compr. Physiol.* doi: 10.1002/cphy.c110045.
2. Krug, S.M., D. Günzel, M.P. Conrad, *et al.* 2012. Charge-selective claudin channels. *Ann. N.Y. Acad. Sci.* **1257:** 20–28.
3. Höber, R. 1910. Eine Methode, die elektrische Leitfähigkeit im Innern von Zellen zu messen. *Pflügers Arch.* **133:** 237–253.
4. McClendon, J.F. 1910. On the dynamics of cell division: II Changes in permeability of developing eggs to electrolytes. *Am. J. Physiol.* **27:** 240–275.
5. Gildemeister, M. 1919. Über elektrischen Widerstand, Kapazität und Polarisation der Haut. I. Versuche an der Froschhaut. *Pflügers. Arch.* **176:** 84–105.
6. McClendon, J.F. 1936. Electric impedance and permeability of living cells. *Science* **84:** 184–185.
7. Fricke, H. 1925. The electric capacity of suspensions of red corpuscles of a dog. *Phys. Rev.* **26:** 682–687.
8. Robertson, J.D. 1957. New observations on the ultrastructure of the membranes of frog peripheral nerve fibers. *J. Biophys. Biochem. Cytol.* **3:** 1043–1047.
9. Mitra K., I. Ubarretxena-Beliandia, T. Taguchi, *et al.* 2004. Modulation of the bilayer thickness of exocytic pathway membranes by membrane proteins rather than cholesterol. *Proc. Natl. Acad. Sci. USA* **101:** 4083–4088.
10. Andersen, O.S. & R.E. Koeppe. 2007. Bilayer thickness and membrane protein function: an energetic perspective. *Annu. Rev. Biophys. Biomol. Struct.* **36:** 107–130.
11. Cole, K.S. 1928. Electric impedance of suspensions of spheres. *J. Gen. Physiol.* **12:** 29–36.
12. Cole, K.S. 1928. Electric impedance of suspensions of Arabica eggs. *J. Gen. Physiol.* **12:** 37–54.
13. Cole, K.S. 1932. Electrical phase angle of cell membranes. *J. Gen. Physiol.* **15:** 641–649.
14. Cole, K.S. & H.J. Curtis. 1938. Electrical impedance of nerve during activity. *Nature* **142:** 209–210.
15. Cole, K.S. & H.J. Curtis. 1938. Electrical impedance of *Nitella* during activity. *J. Gen. Physiol.* **21:** 37–64.
16. Cole, K.S. & H.J. Curtis. 1938. Electrical impedance of single marine eggs. *J. Gen. Physiol.* **21:** 591–599.
17. Curtis, H.J. & K.S. Cole. 1938. Transverse electric impedance of the squid giant axon. *J. Gen. Physiol.* **21,** 757–765.
18. Cole, K.S. & R.H. Cole. 1941. Dispersion and absorption in dielectrics I. Alternating current characteristics. *J. Chem. Phys.* **9:** 341–351.
19. Ivorra, A., M. Genescà, A. Sola, *et al.* 2005. Bioimpedance dispersion width as a parameter to monitor living tissues. *Physiol. Meas.* **26:** S165–S173.
20. Carter, C.W. Jr. 1925. Graphic representation of the impedance of networks containing resistances and two reactances. *Bell Syst. Tech. J.* **4:** 387–401.
21. Bode H.W. 1945. *Network analysis and feedback amplifier design.* Van Nostrand, New York.
22. Cole, K.S. 1972. *Membranes, Ions and Impulses.* University of California Press. Berkeley.
23. Gitter, A.H., M. Fromm & J.D. Schulzke. 1998. Impedance analysis for determination of epithelial and subepithelial resistances in intestinal tissues. *J. Biochem. Biophys. Meth.* **37:** 35–46.
24. Gitter A.H., K. Bendfeldt, J.D. Schulzke & M. Fromm. 2000. Trans/paracellular, surface/crypt, and epithelial/subepithelial resistances of mammalian colonic epithelia. *Pflügers. Arch.* **439:** 477–482.
25. Bürgel, N., C. Bojarski, J. Mankertz, *et al.* 2002. Mechanisms of diarrhea in collagenous colitis. *Gastroenterology* **123:** 433–443.
26. Zeissig, S., N. Bürgel, D. Günzel, *et al.* 2007. Changes in expression and distribution of claudin-2, -5 and -8 lead to discontinuous tight junctions and barrier dysfunction in active Crohn's disease. *Gut* **56:** 61–72.

27. A.J. Kroesen, S. Dullat, J.D. Schulzke, *et al.* 2008. Permanently increased mucosal permeability in patients with backwash ileitis after ileoanal pouch for ulcerative colitis. *Scand. J. Gastroenterol.* **43:** 704–711.

28. Kottra, G. & E. Frömter. 1984. Rapid determination of intraepithelial resistance barriers by alternating current spectroscopy. I. Experimental procedures. *Pflügers Arch.* **402:** 409–420.

29. Schifferdecker, E. & E. Frömter. 1978. The AC impedance of *Necturus* gallbladder epithelium. *Pflügers Arch.* **377:** 125–133.

30. Frömter, E. & J. Diamond. 1972. Route of passive ion permeation in epithelia. *Nat. New Biol.* **235:** 9–13.

31. Gitter, A.H., M. Bertog, J.D. Schulzke & M. Fromm. 1997. Measurement of paracellular epithelial conductivity by conductance scanning. *Pflügers Arch.* **434:** 830–840.

32. Krug, S.M., M. Fromm & D. Günzel. 2009. Two-path impedance spectroscopy for measuring paracellular and transcellular epithelial resistance. *Biophys. J.* **97:** 2202–2211.

33. Martinez-Palomo, A., I. Meza, G. Beaty & M. Cereijido. 1980. Experimental modulation of occluding junctions in a cultured transporting epithelium. *J. Cell Biol.* **87:** 736–745.

34. Schultheiss, G. & H. Martens. 1999. Ca-sensitive Na transport in sheep omasum. *Am. J. Physiol. Gastrointest. Liver Physiol.* **276:** G1331–G1344.

35. Schultz, S.G. 1972. Electrical potential differences and electromotive forces in epithelial tissues. *J. Gen. Physiol.* **59:** 794–798.

36. Krug, S.M., S. Amasheh, J.F. Richter, *et al.* 2009. Tricellulin forms a barrier for small and large solutes depending on its localization within the tight junction. *Molec. Biol. Cell* **20:** 3713–3724.

37. Konishi, Y., K. Hagiwara & M. Shimizu. 2002. Transepithelial transport of fluorescein in Caco-2 cell monolayers and use of such transport in *in vitro* evaluation of phenolic acid availability. *Biosci. Biotechnol. Biochem.* **66:** 2449–2457.

38. Lewis, S.A., D.C. Eaton, C. Clausen & J.M. Diamond. 1977. Nystatin as a probe for investigating the electrical properties of a tight epithelium. *J. Gen. Physiol.* **70:** 427–440.

39. Wills, N.K., S.A. Lewis & D.C. Eaton. 1979. Active and passive properties of rabbit descending colon: a microelectrode and nystatin study. *J. Membr. Biol.* **45:** 81–108.

40. Clausen, C., S.A. Lewis & J.M. Diamond. 1979. Impedance analysis of a tight epithelium using a distributed resistance model. *Biophys. J.* **26:** 291–318.

41. Schmid, T., D. Günzel & M. Bogdan. 2010. Using an artificial neural network to determine electrical properties of epithelia. In *ICANN 2010, Part I, Lect. Notes Comput. Sci.* Diamantaras, K., W. Duch & L.S. Iliadis, Eds., **6352:** 211–216, Springer Verlag. Berlin.

42. Teorell, T. (1946). Application of "square wave analysis" to bioelectric studies. *Acta Physiol. Scand.* **12:** 235–254.

43. Suzuki, K., G. Kottra, L. Kampmann & E. Frömter. 1982. Square wave pulse analysis of cellular and paracellular conductance pathways in *Necturus* gallbladder epithelium. *Pflügers Arch.* **394:** 302–312.

44. Bertrand, C.A., D.M. Durand, G.M. Saidel, *et al.* 1998. System for dynamic measurements of membrane capacitance in intact epithelial monolayers. *Biophys. J.* **75:** 2743–2756.

45. Weber, W.M., H. Cuppens, J.J. Cassiman, *et al.* 1999. Capacitance measurements reveal different pathways for the activation of CFTR. *Pflügers Arch.* **438:** 561–569.

46. Bertrand, C.A., C.L. Laboisse & U. Hopfer. 1999. Purinergic and cholinergic agonists induce exocytosis from the same granule pool in HT29-Cl.16E monolayers. *Am. J. Physiol. Cell Physiol.* **276:** 907–914.

47. Nightingale, E.R. 1959. Phenomenological theory of ion solvation. Effective radii of hydrated ions. *J. Phys. Chem.* **63:**1381–1387.

48. Schultz, S.G. & A.K. Solomon. 1961. Determination of the effective hydrodynamic radii of small molecules by viscometry. *J. Gen. Physiol.* **44:** 1189–1199.

49. Wang, L., Y. Wang, Y. Han, *et al.* 2005. In situ measurement of solute transport in the bone lacunar-canalicular system. *PNAS* **102:** 11911–11916.

50. Tervo, T., F. Joó, A. Palkama & L. Salminen. 1979. Penetration barrier to sodium fluorescein and fluorescein-labelled dextrans of various molecular sizes in brain capillaries. *Experientia* 35: 252–254.

51. Rosenthal, R., S. Milatz, S.M. Krug, *et al.* 2010. Claudin-2, a component of the tight junction, forms a paracellular water channel. *J. Cell Sci.* **123:** 1913–1921.

52. Günzel, D., S. Amasheh, J.F. Richter, *et al.* 2009. Claudin-16 affects transcellular Cl⁻ secretion in MDCK cells. *J. Physiol.* **587:** 3777–3793.

53. Hering, N.A., J.F. Richter, S.M. Krug, *et al.* 2011. *Yersinia enterocolitica* induces barrier dysfunction through regional tight junction changes in colonic HT-29/B6 cell monolayers. *Lab. Invest.* **91:** 310–324.

54. Mankertz, J., M. Amasheh, S.M. Krug, *et al.* 2009. TNFα upregulates claudin-2 expression in epithelial HT-29/B6 cells via phosphatidylinositol-3-kinase signaling. *Cell Tissue Res.* **336:** 67–77.

55. Gitter, A.H., M. Fromm & J.D. Schulzke. 1998. Impedance analysis for determination of epithelial and subepithelial resistance in intestinal tissues. *J. Biochem. Biophys. Meth.* **37:** 35–46.

56. Onnela, N., V. Savolainen, K. Juuti-Uusitalo, *et al.* 2012. Electric impedance of human embryonic stem cell-derived retinal pigment epithelium. *Med. Biol. Eng. Comput.* **50:** 107–116.

Ann. N.Y. Acad. Sci. ISSN 0077-8923

ANNALS OF THE NEW YORK ACADEMY OF SCIENCES
Issue: *Barriers and Channels Formed by Tight Junction Proteins*

Diverse types of junctions containing tight junction proteins in stratified mammalian epithelia

Werner W. Franke[1] and Ulrich-Frank Pape[2]

[1]Helmholtz Group for Cell Biology, German Cancer Research Center (DKFZ), Heidelberg, Germany. [2]Department of Hepatology and Gastroenterology, Charité, University Medicine Berlin, Berlin, Germany

Address for correspondence: Prof. Dr. Werner W. Franke, Helmholtz Group for Cell Biology, German Cancer Research Center (DKFZ), Im Neuenheimer Feld 280, D-69120 Heidelberg, Germany. w.franke@dkfz.de

Molecular compositions and functions of tight junctions (TJs), that is, continuous, cell–cell-connecting zonulae occludentes serving as barrier structures for the paracellular transport of molecules and particles, have hitherto been determined for simple epithelia and for endothelia. In 2002, special TJ structures with barrier functions were identified in the stratum granulosum of mammalian epidermis. In addition, using biochemical and immunocytochemical methods, various types of TJ-type junctions have also been described that also contain claudins and/or occludin as well as typical TJ plaque proteins, in cell layers of all stratified squamous epithelia (e.g., various types of epidermis, gingiva, lingual, and other kinds of oral mucosa, pharynx, esophagus, trachea, vagina, and exocervix), including tissues without a lumen, such as the reticulum and Hassall corpuscles of the thymus, and tumors derived from such epithelia, notably squamous cell carcinomas. Biological and pathological aspects of TJ-related structures in such tissues are discussed.

Keywords: epidermis; stratum granulosum; stratified epithelia; sandwich junctions; stud junctions

Introduction

In the original electron microscopic descriptions, the tight junctions (TJs), as observed in "simple" (monolayer) epithelia and in endothelia, appeared very clearly defined and were brought into connection with one obvious major function: zonulae occludentes. Zonulae occludentes are characterized in cross-sections by maximally tight membrane–membrane contacts (kissing points) forming a near-continuous plasma membrane contact zone around each cell, suggestive of a barrier to paracellular translocations of particles and many types of molecules[1–4] (for an anthology, see Ref. 5). The unraveling of the molecular composition of the TJs led to the discovery of two kinds of tetraspan transmembrane proteins, occludin and cell type–specific combinations of members of the claudin family, accompanied in certain epithelia with occludin-related tricellulins located at sites where three cells meet each other.[6–8] The structure-forming potential of these molecules was convincingly demonstrated by the *de novo* formation of TJ-like structures upon transfection of initially TJ-devoid cells with cDNAs encoding such proteins[9,10] (for recent reviews, see Refs. 11 and 12). And from the beginning, it was also clear that the cytoplasmic portions of these transmembrane proteins are rooted in a dense plaque structure containing proteins ZO-1 and ZO-2, cingulin, and several other proteins[13,14] (for reviews, see Refs. 5, 15, and 16).

The TJ system of the epidermal stratum granulosum

At about the same time when it was reported that the cells of the uppermost living cell layer of the mammalian epidermis, the stratum granulosum, were connected by an extended continuous cell–cell contact TJ structure system, which was immunostained by antibodies to certain claudins, occludin, and several TJ plaque proteins,[17–25] Furuse *et al.* published their fundamental report that the absence of claudin 1 in the epidermis of mice resulted in the loss of the

doi: 10.1111/j.1749-6632.2012.06504.x

Ann. N.Y. Acad. Sci. 1257 (2012) 152–157 © 2012 New York Academy of Sciences.

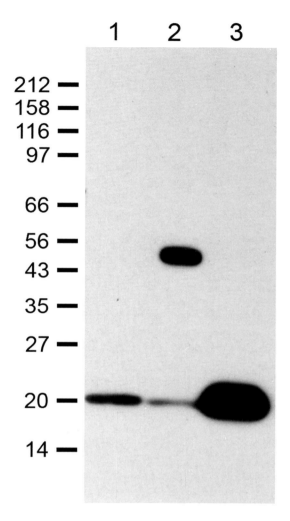

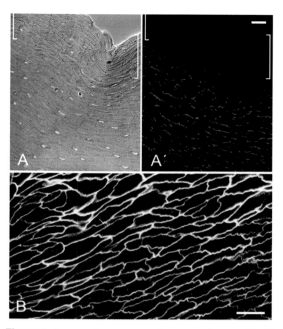

Figure 2. Immunofluorescence microscopy of a cross (A, A′) and a horizontally grazing (B) cryostat section of bovine tongue mucosa, as seen after specific reactions with antibodies to occludin (A′; A, phase contrast optics) or claudin 4 (B). Note that TJ proteins are absent from the dead cell structures, equivalent to the stratum corneum of epidermis (demarcated by brackets in A and A′). Note that punctate or linear structures positive for the TJ protein occludin are seen in the majority of the living cell layers of the tongue mucosa stratum spinosum–equivalent (red, A′) and that in some regions all of the cell–cell contact lines, forming a continuous TJ system, are positive for claudin 4 (B). Bars: 20 μm.

Figure 1. Gel electrophoretic separation (SDS-PAGE) of polypeptides from human tongue epithelium (lane 1) and human epidermis (lane 2) and from a densely grown culture of human epidermal keratinocytes of line HaCaT (lane 3) seen by their specific antibody reactions against claudin 4 using the immunoblot technique. Note the detection of claudin 4 (molecular weight of ∼20 kDa) in all samples. (The additional reaction band seen here at ∼54 kDa results from the reaction of residual immunoglobulin heavy chain, IgG molecular weight of ∼54 kDa.)

barrier structure and thus of body water and led to rapid postnatal death[26] (for reviews, see Refs. 11 and 27). This gene knock-out study, together with the reports of Herrmann *et al.*[28,29] that the disruption of the fatty acid transport protein 4 also resulted in lethal dermopathies, has led to an intense and

ongoing discussion of which structural and molecular elements in the uppermost epidermal cell layers might be essential for mammalian life[25,30,31] (for reviews, see Refs. 32–34).

Biochemical and immunocytochemical demonstrations of TJ proteins in cell–cell contact structures of different cell layers of stratified epithelia

In all stratified mammalian epithelia, we have identified claudin 4, often accompanied by claudin 1, and occludin as well as the TJ plaque proteins ZO-1, ZO-2, and cingulin (see, e.g., Fig. 1).[21,22,24,25,35,36] In systematic studies, however, we have also found, to our great surprise, that in many of these tissues the TJ-related cell–cell junction structures containing TJ marker molecules are not confined to the uppermost living layers but extend, for a variable

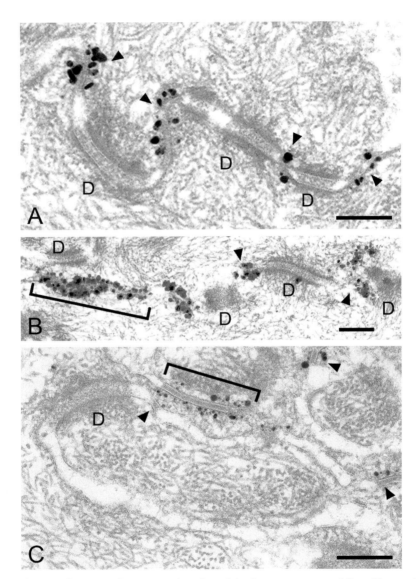

Figure 3. Immunoelectron microscopy of cryostat sections through bovine tongue mucosa, followed by reaction with occludin antibodies and visualization of the immunoglobulins bound by an immunogold reaction plus silver enhancement, showing that the reaction sites (metal grains; arrowheads) are exclusively seen in interdesmosomal (D, desmosomes) spaces. Note that some of these occludin-positive junctions are very small. Extended regions labeled with occludin are also seen (brackets), and a prominent subtype of the occludin-containing junction is characterized by a dense midlayer (sandwich junction (iunctura structa); see the three immunolabeled junctions in C). Bars denote 200 nm.

number of strata, downwards deep into the stratum spinosum equivalents of the specific epithelium, and that the TJ-protein pattern seen in the different cell layers can be highly variable. Figures 2A and A' show, for example, the distribution of occludin-positive junctions in a number of living cell strata of bovine tongue mucosa. Extreme situations of im-

munostaining differences between claudins 1 and 4 as well as between claudins and occludin have been reported for epidermis and other multistratified tissues with drastic differences, for example, between strata with claudin 4–containing TJs on the one hand and occludin-positive ones on the other,[24,35] and in certain regions of bovine gingiva claudin 1–positive TJ-related structures have even been seen in all suprabasal layers, together with protein ZO-1, whereas occludin has been found only in a number (up to 10) of the uppermost strata.[21] Both similar and different TJ protein reactions in regional and cell type–specific variable numbers of layers have been reported for epidermis, snout, and muzzle epithelium, gingiva, lingual, and other portions of oral mucosa, esophagus, trachea, exocervix, and vagina[19,21,24,25,36] (for discussion, see also Ref. 37). Even more surprisingly, we have noted one TJ protein in only some layers, but another TJ protein in many more layers, suggesting that some TJ transmembrane proteins are missing or are markedly underrepresented in the TJ-related structures of some of these layers.

Corresponding observations have also been made with densely grown cultures of keratinocytes, showing complete or regional stratification (see, e.g., also lane 3 in Fig. 1; see also Refs. 18, 19, and 22) and in epithelium-derived cell formations without a lumen, such as hair follicles, with a characteristic prominence of TJ marker molecules in the Henle layer, the reticulum epithelial cells, and the Hassall corpuscles of the thymus.[21,22] Molecular heterogeneities in different cell layers have also been described in cultures of cells from other stratified epithelial systems[19,21,24,25] (for discussion, see also Ref. 37).

While in longitudinal or oblique cross-sections, the immunolocalization results might suggest—or are at least compatible with—a predominance of small or streak-like elongated cell–cell-connecting plasma membrane regions (e.g., Fig. 2A′). Horizontal or grazing sections through the positive strata demonstrate that such occludin- and/or claudin-containing contact zones can circumscribe the entirety of cells in such a layer (for an example of bovine tongue mucosa, see Fig. 2B; for examples of various skin regions, including bovine hoof and muzzle, tonsil-covering pharynx, and vaginal epithelium, see Refs. 19 and 21).

Different subtypes of TJ-related structures distinguished by electron microscopy

The presence of interdesmosomal membrane–membrane contact structures of an ultrastructure similar to, but not identical with, that of typical TJs has also been documented by electron microscopy. Furthermore, immunoelectron

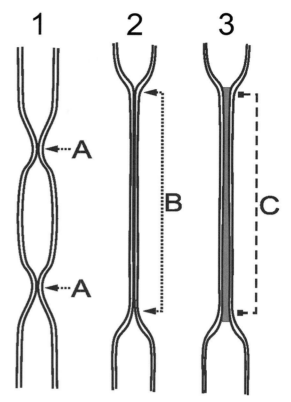

Figure 4. Schematic presentations of different major TJ-type structures frequent in stratified epithelia: typical TJ close cell–cell contacts (kissing points; A); extended, relatively close contacts with a thin, dense middle lamella (B); sandwich junction as an extended junction with an interposed, relatively thick, electron-dense middle lamella (C).[21,38,39]

microscopy has shown that these junctional structures, including the very small ones, and those with relatively thick (10–30 nm), electron-dense middle lamella structures (sandwich junctions), contain occludin (for examples of bovine tongue mucosa, see Fig. 3A–C) and one or more claudins. The precise molecular compositions and configurations of the middle lamellae are not yet known. Three major ultrastructural subtypes are schematically presented in Fig. 4. An additional, very small structure that seems to be related to TJ elements is the "stud junction," best seen in freeze-fractures through the membrane interior (for details, see Ref. 25; for reviews, see Refs. 38 and 39).

Conclusions

Junctions with a TJ-like composition and ultrastructure connecting cells of stratified epithelia have hitherto been the Cinderella of junction research. Now, however, it is clear that TJs of this category do occur in multistratified epithelia and tissues derived therefrom, including tumors. As they display tissue- and cell type–specific, ultrastructural, and compositional differences, it will be important to characterize the molecular composition of the various types of TJs present in the cell layers of normal stratified epithelia to get insights into their functions. It will also be important to determine the compositional pattern of TJ markers in pathological formations, in particular metaplasias and squamous cell carcinomas. Diagnostic information may be valuable in consideration of the accessibility of a given cell–cell connection or of certain groups of cells, for example, certain molecules, including therapeutic agents. Therefore, it is a pressing need to include the TJs of stratified epithelia in cell–cell junction research programs.

Conflicts of interest

The authors report no conflicts of interest.

References

1. Farquhar, M.G. & G.E. Palade. 1963. Junctional complexes in various epithelia. *J. Cell Biol.* **17:** 375–412.
2. Claude, P. & D.A. Goodenough. 1973. Fracture faces of zonulae occludentes from "tight" and "leaky" epithelia. *J. Cell Biol.* **58:** 390–400.
3. Staehelin, L.A. 1974. Structure and function of intercellular junctions. *Int. Rev. Cytol.* **39:** 191–283.
4. Schneeberger, E.E. & M.J. Karnovsky. 1976. Substructure of intercellular junctions in freeze-fractured alveolar-capillary membranes of mouse lung. *Circ. Res.* **38:** 404–411.
5. Cereijido, M. & J. Anderson (eds.) 2001. *Tight Junctions.* 2nd ed. CRC Press. Boca Raton, FL.
6. Furuse, M., T. Hirase, M. Itoh, *et al.* 1993. Occludin: a novel integral membrane protein localizing at tight junctions. *J. Cell Biol.* **123:** 1777–1788.
7. Furuse, M., K. Fujita, T. Hiiragi, *et al.* 1998. Claudin-1 and -2: novel integral membrane proteins localizing at tight junctions with no sequence similarity to occludin. *J. Cell Biol.* **141:** 1539–1550.
8. Ikenouchi, J., M. Furuse, K. Furuse, *et al.* 2005. Tricellulin constitutes a novel barrier at tricellular contacts of epithelial cells. *J. Cell Biol.* **171:** 939–945.
9. Furuse, M., H. Sasaki, K. Fujimoto & S. Tsukita. 1998. A single gene product, claudin-1 or -2, reconstitutes tight junction strands and recruits occludin in fibroblasts. *J. Cell Biol.* **143:** 391–401.
10. Van Itallie, C.M. & J.M. Anderson. 1997. Occludin confers adhesiveness when expressed in fibroblasts. *J. Cell Sci.* **110:** 1113–1121.
11. Furuse, M. 2009. Molecular basis of the core structure of tight junctions. *Cold Spring Harb. Perspect. Biol.* **1:** a002907.
12. Mariano, C., H. Sasaki, D. Brites & M.A. Brito. 2011. A look at tricellulin and its role in tight junction formation and maintenance. *Eur. J. Cell Biol.* **90:** 787–796.
13. Stevenson, B.R., J.D. Siliciano, M.S. Mooseker & D.A. Goodenough. 1986. Identification of ZO-1: a high molecular weight polypeptide associated with the tight junction (zonula occludens) in a variety of epithelia. *J. Cell Biol.* **103:** 755–766.
14. Citi, S., H. Sabanay, R. Jakes, *et al.* 1988. Cingulin, a new peripheral component of tight junctions. *Nature* **333:** 272–276.
15. Citi, S. 2001. The cytoplasmic plaque proteins of the tight junction. In *Tight Junctions.* M. Cereijido & J. Anderson, Eds.: 231–264. CRC Press, Boca Raton, FL.
16. Anderson, J.M. & C.M. Van Itallie. 2009. Physiology and function of the tight junctions. *Cold Spring Harb. Perspect. Biol.* **1:** a002584.
17. Morita, K., M. Itoh, M. Saitou, *et al.* 1998. Subcellular distribution of tight junction-associated proteins (occludin, ZO-1 and ZO-2) in rodent skin changes. *J. Invest. Dermatol.* **110:** 862–866.
18. Brandner, J.M., P. Houdek, C. Kuhn, *et al.* 2001. Tight-junction-systems in mammalian squamous stratified epithelia. *Biol. Cell* **93:** 247.
19. Brandner, J.M., S. Kief, C. Grund, *et al.* 2002. Organization and formation of the tight junction-system in human epidermis and cultured keratinocytes. *Eur. J. Cell Biol.* **81:** 253–263.
20. Pummi, K., M. Malminen, H. Aho, *et al.* 2001. Epidermal tight junctions: ZO-1 and occludin are expressed in mature, developing, and affected skin and in vitro differentiating keratinocytes. *J. Invest. Dermatol.* **117:** 1050–1058.
21. Langbein, L., C. Grund, C. Kuhn, *et al.* 2002. Tight junctions and compositionally related junctional structures in mammalian stratified epithelia and cell cultures derived therefrom. *Eur. J. Cell Biol.* **81:** 419–435.
22. Langbein, L., U.-F. Pape, C. Grund, *et al.* 2003. Tight junction-related structures in the absence of a lumen: occludin, claudins and tight junction plaque proteins in densely packed cell formations of stratified epithelia and squamous cell carcinomas. *Eur. J. Cell Biol.* **82:** 385–400.
23. Turksen, K. & T.-C. Troy. 2002. Junctions gone bad: claudins and loss of the barrier in Cancer. *Biochim. Biophys. Acta* **1816:** 73–79.
24. Schlüter, H., R. Wepf, I. Moll & W.W. Franke. 2004. Sealing the live part of the skin: the integrated meshwork of desmosomes, tight junctions and curvilinear ridge structures in the cells of the uppermost granular layer of the human epidermis. *Eur. J. Cell Biol.* **83:** 655–665.
25. Schlüter, H., I. Moll, H. Wolburg & W.W. Franke. 2007. The different structures containing tight junction proteins in epidermal and other stratified epithelial cells, including squamous cell metaplasia. *Eur. J. Cell Biol.* **86:** 645–655.
26. Furuse, M., M. Hata, K. Furuse, *et al.* 2002. Claudin-based tight junctions are crucial for the mammalian epidermal barrier: a lesson from claudin-1-deficient mice. *J. Cell Biol.* **156:** 1099–1111.

27. Bazzoni, G. & E. Dejana. 2002. Keratinocyte junctions and the epidermal barrier: how to make a skin-tight dress. *J. Cell Biol.* **156:** 947–949.

28. Herrmann, T., F. van der Hoeven, H.-J. Gröne, *et al.* 2003. Mice with targeted disruption of the fatty acid transport protein (Fatp 4, Slc27a4) gene show features of lethal restrictive dermopathy. *J. Cell Biol.* **161:** 1105–1115.

29. Herrmann, T., H.-J. Gröne, L. Langbein, *et al.* 2005. Disturbed epidermal structure in mice with temporally controlled Fatp4 deficiency. *J. Invest. Dermatol.* **125:** 1228–1235.

30. Elias, P.M., N.S. McNutt & D.S. Friend. 1977. Membrane alterations during cornification of mammalian squamous epithelia: a freeze-fracture, tracer, and thin-section study. *Anat. Rec.* **189:** 577–594.

31. Elias, P.M., C. Cullander, T. Mauro, *et al.* 1998. The secretory granular cell: the outermost granular cell as a specialized secretory cell. *J. Invest. Dermatol. Symp. Proc.* **3:** 87–100.

32. Tsuruta, D., K.J. Green, S. Getsios & J.C.R. Jones. 2002. The barrier function of skin: how to keep a tight lid on water loss. *Trends Cell Biol.* **12:** 355–357.

33. Kirschner, N., C. Bohner, S. Rachow & J.M. Brandner. 2010. Tight junctions: is there a role in dermatology? *Arch. Dermatol. Res.* **203:** 483–493.

34. Kirschner, N., P. Houdek, M. Fromm, *et al.* 2010. Tight junctions form a barrier in human epidermis. *Eur. J. Cell Biol.* **89:** 839–842.

35. Franke, W.W. 2009. Discovering the molecular components of intercellular junctions – a historical view. *Cold Spring Harb. Perspect. Biol.* **1:** a003061.

36. Schlüter, H. 2006. Schlussleisten ("Tight Junctions") und biochemisch verwandte Strukturen: Tight Junction-Proteine und -Funktionen in mehrschichtigen Epithelien, mit besonderer Berücksichtigung der bronchialen Plattenepithelmetaplasie. PhD Thesis, Faculty of Biosciences, University of Heidelberg, Germany.

37. Orlando, R.C., E.R. Lacy, N.A. Tobey & Cowart K. 1992. Barriers to paracellular permeability in rabbit esophageal epithelium. *Gastroenterology* **102:** 910–923.

38. Franke, W.W., S. Rickelt, M. Barth & S. Pieperhoff. 2009. The junctions that don't fit the scheme: special symmetrical cell-cell junctions of their own kind. *Cell Tissue Res.* **338:** 1–17.

39. Pieperhoff, S., M. Barth, S. Rickelt & W.W. Franke. 2010. Desmosomal molecules in and out of adhering junctions: normal and diseased states of epidermal, cardiac, and mesenchymally derived cells. *Dermatol. Res. Pract.* **2010:** 139167.

Ann. N.Y. Acad. Sci. ISSN 0077-8923

ANNALS OF THE NEW YORK ACADEMY OF SCIENCES

Issue: *Barriers and Channels Formed by Tight Junction Proteins*

Barriers and more: functions of tight junction proteins in the skin

Nina Kirschner and Johanna M. Brandner

Department of Dermatology and Venerology, University Hospital Hamburg-Eppendorf, Hamburg, Germany

Address for correspondence: Johanna M. Brandner, Department of Dermatology and Venerology, University Hospital Hamburg-Eppendorf, Martinistrasse 52, 20246 Hamburg, Germany. brandner@uke.de

Although the existence of tight junction (TJ) structures (or a secondary epidermal barrier) was postulated for a long time, the first description of TJ proteins in the epidermis (occludin, ZO-1, and ZO-2) was only fairly recent. Since then, a wealth of new insights concerning TJs and TJ proteins, including their functional role in the skin, have been gathered. Of special interest is that the epidermis as a multilayered epithelium exhibits a very complex localization pattern of TJ proteins, which results in different compositions of TJ protein complexes in different layers. In this review, we summarize our current knowledge about the role of TJ proteins in the epidermis in barrier function, cell polarity, vesicle trafficking, differentiation, and proliferation. We hypothesize that TJ proteins fulfill TJ structure-dependent and structure-independent functions and that the specific function of a TJ protein may depend on the epidermal layer where it is expressed.

Keywords: epidermis; claudin; occludin; ZO-1; differentiation; proliferation

Introduction

The multilayered, stratified epithelium of the skin is formed by several keratinocyte layers with different characteristics (Fig. 1). In the basal cell layer, keratinocytes are undifferentiated and able to proliferate. This layer contains stem cells that are indefinitely capable of cell division as well as transient amplifying cells, which show a finite capability of proliferation.[4] On top of this layer, several spinous cell layers reside. This stratum spinosum (SS) is characterized by a high number of desmosomes, which contribute to the appearance of spindle-shaped cells. In addition, the cells express the early differentiation marker cytokeratin 10. However, not all spinous cell layers are identical. Differentiation proceeds from the bottom to the top. This can be seen, for instance, by the presence of the intermediate differentiation marker involucrin in the upper spinous cell layers but not in the lower ones. The subsequent stratum granulosum (SG) consists of 3–5 cell layers. Cells in the SG are characterized by lamellar bodies (LB) as well as keratohyalin granules, which entail the typical granular appearance in light microscopy. SG cells express and process the late differentiation markers filaggrin and loricrin. Finally, the uppermost compartment of the epidermis is the stratum corneum (SC), which consists of corneocytes, that is, dead cells, and intercellular lipids. It forms a very important part of the skin barrier and was thought for decades to be the only barrier present in the epidermis, even though a secondary epidermal barrier was postulated a long time ago.[1,2]

The localization patterns of the tight junction (TJ) proteins reflect the complexity of the epidermis (see Table 1). In human epidermis, claudin-1 (Cldn-1), MUPP-1, and JAM-A are present at the cell–cell borders of all living cell layers (see, e.g., Cldn-1 in Fig. 1B). Cldn-7 is found in all living layers, but only very faintly in the stratum basale (SB). ZO-1 and Cldn-4 are present in the upper layers of the SS and in the SG (see, e.g., ZO-1 in Fig. 1C). Occludin (Ocln), cingulin, and Cldn-3 (very faintly) are restricted to the SG (see, e.g., Ocln in Fig. 1D).[3,5–11] In addition, dependent on immunostaining

doi: 10.1111/j.1749-6632.2012.06554.x

Ann. N.Y. Acad. Sci. 1257 (2012) 158–166 © 2012 New York Academy of Sciences.

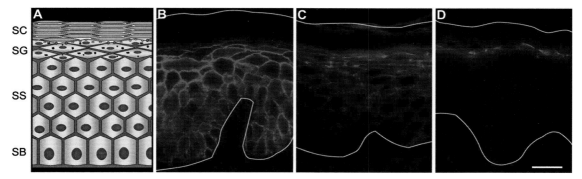

Figure 1. Representative localization patterns of TJ proteins in human epidermis. (A) Schematic overview of human epidermis. Typical TJ structures are shown as green dots. (B–C) Epifluorescence pictures of Cldn-1 (B), ZO-1 (C), and occludin (D) as examples for TJ proteins present at the cell–cell borders in all living layers (B), in the upper layers (C), and in the SG (D); bar 20 µm; white lines indicate the boundaries of the epidermis.

procedures, the proteins are often found in the cytoplasm of cells in layers with and without localization at the cell–cell borders (e.g., Refs. 3, 6–9, 12, and own observations). Cldn-2 and Cldn-5 were up to now only detected in the cytoplasm of the SG.[8,13] Even though the SC is negative for TJ proteins in light microscopy, Cldn-1 and Ocln were detected in the lower layers of SC by electron microscopy (EM).[14–16] Therefore, it is likely that other TJ proteins that have not yet been investigated by EM are present in the SC too. In summary, this means that all TJ proteins are present in the SG, whereas some TJ proteins are also found in few or several deeper layers and/or in the SC. Concerning TJ structures, the typical TJ structures ("kissing points") were found by EM in the SG.[1,5,17,18] In addition, structures containing TJ proteins, but not showing the morphology of typical kissing points, can be found by EM in all layers except for the SB.[18,19] However, TJ proteins are also found outside of distinct membrane structures, for instance, as mentioned above, in the cytoplasm (see Fig. 1), suggesting TJ structure-dependent and structure-independent functions of TJ proteins. Because of varying compositions of TJ protein complexes in the various layers resulting from the different localization patterns described above, it is likely that TJ proteins play distinct roles in the different layers with the SG resembling simple epithelia most, and exhibiting a TJ barrier function. However, while in simple epithelia TJs are the main (paracellular) barrier, in the skin the SC also plays an important role in barrier function. We will summarize here our current knowledge about barrier function as well as other functions of TJ proteins in the epidermis.

Barrier function

When describing barrier function of TJs in the epidermis, several points have to be taken into account. (1) What kind of barrier is meant, that is, which molecules are restricted by the barrier? In general, one has to distinguish between cations, anions, water, small molecules, such as nutrients (or tracers of a similar size), and larger molecules, for example, protein-antigens (or tracers of a similar size). (2) In which layers are these barriers localized? (3) Are these barriers relevant for molecules entering the skin from outside (outside–in barrier) or molecules leaving the skin (inside–out barrier)? (4) And, finally, are there variations of these barriers in different species, different body sites, different ages, and different diseases?

Because of the compact nature of the epidermis and the variability between species, the investigation of these questions is very challenging and the answers not easy to give. The inside–out barrier function of TJs for small molecules, that is, a 557 Da tracer, in newborn murine skin was nicely shown in 2002 by the pioneering work of Furuse et al.[17] They demonstrated that this 557 Da tracer stopped on its way from inside to outside at Ocln-positive spots in the SG, and that this stop was abolished in Cldn-1–deficient mice.[17] Tunggal et al. also showed a loss of this tracer stop in epidermal E-cadherin–deficient mice, which do not express Cldn-1 in the SG.[20] In adult mouse skin, which

Tables 1. Localization patterns of TJ proteins at the cell–cell borders of epidermal layers

Protein	SB	Lower SS	Upper SS	SG	SC	Citations (first description)	Alteration in human diseases/ impact on skin barrier in mouse models
Claudin-1 human	(+)	+	+	+	+[1]	5, 14–16	Psoriasis vulgaris: downregulation in the uppermost and lowermost layers[7] or complete downregulation,[7,10] or no alteration[8] Atopic dermatitis (uninvolved skin): downregulation[35] Impetigo contagiosa: downregulation at the sites of pathogen invasion[32] Hailey-Hailey disease: no alteration[57] Darier's disease: no alteration[57]
Claudin-1 mouse	(+)	+	+	+	−	17	Knockout mice die within one day of birth due to skin barrier defect[17]
Claudin-2 human	−	−	−	+[2]	−	13	
Claudin-3 human	−	−	−	(+)	−	10	Psoriasis vulgaris: downregulation and cytoplasmatic staining[10]
Claudin-4 human	−	−	+	+		12, 58	Psoriasis vulgaris: broader but often cytoplasmic localization[6–8] Hailey-Hailey disease: no alteration[57] Darier's disease: no alteration[57]
Claudin-4 mouse	−	−	+	+	−	17	
Claudin-5 human	−	−	−	(+)[2]	−	8	Psoriasis vulgaris: no major alteration[8]
Claudin-6 mouse	−	−	+	+	−	48	Overexpressing mice die within 48 h of birth due to skin barrier defect[47]
Claudin-7 human	(+)	+	+	+	−	6	Psoriasis vulgaris: downregulation[7]
Claudin-10 Mouse	−	−	−	−	+	48	
Claudin-11 mouse	−	−	−	+	+	48	
Claudin-12 mouse	+	+	+	+	−	48	
Claudin-17 human	−	−	−	+	−	58	
Claudin-18 mouse	−	−	+	+	−	48	
Occludin human	−	−	−	+	+[1]	9, 11, 15, 16	Psoriasis vulgaris: broader localization[6–9,11] Ichthyosis vulgaris: with unknown genotype: broader localization[9] with filaggrin-genotype: downregulation[25] Lichen planus: broader localization[9] Impetigo contagiosa: downregulation at sites of pathogen invasion; upregulation at sites of pathogen colonialization[32] Hailey-Hailey disease: no alteration[57] Darier's disease. no alteration[57]
Occludin rodent	−	−	−	+	−	3	Knockout mice show no obvious skin phenotype[36]
Tricellulin mouse	−	−	−	+	−	23	

Continued

Tables 1. *Continued*

Protein	SB	Lower SS	Upper SS	SG	SC	Citations (first description)	Alteration in human diseases/ impact on skin barrier in mouse models
JAM-A human	+	+	+	+	–	6	Psoriasis vulgaris: slight downregulation uppermost layers[7]
							Systemic sclerosis: downregulation SG[59]
JAM-A mouse	–	+	+	+	–	29	Knockout mice show no obvious skin phenotype[60]
ZO-1 human	–	–	+	+	–	9	Psoriasis vulgaris: broader localization[6–9,11]
							Ichthyosis vulgaris:
							with unknown genotype: broader localization[9]
							with filaggrin-genotype: downregulation[25]
							Lichen planus: broader localization[9]
							Impetigo contagiosa: downregulation at sites of pathogen invasion; upregulation at sites of pathogen colonialization[32]
							Hailey-Hailey disease: broader localization[57]
							Darier's disease: no alteration[57]
ZO-1 rodent	–	–	+	+	–	3	
ZO-2 rodent	–	–	+	+	–	3	
MUPP-1 human	+	+	+	+	–	6	
Cingulin human	–	–	–	+	–	6	

+: positive, (+): faintly positive, –: negative; [1]Only detected by immune electron microscopy; [2]Only cytoplasmatic. SB: stratum basale, SS: stratum spinosum, SG: stratum granulosum, SC: stratum corneum. Additional cytoplasmatic staining was observed for several proteins also in other layers. On the mRNA level, Cldn-3 (human and mouse), Cldn-8 (human and mouse), Cldn-11 (human), Cldn-12 (human), Cldn-18 (human), and Cldn23 (human and mouse) were identified in keratinocytes or skin tissues.[5,10,35,38,61,62] Cldn-17 was identified on mRNA and protein level in human reconstructed epidermis.[63] For more detailed information concerning skin diseases, see Ref. 37.

is much thinner than newborn mouse skin, the inside–out tracer stop was not identified so far. We demonstrated the tracer stop for the first time in the SG of human adult skin,[21] which was then confirmed by others.[22] Hashimoto showed by EM that there is also an inside–out barrier for the electron-dense ion tracer lanthanum at TJ structures in the SG.[1] This barrier was also found by other authors, but was often attributed to LB contents.[23–25]

Concerning the outside–in barrier, it was shown that TJs are a barrier for the 557 Da tracer and for a 45 kDa FITC-ovalbumin in murine skin;[23] however, to perform these experiments, impermeability of the SC had to be overcome first. Therefore, the possibility remains that the outside–in TJ barrier only exists or is relevant when the SC barrier is impaired.

TJ barriers for ions aside from lanthanum, as well as for water, still must be shown in the epidermis. This is very challenging because of the very tight barrier of the SC, which accounts for most of the transepithelial resistance (TER) measurable in the epidermis, leading to the difficulty to distinguish between SC and TJ barriers.[26] However, TER can readily be measured in cultured keratinocytes of mice and humans, showing that at least in cell culture, TJs in keratinocytes are able to form a barrier for ions, even though, strictly speaking, up to now transcellular and paracellular permeability have not been specifically distinguished.[7,27–34] Knockdown of Cldn-1 in human keratinocytes results in decreased TER and increased paracellular tracer flux for sodium fluorescein.[33,35] Further, the knockdown of Ocln decreases TER in human keratinocytes.[33] However, Ocln-deficient mice exhibit no obvious defect in skin barrier function,[36] hinting for a compensation mechanism in these mice.

As described above, the barrier for the 557 Da tracer is localized in healthy skin in the SG. In skin of patients with the skin disease psoriasis vulgaris, the localization of the tracer stop is found in deeper layers of the epidermis, in the upper SS.[37] This fits well with the relocation of the colocalization of all TJ proteins from the SG to the upper SS[7] and to the evidence of TJ structures in this layer in psoriasis vulgaris.[11] For other skin diseases, including the NISCH (neonatal ichthyosis sclerosing cholangitis) syndrome, which is caused by a complete loss of Cldn-1, this barrier function has not yet been investigated. The barrier for lanthanum described in the SG of healthy skin is no longer present in filaggrin-deficient patients with Ichthyosis vulgaris.[25] Interestingly, these patients also show a downregulation of Ocln and ZO-1.[25] However, because they also exhibit differences in LB secretion and other parameters, various factors might contribute to this observation.

There are several hints that the TJ barrier is also important for diseases induced by allergens and pathogens entering the body from outside. This might especially be true when barrier function of the SC is impaired. During skin infection with *Staphylococcus aureus* and *S. epidermidis*, we observed that TJ proteins are upregulated when the skin is colonized with pathogenic and nonpathogenic bacteria, which might be a preventive or compensatory effect to avoid invasion of the pathogens. However, at later time points when skin infection is advancing, TJ proteins are lost, hinting for an interaction of pathogens and TJs.[32] Whether this interaction is direct or indirect still has to be shown. Kubo *et al.* provided evidence that Langerhans cells maintain the TJ barrier when they elongate their dendrites through the granular cell layer to collect antigens from the outside,[23] supporting the hypothesis that intact TJs are important to avoid skin infection. DeBenedetto *et al.* demonstrated a decreased expression of Cldn-1 in uninvolved skin of atopic dermatitis, which might contribute to increased accessibility of this skin for allergens and therefore the triggering of atopic dermatitis lesions.[35] A similar mechanism of skin inflammation because of increased accessibility of the skin for antigens because of Cldn reduction and therefore TJ impairment, was also hypothesized by Yang *et al.* in FGF-receptor knockout mice.[38]

Another kind of barrier: SC compartmentalization

Cldn-1 and Ocln, as well as TJ-related structures, were also detected at the lateral plasma membranes of the lower layers of SC where they neighbor corneodesmosomes.[14–16] This special localization is intriguing because corneodesmosomes—which are important for the orchestrated desquamation of the SC—are first degraded at the apical and basal sides and only later on at the lateral sides of the cells. Therefore, these TJ-related structures, also called TJ remnants, were suggested to form a barrier around corneodesmosomes against degrading enzymes. This could lead to the protection of corneodesmosomes at the lateral plasma membranes from early degradation.[14–16]

TJ barrier regulation/modulation

TJs are highly dynamic complexes, and their function as well as protein expression can be regulated at different levels. Calcium is an important inducer of TJ formation in keratinocytes. An increase of extracellular Ca^{2+} concentration results in a continuous localization of TJ proteins at the cell–cell borders,[5,9,29,34,39] in the establishment of TER, and in reduced paracellular permeability for larger molecules.[29,34,39] Further, the formation of barrier-forming TJs depends on the cell polarity complex Par3/Par6/aPKC.[29] The activation of aPKC is regulated by Tiam1 (T lymphoma invasion and metastasis), an exchange factor for the small GTPase Rac.[39] One putative target molecule of aPKC is Cldn-4, which is phosphorylated during TJ formation in HaCaT cells.[27]

TJ barrier function in keratinocytes (TER) and human skin xenografts or *in vivo* murine skin (557 Da tracer) is impaired by acute UV exposure.[22,40] Chronic UV exposure, as found in sun-exposed skin, results in a downregulation of Cldn-1 in the lowermost layers and a broader expression of Cldn-4, Ocln, and ZO-1 (unpublished data and Ref. 41). In addition, knockdown of TRPV4 (transient receptor potential vanilloid 4 channel), a physiological sensor for mechanical deformation, warm

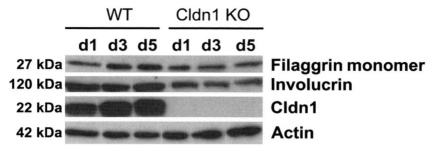

Figure 2. Cldn-1 influences differentiation markers in keratinocytes. Western blot analysis of the differentiation-associated proteins involucrin and filaggrin in wild-type and Cldn-1 knockout keratinocytes one, three, and five days after induction of differentiation by Ca^{2+} switch. Note the decreased protein levels for involucrin at all time points, while filaggrin monomer is only slightly decreased at later time points. Actin was used as gel loading control ($n = 3$).

temperature, and hypoosmolarity, was shown to influence TJ barrier function for the 557 Da tracer in murine skin.[42]

The proinflammatory cytokines TNF-α and IL-1β lead to a biphasic alteration of TER in cultured keratinocytes after induction of TJ assembly by the elevation of calcium concentration. Although there is an upregulation of TER at early time points, there is a downregulation at later time points. When applied to fully developed TJs, only TNF-α exhibited an effect—that is, increased TER.[7] IL-4 shows an increasing effect on TER in keratinocytes,[35] as well as somatostatin, acting via the SST-receptor 3.[43]

Proliferation and differentiation

An influence of TJs or TJ proteins on differentiation (and proliferation) in the skin was first suggested by observations in Cldn-1–deficient mice.[17] Even though no differences for differentiation markers (involucrin, loricrin, transglutaminase-1, and Klf-4) were seen in immunoblotting and Northern blotting directly after birth, a thickened and more compact SC already suggested altered differentiation. When skin of these mice, which die at the first day of birth because of increased transepidermal water loss, was transplanted onto nude mice, it developed hyperplasia. Patients with a complete deficiency of Cldn-1 develop the NISCH syndrome, which includes an ichthyosis-form phenotype that could reflect an abnormal epidermal differentiation (and proliferation) process (OMIM # 607626).[44,45] We were able to show that keratinocytes derived from Cldn-1–deficient mice exhibit downregulation of involucrin and, at later stages of differentiation, also slightly of filaggrin monomers (Fig. 2).[30] Downregulation of involucrin was also

seen in CD44-deficient mice, which display a downregulation of Cldn-1 preceding the downregulation of the differentiation marker.[30] DeBenedetto et al. showed that downregulation of Cldn-1 in human cells results in an increase of keratinocyte proliferation,[35] which might also contribute to the hyperplasia observed in the transplanted skin of Cldn-1–deficient mice. The influence of Cldn-1 on proliferation and differentiation fits nicely to observations made in psoriasis vulgaris, which is characterized by hyperproliferation and impaired differentiation and where a downregulation of Cldn-1 occurs, especially in the lowermost and uppermost layers.[7,10] It might also explain increased proliferation and decrease of involucrin and filaggrin demonstrated in uninvolved skin of atopic dermatitis[46] where a downregulation of Cldn-1 was also observed.[35]

Also for Cldn-6, an influence on differentiation was described. Overexpression of Cldn-6 in homozygous mice leads to a broader localization of cytokeratin 1, filaggrin, loricrin, and transglutaminase-3 in the hyperplastic epidermis and an upregulation on mRNA level of SPRR2D and SPRR2G while there was a downregulation of repetin, SPRR1A, and SPRR2A.[47] In heterozygous mice, which show lower levels of Cldn-6 and therefore survive longer than 48 h, overexpression results in changes of hair follicle differentiation and cycling, namely a shorter anagen phase and altered hair type distribution and length.[48] Also, mice with overexpression of a Cldn-6 cytoplasmatic tail deletion mutant show broader localization of involucrin and cytokeratin 1 (as well as filaggrin), but only two weeks and three months after birth, while expression normalized between these time points.[49]

Whether the effect of TJ proteins on proliferation or differentiation is a direct or indirect effect is not clear yet. For Cldn-1, which is found in the basal cell layer where proliferative cells are present, there might be a direct effect on cell proliferation. However, because this molecule is also located in upper layers of the epidermis, the effect on proliferation can additionally be indirect, for instance via altered ion gradients or an alteration of the skin barrier integrity, which also influence proliferation.[50] The effect on differentiation can occur on several levels. TJ proteins could influence calcium gradients, which are known to be important for cellular differentiation.[51] In addition, they could influence directed LB secretion in the SG (see below), which is important for terminal differentiation. Further, it can be hypothesized that there is a direct effect of TJ proteins on expression or degradation of differentiation-associated proteins. And finally, altered differentiation could be a result of increased proliferation.[26]

Tissue polarity and vesicle trafficking

Even though in multilayered epithelia most cell layers do not show classical polarization (i.e., apical–basal and planar polarity of the cells), there exists, however, tissue polarity, that is, cells from the basal cell layer clearly differ from cells in the SC. In addition, cells in the granular cell layer show polarization. Extrusion of LB and therefore disposal of lipids and specific enzymes is only found at the apical side of the cells. The group of Carien Niessen showed the association of cell polarity complexes with TJs in keratinocytes and their importance for the formation of functional TJs.[20,29,52] Treatment of cultured keratinocytes and skin equivalents with sodium caprate, which is known to open TJs, alters directed LB secretion hinting for an influence of TJs on cell polarization.[53] But sodium caprate increases intracellular DAG, IP$_3$, and Ca^{2+} and is therefore not specific for TJs; rather, it influences several other cellular processes in addition, which might include LB trafficking. In CD44-deficient mice, impairment of cell polarization and loss of directed LB secretion is accompanied by the downregulation of TJ proteins.[30] But, again, CD44 deficiency also influences other cellular processes, including differentiation. Therefore, evidence for an influence of TJs/TJ proteins on cell polarization in keratinocytes is still indirect. In addition, the effect might primarily be on vesicle trafficking—which has been shown to be influenced by TJs in simple epithelia[54,55]—and less on cell polarity. In general, the role of TJs in cell polarity was challenged in the last years. Several authors showed independency of cell polarization from the presence of functional TJs in simple epithelial cells (for review, see Ref. 56).

Conclusions

TJ proteins in the multilayered epithelium of the skin are involved in a multitude of functions, for instance barrier function, cell differentiation, and proliferation. Although barrier function is TJ structure-dependent, other functions may be TJ structure-dependent or structure-independent. The latter might especially be true in the spinous and basal cell layers, where up to now, no typical TJ structures but several TJ proteins have been described. Future experiments dissecting the roles of TJ proteins in the various layers, and the comparison of the results obtained in skin with data from simple epithelia and multilayered epithelia from other origins will contribute to the overall knowledge of TJs and the further elucidation of the significance of TJ proteins in the skin.

Acknowledgment

This work was supported by the Deutsche Forschungsgemeinschaft (FOR 721/2).

Conflicts of interest

The authors declare no conflicts of interest.

References

1. Hashimoto, K. 1971. Intercellular spaces of the human epidermis as demonstrated with lanthanum. *J. Invest. Dermatol.* **57:** 17–31.

2. Rothman, S. & P. Flesch. 1944. The Physiology of the Skin. *Ann. Rev. Physiol.* **6:** 195–224.

3. Morita, K., M. Itoh, M. Saitou, Y. Ando-Akatsuka, *et al.* 1998. Subcellular distribution of tight junction-associated proteins (occludin, ZO-1, ZO-2) in rodent skin. *J. Invest. Dermatol.* **110:** 862–866.

4. Strachan, L.R. & R. Ghadially. 2008. Tiers of clonal organization in the epidermis: the epidermal proliferation unit revisited. *Stem Cell Rev* **4:** 149–157.

5. Brandner, J.M., S. Kief, C. Grund, *et al.* 2002. Organization and formation of the tight junction system in human epidermis and cultured keratinocytes. *Eur. J. Cell. Biol.* **81:** 253–263.

6. Brandner, J.M., S. Kief, E. Wladykowski, *et al.* 2006. Tight junction proteins in the skin. *Skin Pharmacol. Physiol.* **19:** 71–77.

7. Kirschner, N., C. Poetzl, P. von den Driesch, *et al.* 2009. Alteration of tight junction proteins is an early event in psoriasis: putative involvement of proinflammatory cytokines. *Am. J. Pathol.* **175:** 1095–1106.

8. Peltonen, S., J. Riehokainen, K. Pummi & J. Peltonen. 2007. Tight junction components occludin, ZO-1, and claudin-1, -4 and -5 in active and healing psoriasis. *Br. J. Dermatol.* **156:** 466–472.

9. Pummi, K., M. Malminen, H. Aho, *et al.* 2001. Epidermal tight junctions: ZO-1 and occludin are expressed in mature, developing, and affected skin and in vitro differentiating keratinocytes. *J. Invest. Dermatol.* **117:** 1050–1058.

10. Watson, R.E.B., R. Poddar, J.M. Walker, *et al.* 2007. Altered claudin expression is a feature of chronic plaque psoriasis. *J. Pathol.* **212:** 450–458.

11. Yoshida, Y., K. Morita, A. Mizoguchi, *et al.* 2001. Altered expression of occludin and tight junction formation in psoriasis. *Arch. Dermatol. Res.* **293:** 239–244.

12. Morita, K., S. Tsukita & Y. Miyachi. 2004. Tight junction-associated proteins (occludin, ZO-1, claudin-1, claudin-4) in squamous cell carcinoma and Bowen's disease. *Br. J. Dermatol.* **151:** 328–334.

13. Telgenhoff, D., S. Ramsay, S. Hilz, *et al.* 2008. Claudin 2 mRNA and protein are present in human keratinocytes and may be regulated by all-trans-retinoic acid. *Skin Pharmacol. Physiol.* **21:** 211–217.

14. Haftek, M., S. Callejon, Y. Sandjeu, *et al.* 2009. Tight junction-like structures contribute to the lateral contacts between corneocytes and protect corneodesmosomes from premature degradation. *J. Invest. Dermatol.* **129:** S65.

15. Haftek, M., S. Callejon, Y. Sandjeu, *et al.* 2011. Compartmentalization of the human stratum corneum by persistent tight junction-like structures. *Exp. Dermatol.* **20:** 617–621.

16. Igawa, S., M. Kishibe, M. Murakami, *et al.* 2011. Tight junctions in the stratum corneum explain spatial differences in corneodesmosome degradation. *Exp. Dermatol.* **20:** 53–57.

17. Furuse, M., M. Hata, K. Furuse, *et al.* 2002. Claudin-based tight junctions are crucial for the mammalian epidermal barrier: a lesson from claudin-1-deficient mice. *J. Cell Biol.* **156:** 1099–1111.

18. Langbein, L., C. Grund, C. Kuhn, *et al.* 2002. Tight junctions and compositionally related junctional structures in mammalian stratified epithelia and cell cultures derived therefrom. *Eur. J. Cell. Biol.* **81:** 419–435.

19. Schlüter, H., I. Moll, H. Wolburg & W.W. Franke. 2007. The different structures containing tight junction proteins in epidermal and other stratified epithelial cells, including squamous cell metaplasia. *Eur. J. Cell. Biol.* **86:** 645–655.

20. Tunggal, J.A., I. Helfrich, A. Schmitz, *et al.* 2005. E-cadherin is essential for in vivo epidermal barrier function by regulating tight junctions. *EMBO. J.* **24:** 1146–1156.

21. Kirschner, N., P. Houdek, M. Fromm, *et al.* 2010. Tight junctions form a barrier in human epidermis. *Eur. J. Cell. Biol.* **89:** 839–842.

22. Yuki, T., A. Hachiya, A. Kusaka, *et al.* 2011. Characterization of tight junctions and their disruption by UVB in human epidermis and cultured keratinocytes. *J. Invest. Dermatol.* **131:** 744–752.

23. Kubo, A., K. Nagao, M. Yokouchi, *et al.* 2009. External antigen uptake by Langerhans cells with reorganization of epidermal tight junction barriers. *J. Exp. Med.* **206:** 2937–2946.

24. Elias, P.M., N.S. McNutt & D.S. Friend. 1977. Membrane alterations during cornification of mammalian squamous epithelia: a freeze-fracture, tracer, and thin-section study. *Anat. Rec.* **189:** 577–593.

25. Gruber, R., P.M. Elias, D. Crumrine, *et al.* 2011. Filaggrin genotype in ichthyosis vulgaris predicts abnormalities in epidermal structure and function. *Am. J. Pathol.* **178:** 2252–2263.

26. Kirschner, N., R. Rosenthal, D. Günzel, *et al.* 2012. Tight junctions and differentiation—a chicken or the egg question? *Exp. Dermatol.* **21:** 171–175.

27. Aono, S. & Y. Hirai. 2008. Phosphorylation of claudin-4 is required for tight junction formation in a human keratinocyte cell line. *Exp. Cell. Res.* **314:** 3326–3339.

28. Krug, S.M., M. Fromm & D. Gunzel. 2009. Two-path impedance spectroscopy for measuring paracellular and transcellular epithelial resistance. *Biophys J* **97:** 2202–2211.

29. Helfrich, I., A. Schmitz, P. Zigrino, *et al.* 2007. Role of aPKC isoforms and their binding partners Par3 and Par6 in epidermal barrier formation. *J. Invest. Dermatol.* **127:** 782–791.

30. Kirschner, N., M. Haftek, C.M. Niessen, *et al.* 2011. CD44 regulates tight-junction assembly and barrier function. *J. Invest. Dermatol.* **131:** 932–943.

31. Mertens, A.E.E., T.P. Rygiel, C. Olivo, *et al.* 2005. The Rac activator Tiam1 controls tight junction biogenesis in keratinocytes through binding to and activation of the Par polarity complex. *J. Cell Biol.* **170:** 1029–1037.

32. Ohnemus, U., K. Kohrmeyer, P. Houdek, *et al.* 2008. Regulation of epidermal tight-junctions (TJ) during infection with exfoliative toxin-negative Staphylococcus strains. *J. Invest. Dermatol.* **128:** 906–916.

33. Yamamoto, T., Y. Saeki, M. Kurasawa, *et al.* 2008. Effect of RNA interference of tight junction-related molecules on intercellular barrier function in cultured human keratinocytes. *Arch. Dermatol. Res.* **300:** 517–524.

34. Yuki, T., A. Haratake, H. Koishikawa, *et al.* 2007. Tight junction proteins in keratinocytes: localization and contribution to barrier function. *Exp. Dermatol.* **16:** 324–330.

35. De Benedetto, A., N.M. Rafaels, L.Y. McGirt, *et al.* 2011. Tight junction defects in patients with atopic dermatitis. *J. Aller. Clin. Immunol.* **127:** 773–786.

36. Saitou, M., M. Furuse, H. Sasaki, *et al.* 2000. Complex phenotype of mice lacking occludin, a component of tight junction strands. *Mol. Biol. Cell* **11:** 4131–4142.

37. Kirschner, N., C. Bohner, S. Rachow & J. Brandner. 2010. Tight junctions: is there a role in dermatology? *Arch. Dermatol. Res.* **302:** 483–493.

38. Yang, J., M. Meyer, A.-K. Müller, *et al.* 2010. Fibroblast growth factor receptors 1 and 2 in keratinocytes control the epidermal barrier and cutaneous homeostasis. *J. Cell Biol.* **188:** 935–952.

39. Mertens, A.E., T.P. Rygiel, C. Olivo, *et al.* 2005. The Rac activator Tiam1 controls tight junction biogenesis in keratinocytes through binding to and activation of the Par polarity complex. *J. Cell Biol.* **170:** 1029–1037.

40. Yamamoto, T., M. Kurasawa, T. Hattori, *et al.* 2008. Relationship between expression of tight junction-related molecules and perturbed epidermal barrier function in UVB-irradiated hairless mice. *Arch. Dermatol. Res.* **300:** 61–68.

41. Rachow, S., M. Zorn-Kruppa, U. Ohnemus, *et al.* 2011. Common alterations of claudin-1, Claudin-4 and ZO-1 in squamous cell carcinoma, its precursors and sun-exposed skin but specific alterations of Occludin in SCC: relevance for resistance to apoptosis. *J. Investig. Dermatol.* **131:** S2–S26.

42. Sokabe, T., T. Fukumi-Tominaga, S. Yonemura, *et al.* 2011. The TRPV4 channel contributes to intercellular junction formation in keratinocytes. *J. Biol. Chem.* **285:** 18749–18758.

43. Vockel, M., U. Breitenbach, H.-J. Kreienkamp & J.M. Brandner. 2010. Somatostatin regulates tight junction function and composition in human keratinocytes. *Exp. Dermatol.* **19:** 888–894.

44. Feldmeyer, L., M. Huber, F. Fellmann, *et al.* 2006. Confirmation of the origin of NISCH syndrome. *Hum. Mutat.* **27:** 408–410.

45. Hadj-Rabia, S., L. Baala, P. Vabres, *et al.* 2004. Claudin-1 gene mutations in neonatal sclerosing cholangitis associated with ichthyosis: a tight junction disease. *Gastroenterology* **127:** 1386–1390.

46. Egberts, F., M. Heinrich, J.M. Jensen, *et al.* 2004. Cathepsin D is involved in the regulation of transglutaminase 1 and epidermal differentiation. *J. Cell. Sci.* **117:** 2295–2307.

47. Turksen, K. & T.C. Troy. 2002. Permeability barrier dysfunction in transgenic mice overexpressing claudin 6. *Development* **129:** 1775–1784.

48. Troy, T.C., R. Rahbar, A. Arabzadeh, *et al.* 2005. Delayed epidermal permeability barrier formation and hair follicle aberrations in Inv-Cldn6 mice. *Mech. Dev.* **122:** 805–819.

49. Troy, T.-C., A. Arabzadeh, N.M.K. LariviÄře, *et al.* 2009. Dermatitis and aging-related barrier dysfunction in transgenic mice overexpressing an epidermal-targeted claudin 6 tail deletion mutant. *PLoS One* **4:** e7814.

50. Denda, M., J. Sato, T. Tsuchiya, *et al.* 1998. Low humidity stimulates epidermal DNA synthesis and amplifies the hyperproliferative response to barrier disruption: implication for seasonal exacerbations of inflammatory dermatoses. *J. Invest. Dermatol.* **111:** 873–878.

51. Menon, G.K. & S.H. Lee. 2006. Epidermal calcium gradient and the permeability barrier. In *Skin Barrier*. Elias P M & Feingold K R, Eds.: 289–304. New York: Taylor & Francis.

52. Michels, C., S.Y. Aghdam & C.M. Niessen. 2009. Cadherin-mediated regulation of tight junctions in stratifying epithelia. *Ann. N. Y. Acad. Sci.* **1165:** 163–168.

53. Kuroda, S., M. Kurasawa, K. Mizukoshi, *et al.* 2010. Perturbation of lamellar granule secretion by sodium caprate implicates epidermal tight junctions in lamellar granule function. *J. Dermatol. Sci.* **59:** 107–114.

54. Kohler, K. & A. Zahraoui. 2005. Tight junction: a coordinator of cell signalling and membrane trafficking. *Biol. Cell.* **97:** 659–665.

55. Mruk, D.D., A.S. Lau & A.M. Conway. 2005. Crosstalk between Rab GTPases and cell junctions. *Contraception* **72:** 280–290.

56. Giepmans, B.N. & S.C. van Ijzendoorn. 2009. Epithelial cell-cell junctions and plasma membrane domains. *Biochim. Biophys. Acta.* **1788:** 820–831.

57. Raiko, L., P. Leinonen, P.M. Hägg, J. Peltonen, A. Oikarinen & S. Peltonen. 2009. Tight junctions in Hailey-Hailey and Darier's diseases. *Dermatol. Rep.* **1:** e1: 1–5.

58. Brandner, J.M., M. McIntyre, S. Kief, E. Wladykowski & I. Moll. 2003. Expression and localization of tight junction-associated proteins in human hair follicles. *Arch. Dermatol. Res.* **295:** 211–221.

59. Hou, Y., B.J. Rabquer, M.L. Gerber, *et al.* Junctional adhesion molecule-A is abnormally expressed in diffuse cutaneous systemic sclerosis skin and mediates myeloid cell adhesion. *Ann. Rheum. Dis.* **69:** 249–254.

60. Cera, M.R., A. Del Prete, A. Vecchi, *et al.* 2004. Increased DC trafficking to lymph nodes and contact hypersensitivity in junctional adhesion molecule-A-deficient mice. *J. Clin. Invest.* **114:** 729–738.

61. Gareus, R., M. Huth, B. Breiden, *et al.* 2007. Normal epidermal differentiation but impaired skin-barrier formation upon keratinocyte-restricted IKK1 ablation. *Nat. Cell Biol.* **9:** 461–469.

62. Tebbe, B., J. Mankertz, C. Schwarz, *et al.* 2002. Tight junction proteins: a novel class of integral membrane proteins. Expression in human epidermis and in HaCaT keratinocytes. *Arch. Dermatol. Res.* **294:** 14–18.

63. Farwick, M., G. Gauglitz, T. Pavicic, *et al.* 2011. Fifty-kDa hyaluronic acid upregulates some epidermal genes without changing TNF-alpha expression in reconstituted epidermis. *Skin Pharmacol. Physiol.* **24:** 210–217.

Ann. N.Y. Acad. Sci. ISSN 0077-8923

ANNALS OF THE NEW YORK ACADEMY OF SCIENCES
Issue: *Barriers and Channels Formed by Tight Junction Proteins*

Roles for claudins in alveolar epithelial barrier function

Christian E. Overgaard, Leslie A. Mitchell, and Michael Koval

Division of Pulmonary, Allergy and Critical Care Medicine, Department of Medicine, Emory Alcohol and Lung Biology Center and Department of Cell Biology, Emory University, Atlanta, Georgia

Address for correspondence: Michael Koval, Emory University School of Medicine, Division of Pulmonary, Allergy, and Critical Care Medicine, Whitehead Biomedical Research Building, 615 Michael St., Suite 205, Atlanta, GA 30322. mhkoval@emory.edu

Terminal airspaces of the lung, alveoli, are sites of gas exchange that are sensitive to disrupted fluid balance. The alveolar epithelium is a heterogeneous monolayer of cells interconnected by tight junctions at sites of cell–cell contact. Paracellular permeability depends on claudin (cldn)-family tight junction proteins. Of over a dozen alveolar cldns, cldn-3, cldn-4, and cldn-18 are the most highly expressed; other prominent alveolar claudins include cldn-5 and cldn-7. Cldn-3 is primarily expressed by type II alveolar epithelial cells, whereas cldn-4 and cldn-18 are expressed throughout the alveolar epithelium. Lung diseases associated with pulmonary edema, such as alcoholic lung syndrome and acute lung injury, affect alveolar claudin expression, which is frequently associated with impaired fluid clearance due to increased alveolar leak. However, recent studies have identified a role for increased cldn-4 in protecting alveolar barrier function following injury. Thus, alveolar claudins are dynamically regulated, tailoring lung barrier function to control the air–liquid interface.

Keywords: tight junction; acute lung injury; acute respiratory distress syndrome; alcoholic lung disease; sepsis

Introduction

Gas exchange between the lung airspace and the circulatory system is necessary to support respiration in mammals. In order for gas exchange to occur, the lung must maintain a highly specialized barrier between the atmosphere and fluid-filled tissues. The airspace is not completely dry; rather, it is covered by a highly regulated thin layer of fluid known as the air–liquid interface. Lung epithelia maintain the air–liquid interface by providing both a physical barrier to prevent leakage into airspaces and active transport of excess fluid.[1,2] Clinically, during acute respiratory distress syndrome (ARDS), failure of the lung epithelial barrier leads to airspace flooding significantly decreasing the efficiency of gas exchange that exacerbates the severity of acute lung injury.[3,4] The terminal airspaces of the lung, known as alveoli, provide the physical barrier to paracellular fluid permeability.

The alveolar epithelium is heterogeneous and consists of two different cell types: type I and type II alveolar epithelial cells (Fig. 1). Type I cells are a squamous epithelium that cover over 90% of the alveolar surface area.[5] These large thin cells are the primary site of gas exchange between the airspaces and pulmonary capillary vasculature. Type II cells are interspersed throughout the alveoli and serve several functions. Most critically, type II cells produce pulmonary surfactant, lowering surface tension of the air–liquid interface to maintain open airspaces. Type II cells can also differentiate into type I cells in response to injury. As shown in Figure 1, the vast majority of alveolar intercellular junctions are between adjacent type I cells; however, there are also heterocellular interfaces between type I and type II cells.

There are several types of intercellular junctions between alveolar epithelial cells. These include tight junctions, adherens junctions, gap junctions, and desmosomes, each serving a distinct function.[6–10] Tight junctions are the most critical determinant of epithelial barrier function, although evidence suggests that other junctions, particularly adherens junctions, contribute to barrier function by regulating tight junction assembly.[9,11,12]

doi: 10.1111/j.1749-6632.2012.06545.x

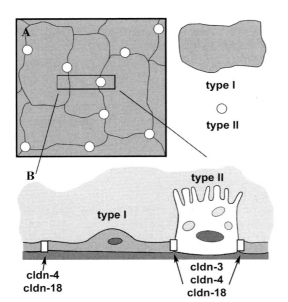

Figure 1. Alveolar epithelial cells. (A) *En face* view of alveolar epithelium, showing the relative size and number of type I and type II cells. (B) Cross-section of the area delineated by the box in A. Shown are type I–type I and type I–type II cell junctions that have distinct claudin composition. The major difference is the presence of high levels of cldn-3 at type I–type II cell tight junctions.

Tight junctions are a network of interwoven strands forming a ring around the lateral plasma membrane. Claudins (cldns), a family of tetraspan transmembrane proteins, form the structural and functional basis for control of tight junction permeability.[13–17] The extracellular domains of claudins interact with extracellular domains of claudins on adjoining cells to form a barrier that restricts ion, solute, and liquid trafficking between cells via the paracellular pathway.[18–21] There are over two dozen mammalian claudins that have tissue-specific patterns of expression that determine tight junction permeability.

Claudins require additional protein components in order to be assembled into tight junctions. They are directly tethered to the actin cytoskeleton via cytosolic scaffold proteins that interact primarily with the C-terminal domain.[22,23] ZO-1 and ZO-2 are the best-characterized scaffold proteins and have been shown to promote and regulate claudin incorporation into tight junctions.[24,25] Tight junctions are also controlled by other transmembrane proteins. For example, the tetraspan transmembrane protein

occludin interacts with claudins to regulate tight junction assembly and barrier function.[19,26] However, occludin is not an absolute requirement for a high-resistance barrier and acts as a proapoptic signal when junctions are disrupted, suggesting an important role for occludin in cell signaling.[27,28] Immunoglobulin-fold transmembrane proteins, including junction adhesion molecule A (JAM-A), also regulate claudin expression and tight junction permeability.[29] Although claudins are one part of the multiprotein complex required to form tight junctions, they nonetheless function as the primary structural component that controls paracellular permeability and the tight junction barrier.

Claudin expression by the alveolus

At least 14 different claudins are expressed at the mRNA and protein level by alveolar epithelial cells.[17,30] The predominant claudins expressed by the alveolar epithelium are cldn-3, cldn-4, and cldn-18.[31] However, other claudins expressed by alveolar epithelium can also influence alveolar barrier function, e.g., cldn-5 and cldn-7, which are associated with decreased and increased alveolar barrier function, respectively (see below).[32,33] Claudin expression is not uniform throughout the alveolus; instead, type II and type I alveolar epithelial cells have distinct patterns of claudin expression[31] (Table 1). The most prominent difference is that type II cells express over 17-fold more cldn-3 than type I cells. By contrast, cldn-4 expression by type II and type I cells is comparable at baseline, although cldn-4 is upregulated during acute lung injury (see below). Type II and type I cells also express comparable levels of cldn-18, which is specifically expressed by alveolar epithelium and is absent from the upper airways. There are two cldn-18 splice variants: cldn-18.1, found primarily in the lung, and cldn-18.2, expressed in the stomach.[34,35]

Resistance to Triton X-100 extraction is a commonly used biochemical assay that correlates with the incorporation of transmembrane proteins into junctional complexes.[36–39] Using this approach, cldn-18 is significantly more insoluble ($\sim$75% insoluble) as compared with cldn-3 ($\sim$40% insoluble) or cldn-4 ($\sim$30% insoluble).[40] Although the basis for enhanced cldn-18 resistance to detergent extraction is not known at present, the C-terminal domain of cldn-18 is roughly twice as large as

Table 1. Alveolar epithelial claudin expression

	Human fetal[77,78]	Human adult[42,69,79]	Rat type II[31,32,65]	Rat type I[31–33,65]	Mouse type II[44,80]
Claudin-3	RP*	P	RP#	RP#	P
Claudin-4	RP	P	RP	RP	P
Claudin-18	RP	P	RP	RP	P
Claudin-5	RP	P	RP	RP	P
Claudin-7	RP		RP	RP	P
Claudin-10b	R		R	R	P
Claudin-12			R	RP	
Claudin-15			R	RP	
Claudin-19			R	R	

*R = mRNA expression detected; P = protein expression detected.
#Rat type II cells express ~17-fold more cldn-3 than type I cells; cldn-4 and cldn-18 expression is comparable.[31,32,65] Other claudin mRNAs expressed by rat alveolar epithelial cells: cldn-9, cldn-11, cldn-20, cldn-22, and cldn-23 (Ref. 31).

that of cldn-3 or cldn-4, which could provide a more effective template for scaffold proteins to bind and crosslink cldn-18 to the cytoskeleton. Whether this is the case remains to be determined; however, this would suggest that association of cldn-18 with cortical actin is an important contributor to alveolar barrier function.[31,41] Consistent with this, proinflammatory hormones decrease cldn-18 expression and assembly into alveolar epithelial tight junctions *in vitro*, which correlates with decreased barrier function[42] (Table 2). Moreover, cldn-18 expression is decreased in murine models of sepsis using cecal ligation and puncture and bleomycin-induced lung injury, which is expected to compromise alveolar barrier function.[43,44]

Alcoholic lung syndrome impairs alveolar barrier function

Chronic alcohol abuse is a clinically significant risk factor for the development of ARDS.[45–47] A key root

cause of this effect is that prolonged ethanol ingestion induces a significant oxidant load driving the alveolar epithelium to produce transforming growth factor beta (TGF-β), which has a deleterious effect on alveolar barrier function[48] and primes the lung for an amplified response to acute lung injury.[49,50] Production of TGF-β in response to alcohol further exacerbates oxidant stress by inhibiting glutathione transport into the airspaces, impairing the antioxidant capacity of the lung.[51,52]

In an otherwise healthy alcoholic, ion channels (e.g., amiloride sensitive sodium channels) can compensate for compromised barrier function and maintain a proper air–liquid interface.[1,53] However, because the alcoholic lung is already under a significant oxidant burden,[17] it is highly susceptible to the effects of a so-called "second hit," such as direct trauma or inflammation due to sepsis. As a result of a second hit, alveolar barrier function in the alcoholic lung is further compromised, overwhelming

Table 2. Changes to alveolar epithelial claudin expression in disease

	Alcoholic lung syndrome[59]	Ventilator induced lung injury[65]	Inflammation[42]	Sepsis-induced ARDS[43]	Pulmonary fibrosis[44]
Claudin-3	decreased	unchanged	unchanged	unchanged	decreased
Claudin-4	unchanged	increased	unchanged	decreased	decreased
Claudin-18	decreased	unchanged	decreased	decreased	decreased
Claudin-5	increased		unchanged	unchanged	decreased
Claudin-7	decreased			unchanged	decreased

control   alcohol

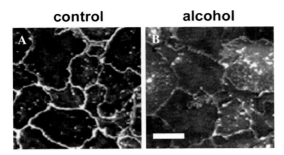

Figure 2. Alcohol impairs assembly of claudins into tight junctions. Model type I alveolar epithelial monolayers were derived from primary cells isolated from either control (A) or alcohol-fed (B) rats that were cultured for six days and then immunolabeled for cldn-7. In contrast with control alveolar epithelial cells, where cldn-7 prominently localized to sites of cell–cell contact (A), cells isolated from alcohol-fed rats had impaired claudin assembly (B), which correlated with impaired barrier function. Bar = 10 μm. Adapted from Fernandez *et al.*[59]

mechanisms of fluid clearance that are already near capacity.

In addition to affecting oxidant load, TGF-β has a direct influence on alveolar epithelial function by promoting epithelial-to-mesenchyme transition (EMT).[54–56] EMT induced by TGF-β directly affects tight junctions through increased expression of transcription factors such as snail, which repress cell polarity and claudin expression.[57,58] In fact, chronic alcohol ingestion decreases expression of several claudins, including cldn-1, cldn-7, and cldn-18.[59] Alcohol also decreases expression of other alveolar tight junction proteins, including occludin and ZO-1, which are likely to contribute to a leaky lung phenotype.[60] In addition to the changes to tight junction protein expression, tight junction formation is also impaired in response to chronic alcohol exposure (Fig. 2). The overall decrease in ZO-1 expression in the alcoholic lung is likely to contribute to decreased tight junction formation, since a decrease in the scaffold impairs the ability of claudins to stably incorporate into tight junctions.[25]

In parallel to alcohol-induced decreases alveolar epithelial cell claudin expression, alcohol surprisingly increases cldn-5 expression.[59] The effect of alcohol on cldn-5 requires an as yet unknown post-translational mechanism of regulation, since alveolar cldn-5 mRNA remains unchanged by alcohol exposure.[59] Several studies have correlated increased cldn-5 with increased lung epithelial paracellular permeability,[32,61,62] suggesting that increased cldn-5

expression is a critical aspect of impaired barrier function in the alcoholic lung. Consistent with this potential mechanism, we have found that transducing normal alveolar epithelial cells with YFP-cldn-5 decreases barrier function *in vitro* (C.E. Overgaard and M. Koval, unpublished results). How cldn-5 could decrease alveolar epithelial barrier function is not known at present. Nonetheless, the ability of cldn-5 to reduce alveolar barrier function is likely to be tissue specific, since cldn-5 is necessary for maintaining the blood–brain barrier, and overexpression of cldn-5 by low-resistance epithelia can increase barrier function.[63,64] Defining roles for cldn-5 in the pathology of alcoholic lung disease will require understanding how cldn-5 interacts with other alveolar epithelial claudins and tight junction proteins.

Cldn-4 expression correlates with increased alveolar fluid clearance

A role for cldn-4 in response to acute lung injury was first implicated in studies where cldn-4 was found to be acutely upregulated by mice in response to ventilator-induced lung injury (VILI).[65] The increase in cldn-4 correlated with decreased severity of injury and was specific, as other claudins, including cldn-3 and cldn-18, were unchanged in response to VILI.[65] Interestingly, alveolar epithelial cldn-4 expression is downregulated in sepsis that is likely to increase the severity of lung injury.[43] Moreover, cultured alveolar epithelial cells show considerable variation in endogenous cldn-4 expression, even within the same monolayer. This suggests that cldn-4 expression is more sensitive to cell phenotype or microenvironment as opposed to other claudins that are more uniformly expressed and regulated, such as cldn-18.[32,66]

A functional role for cldn-4 in lung fluid clearance *in vivo* was confirmed using a peptide fragment derived from *Clostridium perfringens* enterotoxin (CPE), which binds to cldn-3 and cldn-4 with high affinity.[67,68] Intratracheal instillation of a CPE peptide decreased lung cldn-4 content and rendered the lungs more sensitive to VILI.[65] Recently, cldn-4 was found to be associated with increased alveolar fluid clearance rates in *ex vivo* perfused human donor lungs.[69] The extent of lung injury was inversely correlated with cldn-4 expression, supporting a clinically relevant role for cldn-4 in protecting the lung from damage along with improved fluid clearance.

Differential effects of cldn-3 and cldn-4 on alveolar barrier function

Cldn-3 and cldn-4 are the most closely related by amino acid homology.[62] However, since cldn-3 and cldn-4 differ dramatically in their pattern of expression in the alveolus, this suggests that these two claudins serve different roles in alveolar barrier function. To directly test this, alveolar epithelial cells cultured on permeable supports were transduced to specifically increase expression of YFP-tagged versions of either cldn-3 or cldn-4.[40] Expression of YFP-cldn-3 or YFP-cldn-4 has no effect on the expression or localization of other claudins (Fig. 3). However, YPF-cldn-3 and YFP-cldn-4 have differential effects on alveolar epithelial barrier function (Fig. 4). Alveolar epithelial cells transduced with YFP-cldn-4 increased transepithelial resistance (TER) by nearly 50%, providing direct evidence that increased cldn-4 improves alveolar barrier function.[40] By contrast, cells transduced with YFP-cldn-3 show a decrease in TER from $\sim$550 $\Omega \times cm^2$ to $\sim$400 $\Omega \times cm^2$ (Fig. 4). Thus, cldn-3 and cldn-4 have differential effects on alveolar epithelial barrier function.

The context of claudin expression influences their function, since claudin–claudin interactions alter paracellular permeability.[70] For instance, increased cldn-3 augments barrier function of a low-resistance clone of Madin Darby canine kidney (MDCK) epithelial cells from $\sim$50 $\Omega \times cm^2$ to $\sim$100 $\Omega \times cm^2$, most likely by interacting with cldn-2, which acts as a pore forming claudin.[71] This contrasts with the observation that cldn-3 decreases alveolar epithelial barrier function.[40] In this light, it is interesting that cldn-3 is capable of a broad range of heterotypic interactions with other claudins, as opposed to cldn-4, which appears to be restricted to homotypic interactions with cldn-4.[61,62] Whether cldn-3 and cldn-4 have fundamentally different roles in directing tight junction assembly and whether this affects paracellular permeability remains to be determined. However, given that cldn-3 is mainly localized to type II–type I cell interfaces in the alveolus (Fig. 1), it seems likely that type II–type I tight junctions differ from type I–type I cell junctions in paracellular permeability. Determining whether this is the case will require novel methods to measure alveolar tight junction permeability *in situ*.

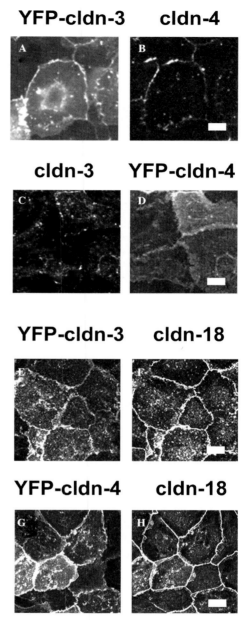

Figure 3. Increasing cldn-3 or cldn-4 expression by alveolar epithelium does not affect tight junction morphology. Model type I alveolar epithelial cells transduced with YFP-cldn-3 (A, B, E, F) or YFP-cldn-4 (C, D, G, H) were fixed and immunostained for cldn-4 (B), cldn-3 (C), or cldn-18 (F, H). YFP-cldn-3 and YFP-cldn-4 localized to the plasma membrane at sites of cell–cell contact. (E–H) Increasing expression of either cldn-3 or cldn-4 had little effect on cldn-18 localization (F, H). Bar, 10 μm. Adapted from Mitchell *et al.*[40]

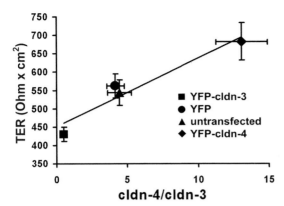

Figure 4. Differential effect of increasing cldn-3 or cldn-4 on alveolar epithelial cell barrier function. Model type I alveolar epithelial cells transduced with YFP-cldn-3 (■), YFP-cldn-4 (♦), YFP-control virus (●), or untransfected controls (▲) were assessed for the effect of altering claudin expression on barrier function, as determined using transepithelial resistance (TER; $\Omega \times cm^2$) (y axis). The expression ratio cldn-4/cldn-3 was determined by immunoblot (x axis), demonstrating that there was a linear relationship between cldn-4/cldn-3 ratio and TER ($r^2 = 0.93$). Cells expressing increased cldn-3 had significantly lower TER than either control cells or cells expressing increased cldn-4 ($P < 0.05$). Increasing cldn-4 also significantly increased barrier function ($P < 0.05$). Adapted from Mitchell *et al.*[40]

Conclusions and perspectives

The strong correlation between cldn-4 and improved lung fluid clearance makes this an appealing therapeutic target for the prevention of ARDS by improving fluid clearance from airspaces. This is particularly appealing for prevention of lung injury during sepsis, where cldn-4 is downregulated.[43] In fact, cldn-4 transcription is under the control of the grainyhead-like 2 (Grhl2) transcription factor, which is activated during development in several epithelia, including lung.[72,73] Whether Grhl2 is expressed by the adult lung and whether it is activated endogenously in response to injury or through a pharmacologically activated pathway remain unknown at present. It is also unclear whether Grhl2 can overcome the effects of transcription factors that are activated in EMT that suppress claudin expression, including snail, slug, and twist.[74–76] This could be a critical issue when attempting to target cldn-4 in alcoholic lung disease, where TGF-β promotes EMT in the alveolus.

Even if cldn-4 is upregulated, is this sufficient to improve fluid clearance when expression of other claudins is impaired? For example, in the context of

decreased cldn-18, which is the major alveolar epithelial claudin, increasing cldn-4 may not augment the alveolar barrier enough to maintain a proper air–liquid balance. Moreover, in the case of alcoholic lung disease, increased cldn-5 may antagonize cldn-4. This could be due to a direct effect on cldn-4, if cldn-5 heteromerically influences cldn-4 assembly or function. Alternatively, cldn-5 may compete with cldn-4, and other claudins, for interaction with scaffold proteins. Cldn-5 may also recruit-specific subclasses of scaffold proteins to tight junctions that could influence assemble as well. In these models, differential affinity for scaffold proteins by claudins dictate the priority, or stability, of claudin integration into tight junction strands through competition for binding to scaffold proteins. In essence, claudin composition could control recruitment of scaffold proteins to tight junctions, the converse of models where scaffold proteins control claudin incorporation junctional strands.[25] If this is the case, then steady state tight junction composition is dictated by coordinated bidirectional interplay between claudins and scaffold proteins.

Acknowledgments

This work was supported by Emory Alcohol and Lung Biology Center/National Institutes of Health (NIH) Grants P50-AA013757 (M.K.), R01-HL083120 (M.K.), AA-013528 (to C.E.O. and L.A.M.), and by the Emory University Research Committee (M.K.).

Conflicts of interest

The authors declare no conflicts of interest.

References

1. Eaton, D.C. *et al.* 2009. The contribution of epithelial sodium channels to alveolar function in health and disease. *Annu. Rev. Physiol.* **71:** 403–423.

2. Van Driessche, W. *et al.* 2007. Interrelations/cross talk between transcellular transport function and paracellular tight junctional properties in lung epithelial and endothelial barriers. *Am. J. Physiol.* **293:** L520–L524.

3. Matthay, M.A. & R.L. Zemans. 2011. The acute respiratory distress syndrome: pathogenesis and treatment. *Annu. Rev. Pathol.* **6:** 147–163.

4. Ware, L.B. & M.A. Matthay. 2000. The acute respiratory distress syndrome. *N. Engl. J. Med.* **342:** 1334–1349.

5. Crapo, J.D. *et al.* 1982. Cell number and cell characteristics of the normal human lung. *Am. Rev. Respir. Dis.* **126:** 332–337.

6. Bartels, H. 1979. The air–blood barrier in the human lung. A freeze-fracture study. *Cell Tissue Res.* **198:** 269–285.

7. Koval, M. 2002. Sharing signals: connecting lung epithelial cells with gap junction channels. *Am. J. Physiol.* **283:** L875–L893.

8. Boitano, S. *et al.* 2004. Cell–cell interactions in regulating lung function. *Am. J. Physiol.* **287:** L455–L459.

9. Komarova, Y.A., D. Mehta & A.B. Malik. 2007. Dual regulation of endothelial junctional permeability. *Sci. STKE* **2007:** re8.

10. Schneeberger, E.E. & R.D. Lynch. 2004. The tight junction: a multifunctional complex. *Am. J. Physiol. Cell Physiol.* **286:** C1213–C1228.

11. Capaldo, C.T. & I.G. Macara. 2007. Depletion of E-cadherin disrupts establishment but not maintenance of cell junctions in MDCK epithelial cells. *Mol. Biol. Cell.* **18:** 189–200.

12. Ivanov, A.I., C.A. Parkos & A. Nusrat. 2010. Cytoskeletal regulation of epithelial barrier function during inflammation. *Am. J. Pathol.* **177:** 512–524.

13. Amasheh, S. *et al.* 2009. Tight junction proteins as channel formers and barrier builders. *Ann. N. Y. Acad. Sci.* **1165:** 211–219.

14. Anderson, J.M. & C.M. Van Itallie. 2009. Physiology and function of the tight junction. *Cold Spring Harb. Perspect. Biol.* **1:** a002584.

15. Angelow, S., R. Ahlstrom & A.S. Yu. 2008. Biology of claudins. *Am. J. Physiol. Renal. Physiol.* **295:** F867–F876.

16. Krause, G. *et al.* 2008. Structure and function of claudins. *Biochimica et Biophysica Acta* **1778:** 631–645.

17. Overgaard, C.E. *et al.* 2011. Claudins: control of barrier function and regulation in response to oxidant stress. *Antioxid. Redox Signal.* **15:** 1179–1193.

18. Piehl, C. *et al.* 2010. Participation of the second extracellular loop of claudin-5 in paracellular tightening against ions, small and large molecules. *Cell Mol. Life Sci.* **67:** 2131–2140.

19. Mrsny, R.J. *et al.* 2008. A key claudin extracellular loop domain is critical for epithelial barrier integrity. *Am. J. Pathol.* **172:** 905–915.

20. Piontek, J. *et al.* 2008. Formation of tight junction: determinants of homophilic interaction between classic claudins. *Faseb J.* **22:** 146–158.

21. Van Itallie, C.M. *et al.* 2006. Two splice variants of claudin-10 in the kidney create paracellular pores with different ion selectivities. *Am. J. Physiol. Renal. Physiol.* **291:** F1288–F1299.

22. Shen, L. & J.R. Turner. 2005. Actin depolymerization disrupts tight junctions via caveolae-mediated endocytosis. *Mol. Biol. Cell* **16:** 3919–3936.

23. Bruewer, M. *et al.* 2004. RhoA, Rac1, and Cdc42 exert distinct effects on epithelial barrier via selective structural and biochemical modulation of junctional proteins and F-actin. *Am. J. Physiol. Cell Physiol.* **287:** C327–C335.

24. Fanning, A.S., C.M. Van Itallie & J.M. Anderson. 2012. Zonula occludens-1 and -2 regulate apical cell structure and the zonula adherens cytoskeleton in polarized epithelia. *Mol. Biol. Cell* **23:** 577–590.

25. Umeda, K. *et al.* 2006. ZO-1 and ZO-2 independently determine where claudins are polymerized in tight-junction strand formation. *Cell* **126:** 741–754.

26. Raleigh, D.R. *et al.* 2011. Occludin S408 phosphorylation regulates tight junction protein interactions and barrier function. *J. Cell Biol.* **193:** 565–582.

27. Yu, A.S. *et al.* 2005. Knockdown of occludin expression leads to diverse phenotypic alterations in epithelial cells. *Am. J. Physiol. Cell Physiol.* **288:** C1231–C1241.

28. Beeman, N., P.G. Webb & H.K. Baumgartner. 2012. Occludin is required for apoptosis when claudin-claudin interactions are disrupted. *Cell Death Dis.* **3:** e273.

29. Laukoetter, M.G. *et al.* 2007. JAM-A regulates permeability and inflammation in the intestine in vivo. *J. Exp. Med.* **204:** 3067–3076.

30. Soini, Y. 2011. Claudins in lung diseases. *Respir. Res.* **12:** 70.

31. Lafemina, M.J. *et al.* 2010. Keratinocyte growth factor enhances barrier function without altering claudin expression in primary alveolar epithelial cells. *Am. J. Physiol.* **299:** L724–L734.

32. Wang, F. *et al.* 2003. Heterogeneity of claudin expression by alveolar epithelial cells. *Am. J. Respir. Cell Mol. Biol.* **29:** 62–70.

33. Chen, S.P. *et al.* 2005. Effects of transdifferentiation and EGF on claudin isoform expression in alveolar epithelial cells. *J. Appl. Physiol.* **98:** 322–328.

34. Niimi, T. *et al.* 2001. Claudin-18, a novel downstream target gene for the T/EBP/NKX2.1 homeodomain transcription factor, encodes lung- and stomach-specific isoforms through alternative splicing. *Mol. Cell. Biol.* **21:** 7380–7390.

35. Hayashi, D. *et al.* 2011. Deficiency of claudin-18 causes paracellular H(+) leakage, up-regulation of interleukin-1beta, and atrophic gastritis in mice. *Gastroenterology* **142:** 292–304.

36. Farshori, P. & B. Kachar. 1999. Redistribution and phosphorylation of occludin during opening and resealing of tight junctions in cultured epithelial cells. *J. Membrane Biol.* **170:** 147–156.

37. Nunbhakdi-Craig, V. *et al.* 2002. Protein phosphatase 2A associates with and regulates atypical PKC and the epithelial tight junction complex. *J. Cell Biol.* **158:** 967–978.

38. Stuart, R.O. & S.K. Nigam. 1995. Regulated assembly of tight junctions by protein kinase C. *Proc. Natl. Acad. Sci. USA* **92:** 6072–6076.

39. Wong, V. & B.M. Gumbiner. 1997. A synthetic peptide corresponding to the extracellular domain of occludin perturbs the tight junction permeability barrier. *J. Cell Biol.* **136:** 399–409.

40. Mitchell, L.A. *et al.* 2011. Differential effects of claudin-3 and claudin-4 on alveolar epithelial barrier function. *Am. J. Physiol.* **301:** L40–L49.

41. Koval, M. 2010. Keratinocyte growth factor improves alveolar barrier function: keeping claudins in line. *Am. J. Physiol.* **299:** L721–L723.

42. Fang, X. *et al.* 2010. Allogeneic human mesenchymal stem cells restore epithelial protein permeability in cultured human alveolar type II cells by secretion of angiopoietin-1. *J. Biol. Chem.* **285:** 26211–26222.

43. Cohen, T.S., G. Gray Lawrence & S.S. Margulies. 2010. Cultured alveolar epithelial cells from septic rats mimic in vivo septic lung. *PLoS One* **5:** e11322.

44. Ohta, H. *et al.* 2011. Altered expression of tight junction molecules in alveolar septa in lung injury and fibrosis. *Am. J. Physiol.* **302:** L193–L205.

45. Joshi, P.C. & D.M. Guidot. 2007. The alcoholic lung: epidemiology, pathophysiology, and potential therapies. *Am. J. Physiol.* **292:** L813–L823.

46. Moss, M. *et al.* 2003. Chronic alcohol abuse is associated with an increased incidence of acute respiratory distress syndrome and severity of multiple organ dysfunction in patients with septic shock. *Crit. Care Med.* **31:** 869–877.

47. Moss, M. *et al.* 2000. The effects of chronic alcohol abuse on pulmonary glutathione homeostasis. *Am. J. Respir. Crit. Care Med.* **161:** 414–419.

48. Bechara, R.I. *et al.* 2004. Transforming growth factor beta1 expression and activation is increased in the alcoholic rat lung. *Am. J. Respir. Crit. Care Med.* **170:** 188–194.

49. Munger, J.S. *et al.* 1999. The integrin alpha v beta 6 binds and activates latent TGF beta 1: a mechanism for regulating pulmonary inflammation and fibrosis. *Cell* **96:** 319–328.

50. Pittet, J.F. *et al.* 2001. TGF-beta is a critical mediator of acute lung injury. *J. Clin. Invest.* **107:** 1537–1544.

51. Arsalane, K. *et al.* 1997. Transforming growth factor-beta1 is a potent inhibitor of glutathione synthesis in the lung epithelial cell line A549: transcriptional effect on the GSH rate-limiting enzyme gamma-glutamylcysteine synthetase. *Am. J. Respir. Cell Mol. Biol.* **17:** 599–607.

52. Jardine, H. *et al.* 2002. Molecular mechanism of transforming growth factor (TGF)-beta1-induced glutathione depletion in alveolar epithelial cells. Involvement of AP-1/ARE and Fra-1. *J. Biol. Chem.* **277:** 21158–21166.

53. Pelaez, A. *et al.* 2004. Granulocyte/macrophage colony-stimulating factor treatment improves alveolar epithelial barrier function in alcoholic rat lung. *Am. J. Physiol.* **286:** L106–L111.

54. Kasai, H. *et al.* 2005. TGF-beta1 induces human alveolar epithelial to mesenchymal cell transition (EMT). *Respir. Res.* **6:** 56.

55. Kim, K.K. *et al.* 2006. Alveolar epithelial cell mesenchymal transition develops in vivo during pulmonary fibrosis and is regulated by the extracellular matrix. *Proc. Natl. Acad. Sci. USA* **103:** 13180–13185.

56. Willis, B.C. & Z. Borok. 2007. TGF-beta-induced EMT: mechanisms and implications for fibrotic lung disease. *Am. J. Physiol.* **293:** L525–L534.

57. Medici, D., E.D. Hay & D.A. Goodenough. 2006. Cooperation between snail and LEF-1 transcription factors is essential for TGF-beta1-induced epithelial-mesenchymal transition. *Mol. Biol. Cell* **17:** 1871–1879.

58. Lee, J.M. *et al.* 2006. The epithelial-mesenchymal transition: new insights in signaling, development, and disease. *J. Cell Biol.* **172:** 973–981.

59. Fernandez, A.L. *et al.* 2007. Chronic alcohol ingestion alters claudin expression in the alveolar epithelium of rats. *Alcohol* **41:** 371–379.

60. Fan, X. *et al.* 2011. Chronic alcohol ingestion exacerbates lung epithelial barrier dysfunction in HIV-1 transgenic rats. *Alcohol. Clin. Exp. Res.* **35:** 1866–1875.

61. Coyne, C.B. *et al.* 2003. Role of claudin interactions in airway tight junctional permeability. *Am. J. Physiol.* **285:** L1166–L1178.

62. Daugherty, B.L. *et al.* 2007. Regulation of heterotypic claudin compatibility. *J. Biol. Chem.* **282:** 30005–30013.

63. Amasheh, S. *et al.* 2005. Contribution of claudin-5 to barrier properties in tight junctions of epithelial cells. *Cell Tissue Res.* **321:** 89–96.

64. Nitta, T. *et al.* 2003. Size-selective loosening of the blood-brain barrier in claudin-5-deficient mice. *J. Cell Biol.* **161:** 653–660.

65. Wray, C. *et al.* 2009. Claudin 4 augments alveolar epithelial barrier function and is induced in acute lung injury. *Am. J. Physiol.* **297:** L219–L227.

66. Koval, M. *et al.* 2010. Extracellular matrix influences alveolar epithelial claudin expression and barrier function. *Am. J. Respir. Cell Mol. Biol.* **42:** 172–180.

67. Veshnyakova, A. *et al.* 2011. Mechanism of clostridium perfringens enterotoxin interaction with claudin-3/-4 suggests structural modifications of the toxin to target specific claudins. *The J. Biol. Chem.* **287:** 1698–1708.

68. Mitchell, L.A. & M. Koval. 2010. Specificity of interaction between clostridium perfringens enterotoxin and claudin-family tight junction proteins. *Toxins* **2:** 1595–1611.

69. Rokkam, D. *et al.* 2011. Claudin-4 levels are associated with intact alveolar fluid clearance in human lungs. *Am. J. Pathol.* **179:** 1081–1087.

70. Van Itallie, C.M., A.S. Fanning & J.M. Anderson. 2003. Reversal of charge selectivity in cation or anion-selective epithelial lines by expression of different claudins. *Am. J. Physiol. Renal Physiol.* **285:** F1078–F1084.

71. Milatz, S. *et al.* 2010. Claudin-3 acts as a sealing component of the tight junction for ions of either charge and uncharged solutes. *Biochim. Biophys. Acta.* **1798:** 2048–2057.

72. Werth, M. *et al.* 2010. The transcription factor grainyhead-like 2 regulates the molecular composition of the epithelial apical junctional complex. *Development* **137:** 3835–3845.

73. Auden, A. *et al.* 2006. Spatial and temporal expression of the Grainyhead-like transcription factor family during murine development. *Gene Expr. Patterns* **6:** 964–970.

74. Carrozzino, F. *et al.* 2005. Inducible expression of Snail selectively increases paracellular ion permeability and differentially modulates tight junction proteins. *Am. J. Physiol. Cell Physiol.* **289:** C1002–C1014.

75. Taube, J. H. *et al.* 2010. Core epithelial-to-mesenchymal transition interactome gene-expression signature is associated with claudin-low and metaplastic breast cancer subtypes. *Proc. Natl. Acad. Sci. USA* **107:** 15449–15454.

76. Martinez-Estrada, O.M. *et al.* 2006. The transcription factors slug and snail act as repressors of claudin-1 expression in epithelial cells. *Biochem. J.* **394:** 449–457.

77. Daugherty, B.L. *et al.* 2004. Developmental regulation of claudin localization by fetal alveolar epithelial cells. *Am. J. Physiol.* **287:** L1266–L1273.

78. Kaarteenaho, R. *et al.* 2010. Divergent expression of claudin -1, -3, -4, -5 and -7 in developing human lung. *Respir. Res.* **11:** 59.

79. Kaarteenaho-Wiik, R. & Y. Soini. 2009. Claudin-1, -2, -3, -4, -5, and -7 in usual interstitial pneumonia and sarcoidosis. *J. Histochem. Cytochem.* **57:** 187–195.

80. Mazzon, E. & S. Cuzzocrea. 2007. Role of TNF-alpha in lung tight junction alteration in mouse model of acute lung inflammation. *Respir. Res.* **8:** 75.

Ann. N.Y. Acad. Sci. ISSN 0077-8923

ANNALS OF THE NEW YORK ACADEMY OF SCIENCES
Issue: *Barriers and Channels Formed by Tight Junction Proteins*

Claudins and alveolar epithelial barrier function in the lung

James A. Frank[1,2,3]

[1]Department of Medicine, Division of Pulmonary and Critical Care Medicine, University of California, San Francisco, California.
[2]Northern California Institute for Research and Education, San Francisco VA Medical Center, San Francisco, California.
[3]Cardiovascular Research Institute, University of California, San Francisco, California

Adress for Coprrespondence: James A. Frank, M.D., Associate Professor, Department of Medicine, Division of Pulmonary and Critical Care Medicine, San Francisco VA Medical Center, 4150 Clement St. Box 111D, San Francisco, California 94121.James.frank@ucsf.edu

The alveolar epithelium of the lung constitutes a unique interface with the outside environment. This thin barrier must maintain a surface for gas transfer while being continuously exposed to potentially hazardous environmental stimuli. Small differences in alveolar epithelial barrier properties could therefore have a large impact on disease susceptibility or outcome. Moreover, recent work has focused attention on the alveolar epithelium as central to several lung diseases, including acute lung injury and idiopathic pulmonary fibrosis. Although relatively little is known about the function and regulation of claudin tight junction proteins in the lung, new evidence suggests that environmental stimuli can influence claudin expression and alveolar barrier function in human disease. This review considers recent advances in the understanding of the role of claudins in the breakdown of the alveolar epithelial barrier in disease and in epithelial repair.

Keywords: alveolar epithelium; tight junction; claudin; acute lung injury; cancer

Barrier function in the lung

As the field of barrierology has emerged over the past 10 years, increasing attention has been directed toward the role of tight junction proteins, including claudins, in determining specific barrier properties of epithelia in health and disease.[1] Significant strides have been made in the understanding of renal and intestinal epithelial barrier regulation by claudins; however, relatively little is known about the nature and regulation of tight junctions in the lung. Consideration of the unique requirements of the epithelial barrier in the airway and alveolar epithelium highlights the importance of additional investigation into the make-up and regulation of lung tight junctions. Most remarkable is the considerable surface area of the alveolar epithelium, which equals approximately 75 m^2. This epithelial barrier is less than 1 micron thick and is exposed daily to more than 8,500 L of air from the outside environment. The alveolar epithelium is made up of type 1 and type 2 cells, the former comprising 95% of the surface area of the lung. Although the alveolar barrier includes both endothelial and epithelial cells, the critical role of the epithelium is highlighted by data demonstrating that changes in epithelial permeability alone are sufficient to cause pulmonary edema.[2] This is because endothelial permeability is relatively high at baseline—for example, the protein concentration of lung lymph fluid is more than 60% of the plasma protein concentration.[3]

In addition to the requirement for low macromolecule permeability in the alveolar epithelium, transepithelial ion transport plays a fundamental role in the regulation of airway surface liquid height and composition, as well as in the removal of edema fluid from the airspaces.[4] Alveolar fluid clearance in the resolution of edema is the energy-dependent removal of water from the airspaces down a sodium concentration gradient. Sodium–potassium ATPase establishes this gradient, and transepithelial sodium transport is regulated at several levels, including by apical epithelial sodium channel (ENaC) expression and conductance.[5] To minimize the energy requirement for this sodium gradient, transepithelial chloride transport is also regulated via channels such as the cystic fibrosis transmembrane conductance regulator (CFTR); in high-transport conditions, sodium transport can become limited by chloride conductance.[6,7] Importantly, a sizable portion of transepithelial chloride conductance occurs via the paracellular route in the alveolar epithelium.[4,8] The precise mechanisms for water transport in the alveolar epithelium remain obscure. Although transcellular water conductance through aquaporins likely plays a role, data from mice genetically deficient in aquaporins suggest that alternate pathways can at least compensate for the absence of these channels;[9] however, the paracellular route may be a key mechanism for transepithelial water transport. Therefore, an idealized conception of alveolar epithelial tight junction properties would include not only low permeability to macromolecules and ions, but also relatively anion-selective paracellular ion transport and potentially water transport.

Disruption of epithelial barrier function is fundamental to many lung diseases, of which the most notable is acute lung injury. This includes the acute respiratory distress syndrome (ARDS), which is primarily characterized by increased alveolar barrier permeability and impaired alveolar fluid clearance. In this syndrome of respiratory failure, preserved epithelial barrier function is inversely associated with patient mortality.[10] Therefore, changes in paracellular ion or macromolecule permeability in the alveolar epithelium may have a direct role in the pathogenesis of this disease. Recent data suggest that differential expression of claudins may play a role in epithelial barrier changes in acute lung injury.

Claudin expression in the lung

The particular claudin expression profiles of human alveolar epithelial cells are not fully known. To begin to determine which claudins are expressed in the alveolar epithelium and how claudins might influence barrier function in the lung, our group used cell type–specific markers and fluorescence activated cell sorting (FACS) to purify populations of type 1 and type 2 cells from rat lungs.[11] Using quantitative real-time PCR, we found that both cell types primarily expressed three claudins: claudin-3, claudin-4, and claudin-18.1. In each cell type, these claudins constituted greater than 97% of claudin transcripts, but the proportion of each of the claudins was different in the two cell types (Fig. 1). In type 1 cells—the cell type covering most of the alveolar epithelial

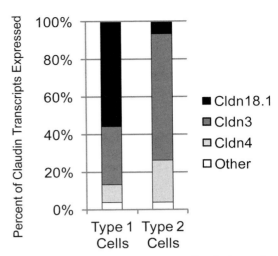

Figure 1. Claudin mRNA expression profiles of primary alveolar epithelial type 1 and type 2 cells. FACS-sorted, freshly isolated primary rat alveolar epithelial cells predominantly express claudin-3, -4, and -18.1, with transcripts for these claudins accounting for 97% of all claudin transcripts in these cells. In type 1 cells, claudin-18.1 is the most abundant transcript, while in type 2 cells claudin-3 is the major transcript. Both cell types express claudin-4. Adapted from Ref. 11.

surface—claudin-18.1 was the most abundantly expressed transcript, while in type 2 cells claudin-3 was the dominant transcript. Both cells expressed relatively high levels of claudin-4. In the healthy lung, type 1 cells form tight junctions with both type 2 cells and other type 1 cells, while type 2 cells form tight junctions primarily with type 1 cells. Of these three claudins expressed in the alveolar epithelium, only claudin-18.1 is unique to the lung,[12] suggesting a lung-specific function. Because claudin-18.1 is the dominant transcript in type 1 cells, it is likely that any unique properties of this protein would partly characterize junctions between type 1 cells. Note that the splice variant claudin-18.2 is expressed in the stomach. Although claudins-1, -5, -7, -12, -15, and -23 are expressed at low levels in alveolar epithelial cells,[11] it is possible that they make important contributions to the alveolar epithelial barrier.

The airway epithelium is made up of more diverse cell types, including Clara cells, ciliated and non-ciliated epithelial cells, goblet cells, neuroendocrine cells, dendritic cells, and others. The specific claudins expressed in each of these cells is not yet known, but transcripts for at least claudins-1, -3, -4, -5, -7, -10, -12, -15, and -18.1 are expressed in whole airway lysates. Studies have identified claudin-1 and

-7 expression in airway dendritic cells within the airway epithelium[13] and claudin-1 in airway smooth muscle cells;[14] claudin-5 is predominantly expressed in endothelial cells.[15] In whole human lung tissue, the most abundant claudin transcripts are claudin-1, -3, -4, -5, -7, -8, -10, -12, -18.1, and -23. This is similar to expression profiles in the rat and mouse, but levels of claudin-15 are probably higher in rodents than in humans based on unpublished data from our laboratory.

Regulation and function of claudins in the alveolar epithelium in acute lung injury

Although claudin expression is not the only mechanism by which cells regulate paracellular permeability, changes in claudin expression in pathological states may provide insight into the contributions of individual proteins to epithelial barrier function. Of the three dominant claudins in the alveolar epithelium, two show notable changes in expression during acute lung injury—claudin-4 and claudin-18.[16–18] Interestingly, regulation of these two proteins is in opposite directions: claudin-4 expression is increased in acute lung injury, while claudin-18 expression may be decreased. The changes in claudin-4 levels appear to be rapid, with an eightfold increase in mRNA levels by 4 h in the ventilator-induced lung injury mouse model. The mechanisms for the increase in claudin-4 expression are not fully known, but data from our group show that protein kinase C (PKC) activation was sufficient to increase claudin-4 mRNA levels and that downstream jun-N-terminal kinase (JNK) inhibition blocked this effect in cultured primary rat and human lung epithelial cells[16] (Fig. 2). Inhibition of ERK1/2 and p38 MAPK did not change PKC-mediated claudin-4 induction in these primary lung cells. It is notable that the claudin-4 promoter contains a conserved, putative AP-1 consensus sequence at the transcription start site, which raises the possibility that JNK may act through AP-1 to increase claudin-4 expression. Previous studies have also shown that SP-1 and the grainyhead transcription factor Ghl2 may be important for basal expression of claudin-4.[19–21] In cultured lung epithelial cells, Madin Darby canine kidney (MDCK) cells, and chick intestinal epithelial cells, the epidermal growth factor (EGF) increases claudin-4 levels.[20,22,23] In the A431 carcinoma cell line, claudin-4 has been reported to be part of a potential protein signature of EGF receptor inhibition; that is, EGFR activation increases claudin-4 expression.[24] Interestingly, the barrier-enhancing properties of EGFR on airway epithelial cells appear to be dependent on JNK activation.[25] In addition, interferon has also been shown to specifically increase claudin-4 expression by a STAT2-dependent mechanism.[26] Others have shown that the flavonoid quercetin specifically induced claudin-4 expression in Caco2 cells; the kinase inhibitors H7 and staurosporine blocked this effect.[27] These inhibitors target PKC and several other kinases. These data constitute groundwork for the mechanistic regulation of claudin-4 expression, but it is not known if any of the pathways identified in these cell culture studies influence the induction of claudin-4 with epithelial cell injury in vivo.

Previous work in MDCK cells has shown that one property of claudin-4 is the formation of a relatively anion-selective paracellular pore pathway.[28] Although knockdown of claudin-4 in this cell type decreases transepithelial electrical resistance, it also results in decreased paracellular anion selectivity as measured by dilution potential. Separate studies have shown that phosphorylation of claudin-4 by WNK-4 in renal epithelial cells enhances membrane localization of claudin-4 and increases transepithelial chloride conductance.[29] This is intriguing because WNK-4 is an activator of chloride conductance in the nephron. Together, these data suggest that claudin-4 limits transepithelial ion transport and confers a greater barrier to sodium than to chloride. In primary cultured rat alveolar type 2 cells, claudin-4 knockdown with siRNA decreased transepithelial electrical resistance without changing paracellular macromolecule permeability.[16] Considering the role of chloride transport in the resolution of pulmonary edema in lung injury, the hypothesis emerges that increased claudin-4 expression may be a part of an adaptive mechanism to enhance the alveolar epithelial barrier and accelerate the resolution of lung edema. In a mouse model of acute lung injury, claudin-4 expression was rapidly and significantly increased. In in vivo studies, peptide inhibition of claudin-4 and -3 with the binding domain of Clostridium perfringens enterotoxin (CPE$_{BD}$) resulted in a significant reduction in alveolar fluid clearance rates without affecting permeability to macromolecules in uninjured mice. CPE$_{BD}$ decreased protein levels of claudin-4, and to

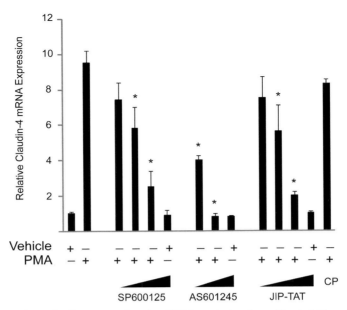

Figure 2. PKC activation induces claudin-4 expression via a JNK-dependent pathway. Phorbol 12-myristate 13-acetate (PMA), an activator of several PKC isoforms, induces a significant increase in claudin-4 expression in primary human type 2–like distal lung epithelial cells at 4 hours. This effect was completely inhibited by the PKC inhibitor Gö6850 (not shown). Inhibition of the JNK MAPK pathway with each of three inhibitors (SP600125, AS601245, and JIP-TAT peptide) blocked the PMA-induced increase in claudin-4 expression in a dose-dependent fashion (*$P < 0.05$, compared with PMA control; data are mean ± SEM; CP = control TAT peptide). Adapted from Ref. 16.

a lesser extent, claudin-3, in the absence of cytotoxicity. Mice treated with CPE$_{BD}$ also developed more severe lung injury and pulmonary edema when exposed to lung injury[16] (Fig. 3). These data further support a protective role for claudin-4 in epithelial barrier function and suggest that one mechanism by which claudin-4 limits edema is by establishing a paracellular sodium barrier to accelerate alveolar fluid clearance.

Claudin-4 and alveolar barrier function in human lungs

Because experimental studies implicate a potential protective role of claudin-4 in the alveolar epithelium, we tested the hypothesis that differences in claudin-4 levels are associated with alveolar barrier function in human lungs rejected for transplantation.[30] Lungs are rejected for transplantation for a variety of reasons, including the presence of clinical lung injury in the donor. Therefore, some lungs rejected for transplant are significantly injured and others are not, but all of the lungs undergo ischemia and reperfusion. In this study, we measured claudin-4 expression in rejected donor lungs using

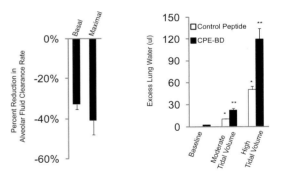

Figure 3. Claudin-3 and -4 knockdown impairs alveolar fluid clearance and increases lung injury severity *in vivo*. *Clostridium perfringens* binding domain peptide (CPE-BD) significantly reduced rates of alveolar fluid clearance by 35–40% in healthy mice. Both basal and beta-adrenergic–stimulated (maximal) rates of alveolar fluid clearance were significantly lower as compared with mice receiving a control peptide (left). Mice given CPE-BD and then exposed to moderate or severe lung injury via escalating tidal volumes on a mechanical ventilator developed more severe pulmonary edema (excess lung water) compared with mice given a control peptide (*$P < 0.05$, compared with baseline lung water; **$P < 0.05$, compared with control peptide-treated mice). Despite reduced rates of fluid clearance at baseline, claudin-3 and -4 knockdown did not induce pulmonary edema in the absence of an additional injurious stimulus. Adapted from Ref. 16.

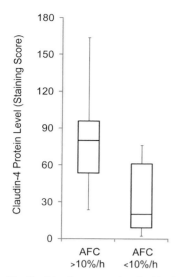

Figure 4. Claudin-4 levels are associated with alveolar fluid clearance rates in human lungs rejected for transplantation. Claudin-4 protein expression levels as assessed by immunostaining were significantly higher in lungs with more preserved alveolar fluid clearance rates. Alveolar fluid clearance was measured in an *ex vivo*–perfused organ system ($P < 0.01$ for this comparison by Mann–Whitney U). Adapted from Ref. 30.

immunostaining and immunoblotting. Claudin-4 levels were quantified on immunostained sections using an automated digital scoring technique that represents, in a single number, the staining intensity and the percentage of cells staining positive. By examining multiple sections from donor lungs, a metric for comparing claudin-4 expression was established. Using an *ex vivo* perfused lung model,[31] alveolar epithelial barrier function was assessed in the donor lungs. Specifically, the rate of alveolar fluid clearance was measured and then compared with claudin-4 staining scores. These data showed a non-normal distribution of claudin-4 levels in the donor lung population. Interestingly, as claudin-4 staining scores increased, alveolar fluid clearance also increased ($r_s = 0.7$, $P < 0.05$ by Spearman rank correlation).[30] Therefore, consistent with the animal and cell culture data, it appears that higher levels of claudin-4 may favor higher rates of alveolar fluid clearance in human lungs (Fig. 4).

To determine if differences in claudin-4 levels were specific—that is, whether claudin-4 was acting as an epithelial cell marker only—levels of other tight junction proteins were also measured. These included claudins-3 and -15, occludin, and ZO-1.

Among these, only claudin-4 levels varied with alveolar fluid clearance rates. In addition, cytokeratin staining showed that claudin-4 levels appeared to vary in epithelial cells and that epithelial cell abundance was comparable in all of the samples studied. These data suggest that claudin-4 levels are dynamic in alveolar epithelial cells during acute lung injury.

To examine if there was an association between claudin-4 expression and surrogate clinical measures of alveolar barrier function, donor lungs were divided into two groups using a clinical lung injury score of 1 to divide the data into two roughly equal sized groups. The four-point clinical lung injury score is derived from measures of oxygenation, mechanical ventilator requirements, and radiographic abnormalities;[32] higher scores indicate more severe injury. These data showed that higher levels of claudin-4 were present in lungs with less severe clinical lung injury.[30] Because levels of claudin-4 were higher in nearly all of the donor lungs than in normal lungs, these data support the hypothesis that lung injury results in a specific induction of claudin-4 and that higher levels of claudin-4 contribute to more preserved alveolar barrier function during injury; however, differences in disease time course, epithelial cell type abundance, and other factors could not be assessed with the available data and remain to be investigated.

A role for claudin-4 in epithelial repair?

Published data to date are consistent with the conclusion that claudin-4–mediated effects on paracellular ion selectivity contribute to the differences in alveolar barrier properties in experimental studies and in human lungs. However, claudin-4 expression patterns in diverse epithelia and in cancer cells raise the possibility that claudin-4 could have an additional unique function in epithelial repair. Notably, claudin-4 is expressed in most epithelial tissues. Although this remains to be determined, it is possible that the effects of claudin-4 on paracellular ion selectivity are less important to barrier function in some organs, such as ovarian surface epithelium, skin, airway epithelium, and intestine. One recent study examined the effect of claudin-4 overexpression on barrier properties in primary rat alveolar epithelial cells in culture. In that study, increased claudin-4 expression did not affect paracellular charge selectivity, suggesting that higher claudin-4 expression levels may not alter paracellular chloride permeability in this cell type.[33] In airway epithelial cells, as opposed to alveolar epithelium, the potential contribution of claudin-4 to ion transport and airway surface liquid height and composition are not certain, but previous studies have shown that airway epithelial cell injury is sufficient to significantly induce claudin-4 expression.[34]

In the naphthalene model of Clara cell injury, naphthalene is administered intraperitoneally to mice, resulting in the selective loss of this cell type from the conducting airways. Cytotoxicity is due to the presence of the cytochrome P450 enzyme CYP2F2 in Clara cells, which converts naphthalene to a toxic metabolite. In the 24–36 h after Clara cell loss, adjacent ciliated epithelial cells squamate and spread to cover the denuded airway epithelium. This is followed by progenitor cell proliferation, and ultimately cells differentiate into the various constituents of the normal airway. Interestingly, at 24 h after naphthalene administration, whole genome array data showed that claudin-4 was among the most highly induced genes in the genome.[34] As the epithelium repaired over the following two weeks, claudin-4 levels returned to normal. When normal epithelial repair was inhibited in transgenic mice expressing herpes simplex virus thymidine kinase in Clara cells exposed to ganciclovir, claudin-4 levels were persistently elevated.[34]

Others have reported that ischemia-reperfusion injury in the gut results in increased claudin-4 expression in migrating epithelial cells at the tips of villi and that claudin-4 levels returned to baseline once epithelial migration was complete.[35] In wounded urothelial cells, claudin-4 expression, but not claudin-8, increased in cells at the leading edge of a repairing wound.[36] In urothelial cells, claudin-8 is a marker of terminal differentiation. Another group found that injured salivary epithelial cells also showed increased claudin-4 expression, and the level of claudin-4 was positively associated with the degree of injury in clinical and experimental studies.[37–39] Still others have found that claudin-4 was expressed in developing mouse embryos and appeared to be required for blastocyst formation.[40] In inner medullary collecting duct (IMCD3) cells, hypertonic stress induced a specific increase in claudin-4.[41]

Together, all of these studies support a role for claudin-4 in epithelial repair. One hypothesis is that claudin-4 facilitates new tight junction formation.

Alternatively, claudin-4 may create a tight junction structure that is favorable to epithelial cell movement during repair. It is also possible that claudin-4 plays a more direct role in epithelial cell movement or differentiation state, a hypothesis suggested by clinical studies and data from tumor cell lines. Claudin-4 expression is increased in several neoplastic tissues, including breast, ovary, pancreas, and prostate, as well as in tumor cell lines.[42–57] In human ovarian surface epithelial (HOSE) cells, high levels of claudin-4 expression were associated with a more invasive phenotype and accelerated cell migration.[42] Knockdown of claudin-4 attenuated this phenotype. The mechanisms that underlie this effect of claudin-4 in HOSE cells remain obscure, but may involve regulation of matrix metalloproteinase expression (MMP) or activity.[42,58] In breast cancer, high claudin-4 expression may be a marker for poor prognosis in certain tumor types.[50,59–61] Data from experimental and human lung studies focused on lung alveolar epithelial barrier function and claudin-4 are entirely consistent with an additional role for claudin-4 in the reestablishment of an intact epithelium after injury, but published studies to date have not addressed this possibility.

Claudins-3 and -18.1

The functions of claudins-3 and -18.1 in the alveolar epithelium are largely unexplored. Claudin-3 is the most abundant claudin sequence in rat type 2 alveolar epithelial cells. Recent work on claudin-3 found that, despite significant sequence homology with claudin-4, claudin-3 conferred no ion selectivity to the paracellular pathway.[33,62] In cultured primary rat alveolar epithelial cells, overexpression of claudin-3 via adenoviral transduction decreased transepithelial electrical resistance and increased paracellular permeability to macromolecules.[33] In MDCK cells, increased claudin-3 expression conferred no paracellular charge selectivity to ion transport but decreased macromolecule permeability without affecting water permeability.[62] It is possible that claudin-3 has a general sealing function, as previous studies have shown that claudin-3, unlike claudin-4, can bind with several other claudins on opposing cells.[63] It is also possible that claudin-3 contributes to paracellular water transport, but this has not been fully elucidated. Claudin-3 expression levels were unchanged in a short-term mouse model of acute lung injury.[16]

Claudin-18.1 is the only known lung-specific tight junction protein, but whether this protein has a lung-specific function is unknown. Recent work in a claudin-18.2 knockout mouse showed that claudin-18.2 formed distinct tight junction strands in the gastric epithelium that were important to limiting paracellular proton transport.[64] Claudin-18.1 (lung) and -18.2 (stomach) differ in sequence in the first extracellular loop, which likely determines ion selectivity, but they share a common sequence in the second extracellular loop and carboxyl-terminal domain. This raises the hypothesis that claudin-18.1 forms distinct strands in the alveolar epithelium, but details regarding claudin-18.1 function, including ion selectivity, have not been reported. Prior studies of experimental lung injury indicate that claudin-18 levels are decreased in the days following bleomycin treatment in mice,[18] but it is uncertain whether this is type 1 cell loss or a specific downregulation of the protein. Interestingly, in cultured primary human type 2 cells, inflammatory stimuli result in a loss of claudin-18 expression by 24 h.[17]

Conclusions and outlook

Recent work has begun to define the claudin expression profiles in lung cells, an important step in determining how tight junction protein regulation influences paracellular transport. In rodent alveolar epithelium, claudins-3, -4, and -18 predominate. Studies to date have demonstrated a specific induction of claudin-4 in acute lung injury. Available human data and functional studies in animals point to a barrier-enhancing role for claudin-4 in the alveolar epithelium. Although a beneficial effect of claudin-4 on paracellular ion transport may be part of the mechanism for this barrier-promoting function in the lung, circumstantial data from other tissues and disease states raise the possibility that claudin-4 may have additional roles in the reestablishment of the epithelial barrier following injury. Because defects in epithelial barrier function in the lung contribute to a variety of lung diseases, future studies of the function and regulation of claudins may provide new therapeutic insights into lung diseases, including acute lung injury, asthma, and pulmonary fibrosis.

Conflicts of interest

The author declares no conflicts of interest.

References

1. Tsukita, S., Y. Yamazaki, T. Katsuno & A. Tamura. 2008. Tight junction-based epithelial microenvironment and cell proliferation. *Oncogene* **27:** 6930–6938.
2. Gorin, A.B. & P.A. Stewart. 1979. Differential permeability of endothelial and epithelial barriers to albumin flux. *J. Appl. Physiol.* **47:** 1315–1324.
3. Vreim, C.R., P.D. Snashall, R.H. Demling & N.C. Staub. 1976. Lung lymph and free interstitial fluid protein composition in sheep with edema. *Am. J. Physiol.* **230:** 1650–1653.
4. Matthay, M.A., H.G. Folkesson & C. Clerici. 2002. Lung epithelial fluid transport and the resolution of pulmonary edema. *Physiol. Rev.* **82:** 569–600.
5. Matthay, M.A., L. Robriquet & X. Fang. 2005. Alveolar epithelium: role in lung fluid balance and acute lung injury. *Proc. Am. Thorac. Soc.* **2:** 206–213.
6. Fang, X., Y. Song, J. Hirsch, *et al.* 2006. Contribution of CFTR to apical-basolateral fluid transport in cultured human alveolar epithelial type II cells. *Am. J. Physiol. Lung Cell. Mol. Physiol.* **290:** L242–L249.
7. Fang, X., N. Fukuda, P. Barbry, *et al.* 2002. Novel role for CFTR in fluid absorption from the distal airspaces of the lung. *J. Gen. Physiol.* **119:** 199–207.
8. Kim, K.J., J.M. Cheek & E.D. Crandall. 1991. Contribution of active Na+ and Cl- fluxes to net ion transport by alveolar epithelium. *Respir. Physiol.* **85:** 245–256.
9. Verkman, A.S., M.A. Matthay & Y. Song. 2000. Aquaporin water channels and lung physiology. *Am. J. Physiol. Lung Cell. Mol. Physiol.* **278:** L867–L879.
10. Ware, L.B. & M.A. Matthay. 2001. Alveolar fluid clearance is impaired in the majority of patients with acute lung injury and the acute respiratory distress syndrome. *Am. J. Respir. Crit. Care Med.* **163:** 1376–1383.
11. LaFemina, M.J., D. Rokkam, A. Chandrasena, *et al.* 2010. Keratinocyte growth factor enhances barrier function without altering claudin expression in primary alveolar epithelial cells. *Am. J. Physiol. Lung Cell. Mol. Physiol.* **299:** L724–L734.
12. Tureci, O., M. Koslowski, G. Helftenbein, *et al.* 2011. Claudin-18 gene structure, regulation, and expression is evolutionary conserved in mammals. *Gene* **481:** 83–92.
13. Sung, S.S., S.M. Fu, C.E. Rose, Jr., *et al.* 2006. A major lung CD103 (alphaE)-beta7 integrin-positive epithelial dendritic cell population expressing Langerin and tight junction proteins. *J. Immunol.* **176:** 2161–2172.
14. Fujita, H., M. Chalubinski, C. Rhyner, *et al.* 2011. Claudin-1 expression in airway smooth muscle exacerbates airway remodeling in asthmatic subjects. *J. Allergy Clin. Immunol.* **127:** 1612–1621.
15. Morita, K., H. Sasaki, M. Furuse & S. Tsukita. 1999. Endothelial claudin: claudin-5/TMVCF constitutes tight junction strands in endothelial cells. *J. Cell. Biol.* **147:** 185–194.
16. Wray, C., Y. Mao, J. Pan, *et al.* 2009. Claudin-4 augments alveolar epithelial barrier function and is induced in acute lung injury. *Am. J. Physiol. Lung Cell. Mol. Physiol.* **297:** L219–227.

17. Fang, X., A.P. Neyrinck, M.A. Matthay & J.W. Lee. 2010. Allogeneic human mesenchymal stem cells restore epithelial protein permeability in cultured human alveolar type II cells by secretion of angiopoietin-1. *J. Biol. Chem.* **285:** 26211–26222.

18. Ohta, H., S. Chiba, M. Ebina, M. Furuse & T. Nukiwa. 2012. Altered expression of tight junction molecules in alveolar septa in lung injury and fibrosis. *Am. J. Physiol. Lung Cell. Mol. Physiol.* **302:** L193–L205.

19. Werth, M., K. Walentin, A. Aue, *et al.* 2010. The transcription factor grainyhead-like 2 regulates the molecular composition of the epithelial apical junctional complex. *Development* **137:** 3835–3845.

20. Ikari, A., K. Atomi, A. Takiguchi, *et al.* 2009. Epidermal growth factor increases claudin-4 expression mediated by Sp1 elevation in MDCK cells. *Biochem. Biophys. Res. Commun.* **384:** 306–310.

21. Honda, H., M.J. Pazin, H. Ji, R.P. Wernyj & P.J. Morin. 2006. Crucial roles of Sp1 and epigenetic modifications in the regulation of the CLDN4 promoter in ovarian cancer cells. *J. Biol. Chem.* **281:** 21433–21444.

22. Singh, A.B. & R.C. Harris. 2004. Epidermal growth factor receptor activation differentially regulates claudin expression and enhances transepithelial resistance in Madin-Darby canine kidney cells. *J. Biol. Chem.* **279:** 3543–3552.

23. Lamb-Rosteski, J.M., L.D. Kalischuk, G.D. Inglis & A.G. Buret. 2008. Epidermal growth factor inhibits Campylobacter jejuni-induced claudin-4 disruption, loss of epithelial barrier function, and Escherichia coli translocation. *Infect. Immun.* **76:** 3390–3398.

24. Myers, M.V., H.C. Manning, R.J. Coffey & D.C. Liebler. 2012. Protein expression signatures for inhibition of epidermal growth factor receptor mediated signaling. *Mol. Cell. Proteomics* **11:** M111.015222.

25. Terakado, M., Y. Gon, A. Sekiyama, *et al.* 2011. The Rac1/JNK pathway is critical for EGFR-dependent barrier formation in human airway epithelial cells. *Am. J. Physiol. Lung Cell. Mol. Physiol.* **300:** L56–L63.

26. Jia, D., R. Rahbar & E.N. Fish. 2007. Interferon-inducible Stat2 activation of JUND and CLDN4: mediators of IFN responses. *J. Interferon. Cytokine. Res.* **27:** 559–565.

27. Amasheh, M., S. Schlichter, S. Amasheh, *et al.* 2008. Quercetin enhances epithelial barrier function and increases claudin-4 expression in caco-2 cells. *J. Nutr.* **138:** 1067–1073.

28. Van Itallie, C., C. Rahner & J.M. Anderson. 2001. Regulated expression of claudin-4 decreases paracellular conductance through a selective decrease in sodium permeability. *J. Clin. Invest.* **107:** 1319–1327.

29. Ohta, A., S.S. Yang, T. Rai, *et al.* 2006. Overexpression of human WNK1 increases paracellular chloride permeability and phosphorylation of claudin-4 in MDCKII cells. *Biochem. Biophys. Res. Commun.* **349:** 804–808.

30. Rokkam, D., M.J. Lafemina, J.W. Lee, M.A. Matthay & J.A. Frank. 2011. Claudin-4 levels are associated with intact alveolar fluid clearance in human lungs. *Am. J. Pathol.* **179:** 1081–1087.

31. Frank, J.A., R. Briot, J.W. Lee, *et al.* 2007. Physiological and biochemical markers of alveolar epithelial barrier dysfunction in perfused human lungs. *Am. J. Physiol. Lung Cell. Mol. Physiol.* **293:** L52–L59.

32. Murray, J.F., M.A. Matthay, J.M. Luce & M.R. Flick. 1988. An expanded definition of the adult respiratory distress syndrome. *Am. Rev. Respir. Dis.* **138:** 720–723.

33. Mitchell, L.A., C.E. Overgaard, C. Ward, S.S. Margulies & M. Koval. 2011. Differential effects of claudin-3 and claudin-4 on alveolar epithelial barrier function. *Am. J. Physiol. Lung Cell. Mol. Physiol.* **301:** L40–L49.

34. Snyder, J.C., A.C. Zemke & B.R. Stripp. 2009. Reparative capacity of airway epithelium impacts deposition and remodeling of extracellular matrix. *Am. J. Respir. Cell. Mol. Biol.* **40:** 633–642.

35. Inoue, K., M. Oyamada, S. Mitsufuji, T. Okanoue & T. Takamatsu. 2006. Different changes in the expression of multiple kinds of tight-junction proteins during ischemia-reperfusion injury of the rat ileum. *Acta. Histochem. Cytochem.* **39:** 35–45.

36. Kreft, M.E., M. Sterle & K. Jezernik. 2006. Distribution of junction- and differentiation-related proteins in urothelial cells at the leading edge of primary explant outgrowths. *Histochem. Cell. Biol.* **125:** 475–485.

37. Ewert, P., S. Aguilera, C. Alliende, *et al.* 2010. Disruption of tight junction structure in salivary glands from Sjogren's syndrome patients is linked to proinflammatory cytokine exposure. *Arthritis Rheum.* **62:** 1280–1289.

38. Michikawa, H., J. Fujita-Yoshigaki & H. Sugiya. 2008. Enhancement of barrier function by overexpression of claudin-4 in tight junctions of submandibular gland cells. *Cell Tissue Res.* **334:** 255–264.

39. Fujita-Yoshigaki, J. 2011. Analysis of changes in the expression pattern of claudins using salivary acinar cells in primary culture. *Methods Mol. Biol.* **762:** 245–258.

40. Moriwaki, K., S. Tsukita & M. Furuse. 2007. Tight junctions containing claudin 4 and 6 are essential for blastocyst formation in preimplantation mouse embryos. *Dev. Biol.* **312:** 509–522.

41. Lanaspa, M.A., A. Andres-Hernando, C.J. Rivard, Y. Dai & T. Berl. 2008. Hypertonic stress increases claudin-4 expression and tight junction integrity in association with MUPP1 in IMCD3 cells. *Proc. Natl. Acad. Sci. USA* **105:** 15797–15802.

42. Agarwal, R., T. D'Souza & P.J. Morin. 2005. Claudin-3 and claudin-4 expression in ovarian epithelial cells enhances invasion and is associated with increased matrix metalloproteinase-2 activity. *Cancer Res.* **65:** 7378–7385.

43. Boireau, S., M. Buchert, M.S. Samuel, *et al.* 2007. DNA-methylation-dependent alterations of claudin-4 expression in human bladder carcinoma. *Carcinogenesis* **28:** 246–258.

44. Cunningham, S.C., F. Kamangar, M.P. Kim, *et al.* 2006. Claudin-4, mitogen-activated protein kinase kinase 4, and stratifin are markers of gastric adenocarcinoma precursor lesions. *Cancer Epidemiol. Biomarkers Prev.* **15:** 281–287.

45. Halder, S.K., G. Rachakonda, N.G. Deane & P.K. Datta. 2008. Smad7 induces hepatic metastasis in colorectal cancer. *Br. J. Cancer* **99:** 957–965.

46. Hanada, S., A. Maeshima, Y. Matsuno, *et al.* 2008. Expression profile of early lung adenocarcinoma: identification of MRP3 as a molecular marker for early progression. *J. Pathol.* **216:** 75–82.

47. Hough, C.D., C.A. Sherman-Baust, E.S. Pizer, *et al.* 2000. Large-scale serial analysis of gene expression reveals genes differentially expressed in ovarian cancer. *Cancer Res.* **60:** 6281–6287.

48. Kleinberg, L., A. Holth, C.G. Trope, R. Reich & B. Davidson. 2008. Claudin upregulation in ovarian carcinoma effusions is associated with poor survival. *Hum. Pathol.* **39:** 747–757.

49. Landers, K.A., H. Samaratunga, L. Teng, *et al.* 2008. Identification of claudin-4 as a marker highly overexpressed in both primary and metastatic prostate cancer. *Br. J. Cancer* **99:** 491–501.

50. Lanigan, F., E. McKiernan, D.J. Brennan, *et al.* 2009. Increased claudin-4 expression is associated with poor prognosis and high tumour grade in breast cancer. *Int. J. Cancer* **124:** 2088–2097.

51. Li, J., S. Chigurupati, R. Agarwal, *et al.* 2009. Possible angiogenic roles for claudin-4 in ovarian cancer. *Cancer Biol. Ther.* **8:** 1806–1814.

52. Michl, P., C. Barth, M. Buchholz, *et al.* 2003. Claudin-4 expression decreases invasiveness and metastatic potential of pancreatic cancer. *Cancer Res.* **63:** 6265–6271.

53. Seckin, Y., S. Arici, M. Harputluoglu, *et al.* 2009. Expression of claudin-4 and beta-catenin in gastric premalignant lesions. *Acta. Gastroenterol. Belg.* **72:** 407–412.

54. Tsutsumi, K., N. Sato, L. Cui, *et al.* 2011. Expression of claudin-4 (CLDN4) mRNA in intraductal papillary mucinous neoplasms of the pancreas. *Mod. Pathol.* **24:** 533–541.

55. Lonardi, S., C. Manera, R. Marucci, *et al.* 2011. Usefulness of claudin 4 in the cytological diagnosis of serosal effusions. *Diagn. Cytopathol.* **39:** 313–317.

56. Facchetti, F., S. Lonardi, F. Gentili, *et al.* 2007. Claudin 4 identifies a wide spectrum of epithelial neoplasms and represents a very useful marker for carcinoma versus mesothelioma diagnosis in pleural and peritoneal biopsies and effusions. *Virchows. Arch.* **451:** 669–680.

57. Konecny, G.E., R. Agarwal, G.A. Keeney, *et al.* 2008. Claudin-3 and claudin-4 expression in serous papillary, clear-cell, and endometrioid endometrial cancer. *Gynecol. Oncol.* **109:** 263–269.

58. Lee, L.Y., C.M. Wu, C.C. Wang, *et al.* 2008. Expression of matrix metalloproteinases MMP-2 and MMP-9 in gastric cancer and their relation to claudin-4 expression. *Histol. Histopathol.* **23:** 515–521.

59. Blanchard, A.A., G.P. Skliris, P.H. Watson, *et al.* 2009. Claudins 1, 3, and 4 protein expression in ER negative breast cancer correlates with markers of the basal phenotype. *Virchows. Arch.* **454:** 647–656.

60. Kulka, J., A.M. Szasz, Z. Nemeth, *et al.* 2009. Expression of tight junction protein claudin-4 in basal-like breast carcinomas. *Pathol. Oncol. Res.* **15:** 59–64.

61. Szasz, A.M., Z. Nemeth, B. Gyorffy, *et al.* 2011. Identification of a claudin-4 and E-cadherin score to predict prognosis in breast cancer. *Cancer Sci.* **102:** 2248–2254.

62. Milatz, S., S.M. Krug, R. Rosenthal, *et al.* 2010. Claudin-3 acts as a sealing component of the tight junction for ions of either charge and uncharged solutes. *Biochim. Biophys. Acta.* **1798:** 2048–2057.

63. Daugherty, B.L., C. Ward, T. Smith, *et al.* 2007. Regulation of heterotypic claudin compatibility. *J. Biol. Chem.* **282:** 30005–30013.

64. Hayashi, D., A. Tamura, H. Tanaka, *et al.* 2012. Deficiency of claudin-18 causes paracellular H(+) leakage, up-regulation of interleukin-1beta, and atrophic gastritis in mice. *Gastroenterology* **142:** 292–304.

Ann. N.Y. Acad. Sci. ISSN 0077-8923

ANNALS OF THE NEW YORK ACADEMY OF SCIENCES
Issue: *Barriers and Channels Formed by Tight Junction Proteins*

Relevance of endothelial junctions in leukocyte extravasation and vascular permeability

Dietmar Vestweber

Max Planck Institute of Molecular Biomedicine, Münster, Germany

Address for correspondence: Dietmar Vestweber, Max Planck Institute of Molecular Biomedicine, Röntgenstr. 20, D-48149 Münster, Germany. vestweb@mpi-muenster.mpg.de

Inflammation and immune surveillance rely on the ability of leukocytes to leave the blood stream and enter tissue. Cytokines and chemokines regulate expression and the activation state of adhesion molecules that enable leukocytes to adhere and arrest at sites of leukocyte exit. Capturing and arrest is followed by the transmigration of leukocytes through the vessel wall—a process called diapedesis. The review will focus on recently published novel approaches to determine the route that leukocytes take *in vivo* when they migrate through the endothelial layer of blood vessels. This work has revealed the dominant importance of the junctional pathway between endothelial cells *in vivo*. In addition, recent progress has improved our understanding of the molecular mechanisms that regulate junctional stability, the opening of endothelial junctions during leukocyte extravasation, and the induction of vascular permeability.

Keywords: endothelial junctions; inflammation; vascular permeability; VE-cadherin

Introduction

Leukocyte extravasation and entry into tissue is essential for initiation and maintenance of the inflammatory process as well as for immune surveillance by lymphocytes. Recruitment of leukocytes is initiated by signaling factors such as cytokines and chemokines that act in concert with selectins, leukocyte-integrins, and members of the Ig-superfamily, which, in combination, mediate capturing, rolling, adhesion, and eventually migration of leukocytes at the luminal surface of the endothelium of postcapillary venules.[1]

This complex process is the prelude for diapedesis, the transmigration of leukocytes through the barrier of the blood vessel wall composed of endothelial cells, the basement membrane, and perivascular cells. The transmigration process is mechanistically not yet well understood. Leukocytes can principally use two different routes to overcome the endothelium: they can move through the junctions between adjacent endothelial cells (paracellular) or they can move directly through the body of single endothelial cells (transcellular; Fig. 1). Because the stability of endothelial cell contacts is maintained by the function of VE-cadherin and not by tight junction proteins, the paracellular diapedesis process will be mainly discussed from the perspective of the adherens junctions, though it is known that tight junction–associated proteins such as the JAMs and ESAM do also contribute to the regulation of leukocyte diapedesis. This review will focus on recently published novel approaches in analyzing leukocyte diapedesis *in vivo*.

The two routes for leukocyte diapedesis

The paracellular route and the transcellular route have both clearly been documented in *in vitro* transmigration assays with cultured endothelial cells and various leukocyte primary isolates.[2–9] At first glance, the transcellular route might seem to require a more complex machinery than the simple opening of junctions for the junctional route. However, *in vitro* and *in vivo* endothelial cells are often only a few micrometers thick, especially in areas close to intercellular junctions. If indentation of the apical plasma membrane of an endothelial cell would lead to fusion with the basal plasma membrane, possibly assisted by intracellular vesicular structures that could serve by bridging the distance between apical

doi: 10.1111/j.1749-6632.2012.06558.x

Two routes of leukocyte extravasation

Figure 1. Leukocytes can overcome the endothelial barrier *in vitro* and *in vivo* by either moving directly through the body of an endothelial cell (transcellular) or by passing through the junctions between endothelial cells (paracellular).

and basal cell surface, a channel could be formed that allows a leukocyte to traverse through the thin cell body.[10]

Careful quantification revealed that about 7–11% of neutrophils, monocytes, or lymphocytes indeed migrate via a transcellular route through the monolayer of human umbilical vein endothelial cells (HUVECs), whereas this percentage increased to about 30% if microvascular endothelial cells[6] or to about 50% if ICAM-1–overexpressing HUVEC[8] were analyzed. The importance of the paracellular pathway was also indirectly supported by results showing that stabilization of endothelial junctions by activating Rap1 strongly inhibited leukocyte transmigration.[11] Mechanistically, it is not clear how the diapedesis process is regulated, though several endothelial cell adhesion receptors, such as PECAM-1, the JAMs, ESAM, CD99, and CD99L2, have been described to participate because antibodies against them inhibit the transmigration process.[12,13] Because these proteins are found strongly enriched at endothelial cell contacts, with the JAMs and ESAM even being confined to tight junctions, it was initially thought that they participate exclusively in paracellular diapedesis. However, some of them such as PECAM-1, JAM and CD99 were recently also found to surround leukocytes that move through a transcellular route,[6,14] again leaving room for a physiological role of the transcellular route. To what extent *in vitro* transmigration assays through endothelial cell layers in the absence of a proper basement membrane can indeed represent the physiological *in vivo* situation is unknown. In fact, it is likely that the lack of a basement membrane would also affect the stability of endothelial junctions and, therefore, it could

be that results from *in vitro* transmigration assays might lead to an overestimation of the importance of the paracellular migration pathway.

In vivo, it is much more difficult to determine with certainty which route an extravasating leukocyte takes. This has been intensively studied over the last five decades in numerous reports using electron microscopy (EM). The challenge with these types of studies is that only a large number of serial sections through the whole body of a diapedesing leukocyte allows for the unambiguous determination of which route this leukocyte takes. This has been most impressively demonstrated by Schoefl analyzing diapedesing lymphocytes in high endothelial venules (HEV) in lymph nodes.[15] He showed that a lymphocyte that appeared in 23 serial sections as being located inside an endothelial cell, was found to be in contact with a second endothelial cell in the following serial sections, revealing that it did not actually transmigrate through a single endothelial cell but instead between endothelial cells, thereby forcing a large inversion into the body of the first endothelial cell. Due to such difficulties, the many studies that attempt to determine the route of transmigration without analyzing serial sections are not informative. Because the preparation of serial sections is rather laborious and technically demanding, most studies using this technique only analyze about a handful of leukocytes. This may explain why, until today, even very thorough studies support contradictory conclusions, either claiming that leukocytes use exclusively the paracellular diapedesis route,[15–17] or exit exclusively via the transcellular route.[18–20] An exception is the study by Marchesi and Gowan[21] who reported that naive lymphocytes enter lymph nodes via the transcellular route whereas neutrophils seemed to use the paracellular route to enter inflamed lymph nodes. Interestingly, although studies on lymphocyte homing into lymph nodes through the specialized HEV provided controversial results, a majority of studies of other tissues and various inflammatory situations claim that extravasation in different tissues, such as skin, pancreas, brain, and bone marrow, is transcellular.[18,19,22–25]

Recently, high resolution, live, intravital microscopy using a laser-scanning confocal microscope was applied as an alternative method to study neutrophil extravasation in the cremaster after applying different inflammatory stimuli.[26] This technique allowed analysis of the extravasation of 100 leukocytes per condition and thereby a more meaningful evaluation of this process than with EM based on serial sections. Importantly, this study revealed that no matter which stimulus was used (IL-1β, fMLP, or ischemia-reperfusion), about 90% of the extravasated neutrophils used the paracellular route.[26] This is presently the most convincing imaging study establishing the dominant physiological relevance of the paracellular extravasation route for neutrophils in inflammation. However, this result for the cremaster cannot necessarily be generalized to other tissues and it is, at present, quite difficult to employ this technique with similar optical resolution in tissues that are less accessible than the cremaster tissue. Furthermore, despite the obvious advantages over EM, confocal microscopy has limitations in the resolution along the z-axis. This is the reason why it cannot always be determined with certainty whether a leukocyte is indeed extravasating through junctions or simply very close to a junction. A recent combination of optical microscopy and atomic force microscopy in an *in vitro* study revealed that a leukocyte that transmigrated in direct contact with junctions did not dissociate the cell contact and diapedesed transcellularly very close to the junctions.[27]

A genetic approach to determine the relevance of endothelial junctions for leukocyte extravasation *in vivo*

To determine the physiological relevance of the junctional diapedesis route independent of the technical limitations of imaging methods, a genetic approach was recently employed by generating mice with highly stabilized or even "locked" endothelial junctions.[28] This approach was based on the assumption that extravasating leukocytes that traverse the endothelial barrier through the junctional pathway would have to dissociate adhesive mechanisms that provide endothelial cell contact integrity. Because VE-cadherin is of dominant importance for the stability of endothelial junctions *in vivo*,[29,30] it was hypothesized that VE-cadherin homophilic interactions need to be disengaged during paracellular diapedesis of leukocytes. Thus, modifying VE-cadherin in a way that would make it constitutively active should render junctions resistant to opening. VE-cadherin needs to associate via β-catenin (or plakoglobin) with α-catenin for proper support of stable endothelial junctions, and it was suggested[31] that the dissociation of this complex or the detachment of this complex from the actin cytoskeleton allows the downregulation of adhesive activities. Indeed, an E-cadherin-α-catenin fusion protein has been demonstrated *in vitro* to be more efficient in stabilizing contacts between transfected cells than unmodified E-cadherin.[32,33] Following this example, an analogous VE-cadherin-α-catenin fusion construct, lacking the β-catenin binding site, was used to generate a knock-in mouse by replacing endogenous VE-cadherin by this construct.[28]

The resulting homozygous knock-in mice were fertile; however, breeding resulted in only 40–50% of viable pups. The developmental defects leading to embryonic lethality are still under investigation. Fortunately, viable mice of homozygous mutant genotype showed no obvious phenotype and were healthy under special pathogen free conditions. The skin of these mice turned out to be resistant to the induction of vascular permeability by potent reagents, such as VEGF and histamine.[28] These results revealed, first, that both stimuli obviously induce vascular permeability by the paracellular route. Second, these results established that the generated knock-in mice indeed possessed highly stabilized endothelial junctions. Reduced leukocyte extravasation in these mice would, therefore, be a strong indication for the prominent role of the paracellular diapedesis route.

Neutrophil recruitment into the IL-1β–inflamed cremaster and the LPS-stimulated lung was reduced in these mice by up to 74% and 63%, respectively. Hapten-induced recruitment of activated lymphocytes into the skin in a delayed type

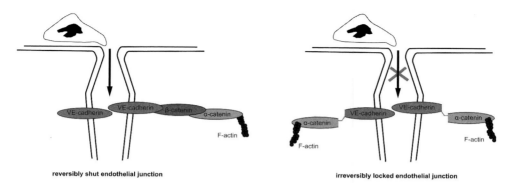

Figure 2. Genetically modified knock-in mice were generated by replacing VE-cadherin by a VE-cadherin–α-catenin fusion construct. These mice had irreversibly "locked" endothelial contacts that led to strong inhibition of leukocyte extravasation into various inflamed tissues. This approach demonstrated the importance of the paracellular route for leukocyte extravasation *in vivo*.[28]

hypersensitivity reaction (DTH) was reduced by 57%[28] (Fig. 2). Assuming that VE-cadherin-α-catenin may not completely lock endothelial junctions, these numbers are minimum estimates for the relevance of the junctional extravasation pathway. The results obtained in the cremaster are in good agreement with life imaging results by Woodfin *et al.*[26]

Unexpectedly, and in strong contrast to the inflammation models, no inhibition was observed for lymphocyte homing in these mice. It can, of course, not be excluded that HEV might allow transcellular diapedesis of naive lymphocytes, as some publications have suggested. However, it is also possible that the junctions of these specialized vessels that represent sites of continuous, rapid, and very efficient lymphocyte recruitment might be regulated by different mechanisms than those of other vessels. It is possible that the junctions of HEV are endued with additional molecular mechanisms that enable them to open their junctions despite of expressing VE-cadherin-α-catenin. In addition, it has been reported that lymphocytes are often found in groups, being located along the same contact between endothelial cells.[15] Thus, if one lymphocyte has managed to open the junctions, several might rapidly follow behind without giving the junction a chance to close again. This would be a mechanism that would reduce inhibitory effects of a VE-cadherin-α-catenin fusion protein.

It is not yet fully understood how VE-cadherin-α-catenin stabilizes endothelial cell contacts. Several indirect mechanisms that would not act via a direct effect on the adhesive activity of VE-

cadherin, could be excluded. Endocytosis of VE-cadherin-α-catenin was similar in kinetics and efficiency as VE-cadherin. The lack of β-catenin association had no effect, since an alternative VE-cadherin-α-catenin fusion protein containing full length VE-cadherin had the same inhibitory effect on neutrophil transmigration. In addition, classical β-catenin–dependent Wnt signaling was not affected in endothelial cells expressing VE-cadherin-α-catenin. VEGF-R2 signaling was also not affected, and expression levels of other endothelial adhesion molecules known to play a role in leukoyte extravasation were normal. Similarly, the association of VE-cadherin with the tyrosine phosphatase VE-PTP (see below) was not affected.

It has been argued that α-catenin cannot directly connect β-catenin with actin and that only dimeric α-catenin binds actin, thereby competing with Arp2/3 binding and affecting actin dynamics.[34,35] This pool of "free" dimeric α-catenin is probably in equilibrium with the pool of VE-cadherin-associated "junction-bound" α-catenin, however, it is not known how much of this free α-catenin is present at endothelial junctions. In this context, it is noteworthy that the overall amount of α-catenin detected by antibody staining at endothelial junctions was not different between endothelial cells that exclusively expressed VE-cadherin–α-catenin or VE-cadherin. In addition, a pool of α-catenin was still detectable in VE-cadherin-α-catenin–expressing endothelial cells.

Thus, VE-cadherin-α-catenin and VE-cadherin behaved similar in any respect that was tested, except in their association with the actin cytoskeleton.[28] The VE-cadherin–α-catenin was much less detergent extractable than VE-cadherin and fluorescence recovery after photobleaching (FRAP) analysis revealed that the membrane mobility of the fusion construct was strongly reduced compared to VE-cadherin. The mobility of the fusion construct increased upon interfering with the formation of the actin cytoskeleton. Thus, VE-cadherin–α-catenin associates more efficiently with the actin cytoskeleton, which may be the reason why this modified VE-cadherin stabilizes endothelial junctions.

Mechanisms that control the stability and the opening of endothelial junctions

The finding that the VE-cadherin–α-catenin fusion protein strongly impairs leukocyte extravasation *in vivo* in various tissues under various inflammatory situations clearly demonstrates that the VE-cadherin–catenin complex is a central target for the opening of endothelial junctions during leukocyte extravasation in the living animal. This is in agreement with *in vitro* studies that showed that certain tyrosine/phenylalanine point mutations of VE-cadherin could inhibit the transmigration of leukocytes through cultured endothelial cell layers.[36,37] Although these results suggest that tyrosine phosphorylation of VE-cadherin is involved in opening endothelial cell contacts, it was shown that this alone was not sufficient to destabilize endothelial junctions.[38] Interestingly, other *in vitro* studies reported that even adhesion blocking antibodies against VE-cadherin would only dissociate endothelial cell contacts if the production of reactive oxygen species (ROS) was not impaired.[39] Beside ROS production, which required activated Rac, this study also suggested an active role for the kinase Pyk in the destabilization of endothelial junctions. Because the stimulation of VCAM-1 by leukocytes can trigger the activation of each of these signaling steps,[40] including the phosphorylation of components of the VE-cadherin–catenin complex, this would provide a potential framework for a leukocyte-driven mechanism that could be involved in opening of endothelial junctions. ICAM-1 is another endothelial receptor for leukocytes, whose crosslinking was shown to trigger VE-cadherin, phosphorylation,[37] and eNOS, Src, and Rho GTPase activation were reported to be involved in this signaling mechanism.[41] In agreement with this, it was shown that

VE-cadherin was rearranged at junctions during *in vitro* transmigration,[42] and *in vivo*, blocking of ICAM-1 signaling affected VE-cadherin rearrangement.[43] In addition, it was shown *in vitro* that endothelial cells require RhoA, Rho kinase, myosin light chain phosphorylation, and actomyosin contraction for leukocyte diapedesis, which probably provide the force that disrupts endothelial junctions,[44–48] once the adhesive mechanisms between endothelial cells are weakened.

There is no doubt that released factors, such as oxygen metabolites or proteolytic enzymes from neutrophils, can be important inducers of vascular permeability in certain inflammatory settings.[49,50] However, there is also increasing evidence that sites of leukocyte extravasation and sites of permeability induction can be spatially and temporally distinct.[51–53] Thus, leukocyte-induced opening of endothelial junctions does not necessarily lead to an increase in permeability. Thus, endothelial cells must be endued with mechanisms that prevent leakage despite leukocytes moving through opened junctions. Whether this is achieved by leukocytes plugging the junctions, or by endothelial protrusions locking the diapedesis pores behind the diapedesing leukocytes as so called "domes"[54] will be interesting to investigate in the future.

The relevance of the endothelial specific tyrosine phosphatase VE-PTP for the control of endothelial junctions in leukocyte diapedesis

Leukocyte transmigration and tyrosine phosphorylation of components of the VE-cadherin–catenin complex is supported by kinases, such as Pyk and AMP-activated protein kinase,[39,41] and also controlled by the vascular endothelial protein tyrosine phosphatase (VE-PTP). VE-PTP is the only known endothelial specific receptor-type tyrosine phosphatase,[55,56] and it interacts with VE-cadherin via intracellular, but mainly extracellular, domains.[57] VE-PTP is required for optimal adhesive function of VE-cadherin and for endothelial cell contact integrity.[58] Major substrates of VE-PTP, which are relevant for the adhesive function of VE-cadherin, are VE-cadherin itself and the catenin plakoglobin.[58] Other tyrosine phosphatases that affect endothelial cell contact integrity and VE-cadherin–catenin phosphorylation are SHP2,[59,60] and the receptor type tyrosine phosphatase RPTP-μ.[61] However,

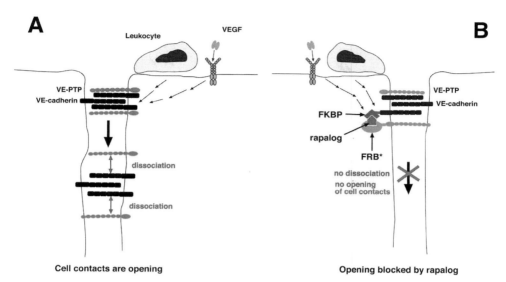

Figure 3. (A) The phosphatase VE-PTP associates with VE-cadherin and is required for optimal VE-cadherin function and endothelial cell contact integrity. The docking of leukocytes to endothelial cells as well as the stimulation by the permeability-stimulating factor VEGF trigger the dissociation of VE-PTP from VE-cadherin, a prerequisite for the opening of endothelial junctions.[58] (B) Knock-in mice were generated expressing VE-cadherin and VE-PTP fusion proteins containing each another, additional protein domain (FKBP and FRB*, see text for details). These additional domains offer two different binding sites for a small molecular weight compound (rapalog). This approach allowed us to show that irreversible stabilization of the association of VE-cadherin–FKBP and VE-PTP–FRB* by rapalog dramatically inhibited leukocyte extravasation and VEGF-induced permeability *in vivo*.[62] Thus, the dissociation of VE-PTP from VE-cadherin is *in vivo* necessary for the opening of endothelial cell contacts in both processes.

neither of these phophatases is specific for endothelial cells and whether they play a role in leukocyte diapedesis is unknown. In contrast, VE-PTP was proposed to participate in leukocyte transmigration, since docking of leukocytes to the apical surface of endothelial cells triggered, within minutes, dissociation of VE-PTP from VE-cadherin.[58] Importantly, this dissociation was also observed upon stimulation of endothelial cells with VEGF, a factor that induces vascular permeability (Fig. 3A). Because VE-PTP enhances the adhesive activity of VE-cadherin, its dissociation from VE-cadherin triggered by different cell contact-destabilizing stimuli was considered a prerequisite for the opening of endothelial junctions.

The relevance of the dissociation of VE-PTP from VE-cadherin for the initiation of leukocyte diapedesis was recently directly analyzed *in vivo*. First, it was demonstrated that LPS-triggered neutrophil recruitment into the lung, as well as intravenous administration of VEGF, stimulated the dissociation of VE-PTP from VE-cadherin *in vivo*.[62] To determine whether this dissociation was indeed necessary for

leukocyte extravasation and the induction of vascular permeability, knock-in mice were generated so that dissociation of these two proteins could be specifically and inducibly prevented. To this end, the C-termini of VE-cadherin and VE-PTP were each fused to a different protein domain that also contained different binding sites for a small molecular weight compound, a derivative of rapamycin called rapalog. The two additional protein domains were FK506-binding protein (FKBP) and FRB*, a mutated form of the rapamycin-binding domain of mammalian target of rapamycin (mTOR). The rapalog compound was not able to bind to mTOR itself and had lost the immunosuppressive activity of rapamycin. To make use of this system *in vivo*, the two respective cDNAs coding for VE-cadherin–FKBP and VE-PTP–FRB*, respectively, separated by an internal ribosomal entry site (IRES), were recombined into the VE-cadherin locus of knock-in mice, thereby replacing endogenous VE-cadherin. The resulting mice were viable and fertile, and VE-cadherin-FKBP expression levels were similar to those of VE-cadherin in wild-type (WT) mice. The expression level of VE-PTP–FRB* was about a third of that of endogenous VE-PTP. Because both constructs had been inserted into the VE-cadherin locus, the fusion proteins were specifically expressed in endothelial cells.[62]

Testing the effects of the rapalog compound in these mice revealed that intravenous administration of this compound indeed inhibited VEGF-induced dissociation of VE-cadherin–FKBP from VE-PTP–FRB* and blocked almost completely the induction of vascular permeability by VEGF in the skin and by LPS in the lung. No such effect was seen when the rapalog was administered to wt mice. Thus, preventing the dissociation of VE-PTP from VE-cadherin indeed prevented vascular leak induction. A similar situation was found for leukocyte extravasation. The rapalog compound strongly inhibited neutrophil extravasation in the IL-1β–stimulated cremaster by about 50% as well as neutrophil recruitment into the lungs of mice exposed to nebulized LPS- by almost 60%. Again, such effects were only seen in the knock-in mice, not in WT mice.

This study established that the endothelial specific phosphatase VE-PTP needs to be dissociated from VE-cadherin to allow the opening of endothelial contacts during leukocyte extravasation and induction of vascular permeability. This demonstrates that tyrosine phosphorylation of the VE-cadherin–catenin complex, or of proteins associated with this complex, is necessary *in vivo* to disconnect endothelial junctions. In addition, this study presents additional and independent evidence for the importance of the paracellular route of leukocyte diapedesis *in vivo* (Fig. 3B).[62]

Conflicts of interest

The author declares no conflicts of interest.

References

1. Ley, K. *et al.* 2007. Getting to the site of inflammation: the leukocyte adhesion cascade updated. *Nat. Rev. Immunol.* **7:** 678–689.
2. Kvietys, P.R. & M. Sandig. 2001. Neutrophil diapedesis: paracellular or transcellular. *News Physiol. Sci.* **16:** 15–19.
3. Muller, W.A. 2001. Migration of leukocytes across endothelial junctions: some concepts and controversies. *Microcirculation* **8:** 181–193.
4. Muller, W.A. 2003. Leukocyte-endothelial-cell interactions in leukocyte transmigration and the inflammatory response. *Trends Immunol.* **24:** 326–333.
5. Carman, C.V. & T.A. Springer. 2004. A transmigratory cup in leukocyte diapedesis both through individual vascular endothelial cells and between them. *J. Cell Biol.* **167:** 377–388.
6. Carman, C.V. *et al.* 2007. Transcellular diapedesis is initiated by invasive podosomes. *Immunity* **26:** 784–797.
7. Carman, C.V. & T.A. Springer. 2008. Trans-cellular migration: cell-cell contacts get intimate. *Curr. Opin. Cell Biol.* **20:** 1–8.
8. Yang, L. *et al.* 2005. ICAM-1 regulates neutrophil adhesion and transcellular migration of TNF-alpha-activated vascular endothelium under flow. *Blood* **106:** 584–592.
9. Millan, J.L. *et al.* 2006. Lymphocyte transcellular migration occurs through recruitment of endothelial ICAM-1 to caveola- and F-actin-rich domains. *Nat. Cell Biol.* **8:** 113–123.
10. Feng, D. *et al.* 2002. Ultrastructural studies define soluble macromolecular, particulate, and cellular transendothelial cell pathways in venules, lymphatic vessels, and tumor-associated microvessels in man and animals. *Microsc. Res. Tech.* **57:** 289–326.
11. Wittchen, E.S. *et al.* 2005. Rap1 GTPase inhibits leukocyte transmigration by promoting endothelial barrier function. *J. Biol. Chem.* **280:** 11675–11682.
12. Vestweber, D. 2007. Adhesion and signaling molecules controlling the transmigration of leukocytes through endothelium. *Immunol. Rev.* **218:** 178–196.
13. Muller, W.A. 2011. Mechanisms of leukocyte transendothelial migration. *Annu. Rev. Pathol.* **6:** 323–344.
14. Mamdouh, Z., A. Mikhailov & W.A. Muller. 2009. Transcellular migration of leukocytes is mediated by the endothelial lateral border recycling compartment. *J. Exp. Med.* **206:** 2795–2808.
15. Schoefl, G.I. 1972. The migration of lymphocytes across the vascular endothelium in lymphoid tissue. A reexamination. *J. Exp. Med.* **136:** 568–588.
16. Marchesi, V. 1961. The site of leukocyte emigration during inflammation. *Q. J. Exp. Physiol. Cogn. Med. Sci.* **46:** 115–118.
17. Anderson, A.O. & N.D. Anderson. 1976. Lymphocyte emigration from high endothelial venules in rat lymph nodes. *Immunology* **31:** 731–748.
18. Feng, D. *et al.* 1998. Neutrophils emigrate from venules by a transendothelial cell pathway in response to FMLP. *J. Exp. Med.* **187:** 903–915.
19. Hoshi, O. & T. Ushiki. 1999. Scanning electron microscopic studies on the route of neutrophil extravasation in the mouse after exposure to the chemotactic peptide N-formyl-methionyl-leucyl-phenylalanine (fMLP). *Arch. Histol. Cytol.* **62:** 253–260.
20. Farr, A.G. & P.P.H. DeBruyn. 1975. The mode of lymphocyte migration through postcapillary venule endothelium in lymph node. *Am. J. Anat.* **143:** 55–92.
21. Marchesi, V.T. & J.L. Gowans. 1964. The migration of lymphocytes through the endothelium of venules in lymph nodes: an electron microscope study. *Proc. R. Soc. Lond. B. Biol. Sci.* **159:** 283–290.
22. Fujita, S. *et al.* 1991. An ultrastructural study of *in vivo* interactions between lymphocytes and endothelial cells in the pathogenesis of the vascular leak syndrome induced by interleukin-2. *Cancer* **68:** 2169–2174.

23. Wolburg, H., K. Wolburg-Buchholz & B. Engelhardt. 2005. Diapedesis of mononuclear cells across cerebral venules during experimental autoimmune encephalomyelitis leaves tight junctions intact. *Acta Neuropathol. (Berl).* **109:** 181–190.

24. Hurley, J. & N. Xeros. 1961. Electron microscopic observation on the emigration of leukocytes. *Austral. J. Exp. Biol. Med. Sci.* **39:** 609–624.

25. Chamberlain, J.K. & M.A. Lichtman. 1978. Marrow cell egress: specificity of the site of penetration into the sinus. *Blood* **52:** 959–968.

26. Woodfin, A. *et al.* 2011. The junctional adhesion molecule JAM-C regulates polarized transendothelial migration of neutrophils *in vivo*. *Nat. Immunol.* **12:** 761–769.

27. Riethmuller, C., I. Nasdala & D. Vestweber. 2008. Nano-surgery at the leukocyte-endothelial docking site. *Pflugers Arch. (Eur. J. Physiol.).* **456:** 71–81.

28. Schulte, D. *et al.* 2011. Stabilizing the VE-cadherin-catenin complex blocks leukocyte extravasation and vascular permeability. *EMBO J* **30:** 4157–4170.

29. Gotsch, U. *et al.* 1997. VE-cadherin antibody accelerates neutrophil recruiment *in vivo*. *J. Cell Sci.* **110:** 583–588.

30. Corada, M. *et al.* 1999. Vascular endothelial-cadherin is an important determinant of microvascular integrity *in vivo*. *Proc. Natl. Acad. Sci. USA* **96:** 9815–9820.

31. Monaghan-Benson, E. & K. Burridge. 2009. The regulation of vascular endothelial growth factor-induced microvascular permeability requires Rac and reactive oxygen species. *J. Biol. Chem.* **284:** 25602–25611.

32. Nagafuchi, A., S. Ishihara & S. Tsukita. 1994. The roles of catenins in the cadherin-mediated cell adhesion: functional analysis of E-cadherin-alpha catenin fusion molecules. *J. Cell Biol.* **127:** 235–245.

33. Ozawa, M. & R. Kemler. 1998. Altered cell adhesion activity by pervanadate due to the dissociation of alpha-catenin from the E-cadherin-catenin complex. *J. Biol. Chem.* **273:** 6166–6170.

34. Yamada, S. *et al.* 2005. Deconstructing the cadherin-catenin-actin complex. *Cell* **123:** 889–901.

35. Drees, F. *et al.* 2005. Alpha-catenin is a molecular switch that binds E-cadherin-beta-catenin and regulates actin-filament assembly. *Cell* **123:** 903–915.

36. Allingham, M.J., J.D. van Buul & K. Burridge. 2007. ICAM-1-mediated, Src- and Pyk2-dependent vascular endothelial cadherin tyrosine phosphorylation is required for leukocyte transendothelial migration. *J. Immunol.* **179:** 4053–4064.

37. Turowski, P. *et al.* 2008. Phosphorylation of vascular endothelial cadherin controls lymphocyte emigration. *J. Cell Sci.* **121:** 29–37.

38. Adam, A.P. *et al.* 2010. SRC-induced tyrosine phosphorylation of VE-cadherin is not sufficient to decrease barrier function of endothelial monolayers. *J. Biol. Chem.* **285:** 7045–7055.

39. van Buul, J.D. *et al.* 2005. Proline-rich tyrosine kinase 2 (Pyk2) mediates vascular endothelial-cadherin-based cell-cell adhesion by regulating beta-catenin tyrosine phosphorylation. *J. Biol. Chem.* **280:** 21129–21136.

40. van Wetering, S. *et al.* 2003. VCAM-1-mediated Rac signaling controls endothelial cell-cell contacts and leukocyte

41. Martinelli, R. *et al.* 2009. ICAM-1-mediated endothelial nitric oxide synthase activation via calcium and AMP-activated protein kinase is required for transendothelial lymphocyte migration. *Mol. Biol. Cell* **20:** 995–1005.

42. Shaw, S.K. *et al.* 2001. Real-time imaging of vascular endothelial-cadherin during leukocyte transmigration across endothelium. *J. Immunol.* **167:** 2323–2330.

43. Sumagin, R. & I.H. Sarelius. 2010. Intercellular adhesion molecule-1 enrichment near tricellular endothelial junctions is preferentially associated with leukocyte transmigration and signals for reorganization of these junctions to accommodate leukocyte passage. *J. Immunol.* **184:** 5242–5252.

44. Strey, A. *et al.* 2002. Endothelial Rho signaling is required for monocyte transendothelial migration. *FEBS Letters* **517:** 261–266.

45. Saito, H. *et al.* 1998. Endothelial myosin light chain kinase regulates neutrophil migration across human umbilical vein endothelial cell monolayer. *J. Immunol.* **161:** 1533–1540.

46. Saito, H. *et al.* 2002. Endothelial Rho and Rho kinase regulate neutrophil migration via endothelial myosin light chain phosphorylation. *J. Leuk. Biol.* **72:** 829–836.

47. Garcia, J.G. *et al.* 1998. Adherent neutrophils activate endothelial myosin light chain kinase: role in transendothelial migration. *J. Appl. Physiol.* **84:** 1817–1821.

48. Stroka, K.M. & H. Aranda-Espinoza. 2011. Endothelial cell substrate stiffness influences neutrophil transmigration via myosin light chain kinase-dependent cell contraction. *Blood* **118:** 1632–1640.

49. Bertuglia, S. & A. Colantuoni. 2000. Protective effects of leukopenia and tissue plasminogen activator in microvascular ischemia-reperfusion injury. *Am. J. Physiol. Heart Circ. Physiol.* **278:** H755-H761.

50. Sumagin, R., E. Lomakina & I.H. Sarelius. 2008. Leukocyte-endothelial cell intercations are linked to vascular permeability via ICAM-1-mediated signaling. *Am. J. Physiol. Heart Circ. Physiol.* **295.**

51. Baluk, P. *et al.* 1998. Endothelial gaps and adherent leukocytes in allergen-induced early- and late-phase plasma leakage in rat airways. *Am. J. Pathol.* **152:** 1463–1476.

52. Kim, M.H., F.R. Curry & S.I. Simon. 2009. Dynamics of neutrophil extravasation and vascular permeability are uncoupled during aseptic cutaneous wounding. *Am. J. Physiol. Cell Physiol.* **296:** C848-C856.

53. He, P. 2010. Leukocyte/endothelium intercations and microvessel permeability: coupled or uncoupled? *Cardiovasc. Res.* **87:** 281–290.

54. Phillipson, M. *et al.* 2008. Endothelial domes encapsulate adherent neutrophils and minimize increases in vascular permeability in paracellular and transcellular emigration. *PLoS ONE* **3:** e1649.

55. Fachinger, G., U. Deutsch & W. Risau. 1999. Functional interaction of vascular endothelial-protein-tyrosine phosphatase with the angiopoietin receptor Tie-2. *Oncogene* **18:** 5948–5953.

56. Baumer, S. *et al.* 2006. Vascular endothelial cell specific phospho-tyrosine phosphatase (VE-PTP) activity is required for blood vessel development. *Blood* **107:** 4754–4762.

transmigration. *Am. J. Physiol. Cell Physiol.* **285:** C343–352.

57. Nawroth, R. *et al.* 2002. VE-PTP and VE-cadherin ectodomains interact to facilitate regulation of phosphorylation and cell contacts. *EMBO J.* **21:** 4885–4895.

58. Nottebaum, A.F. *et al.* 2008. VE-PTP maintains the endothelial barrier via plakoglobin and becomes dissociated from VE-cadherin by leukocytes and by VEGF. *J. Exp. Med.* **205:** 2929–2945.

59. Ukropec, J.A. *et al.* 2000. SHP2 association with VE-cadherin complexes in human endothelial cells is regulated by thrombin. *J. Biol. Chem.* **275:** 5983–5986.

60. Grinnell, K.L., B. Casserly & E.O. Harrington. 2010. Role of protein tyrosine phosphatase SHP2 in barrier function of pulmonary endothelium. *Am. J. Physiol. Lung Cell Mol. Physiol.* **298:** L361–370.

61. Sui, X.F. *et al.* 2005. Receptor protein tyrosine phosphatase micro regulates the paracellular pathway in human lung microvascular endothelia. *Am. J. Pathol.* **166:** 1247–1258.

62. Broermann, A. *et al.* 2011. Dissociation of VE-PTP from VE-cadherin is required for leukocyte extravasation and for VEGF-induced vascular permeability *in vivo. J. Exp. Med.* **208:** 2393–2401.

Ann. N.Y. Acad. Sci. ISSN 0077-8923

Involvement of claudins in zebrafish brain ventricle morphogenesis

Jingjing Zhang,[1,2] Martin Liss,[1] Hartwig Wolburg,[3] Ingolf E. Blasig,[4] and Salim Abdelilah-Seyfried[1]

[1]Max Delbrück Center for Molecular Medicine, Berlin, Germany. [2]Institute of Neurology, Affiliated Hospital of Guangdong Medical College, Zhanjiang, China. [3]Institute of Pathology, University of Tübingen, Tübingen, Germany. [4]Leibniz-Institute for Molecular Pharmacology, Berlin, Germany

Address for correspondence: Salim Abdelilah-Seyfried, Max Delbrück Center for Molecular Medicine, Robert-Rössle Str. 10, 13125, Berlin, Germany. seyfried@mdc-berlin.de

Zebrafish brain ventricle morphogenesis involves an initial circulation-independent opening followed by a blood flow– and circulation-dependent expansion process. Zebrafish claudin-5a is required for the establishment of a neuroepithelial–ventricular barrier, which maintains the hydrostatic pressure within the ventricular cavity, thereby contributing to brain ventricle opening and expansion. In mammalia, several claudin family members, including claudin-3 and claudin-5, are expressed within microvessel endothelial cells of the blood–brain barrier. Whether zebrafish brain ventricle morphogenesis provides a model for studying these claudins during early embryonic development was unknown. This review focuses on the expression and function of these zebrafish claudins during brain ventricle morphogenesis.

Keywords: brain ventricle morphogenesis; tight junction; claudin-5a; claudin-3

Introduction

As an essential structural and functional part of the brain, the ventricular system consists of several connected cavities that are filled with cerebrospinal fluid, which accumulates within and widens these luminal spaces.[1–4] Eventually, the ventricular system develops into fore-, mid-, and hindbrain vesicles, which contribute to brain morphogenesis. Zebrafish has been an important model in developmental biology for more than two decades. Neural tube formation in this organism involves some mechanisms similar to amniote or amphibian neurulation (reviewed in Ref. 5). This supports the use of zebrafish as a model to study brain morphogenesis. Recent work has demonstrated that primary zebrafish brain ventricle formation occurs rapidly, involving two main steps: brain ventricle opening and brain ventricle expansion. Initial brain ventricle opening occurs over a four-hour period, between 17 and 21 hours postfertilization (hpf), during which the neural tube opens into the three primary brain ventricles: the fore-, mid-, and hindbrain ventricles. This step is circulation independent and depends on the morphogenesis of an intact polarized neuroepithelium, the activity of Na$^+$, K$^+$-ATPase (Atp1α1) and localized cell proliferation.[6] The later brain ventricle expansion process, which occurs between 21 and 36 hpf, depends on heartbeat and circulation, because in *troponin T2a* mutants, which lack heart beat and circulation, brain ventricle expansion is impaired.[6]

During vertebrate brain ventricular expansion, neuroepithelial cells are connected by adhesion and tight junctions. These junctions attach cells with each other and form a functional unit between the inside and outside of the neural tube.[7–9] During initial zebrafish brain ventricle opening, the neuroepithelial cell layer functions as a cerebral–ventricular barrier system that contributes to the initial zebrafish brain ventricle opening and expansion processes[10] (reviewed in Ref. 11). Tight junctions are the main cell–cell contact structures between neuroepithelial cells at these stages. Antisense

doi: 10.1111/j.1749-6632.2012.06507.x

oligonucleotide morpholino (MO)–mediated knockdown of zebrafish claudin-5a revealed a reduction of tight junction permeability of the neuroepithelial cell layer as assayed by the paracellular diffusion of the electron-dense molecule lanthanum nitrate. As a consequence of disruption of the neuroepithelial–ventricular barrier, zebrafish brain ventricles failed to inflate during early embryogenesis at 20 hpf.[10]

Different claudins contribute to diverse barriers within various epithelial and endothelial cell types.[12–15] In the central nervous system, the blood–brain barrier (BBB) is formed by tight junctions and preserves a restrictive paracellular diffusion function. In mouse and human, both claudin-3 and claudin-5 are detected within endothelial cell junctions of the BBB.[13,16–18] Loss of murine claudin-5 caused a size selective loosening of the BBB, even though no developmental or morphological abnormalities were evident in the blood vessels where claudin-5 is normally expressed.[13] Tracer experiments and magnetic resonance imaging showed that the BBB was strongly affected in this mutant and leaky for small molecules of less than 800 Da, whereas barrier properties were maintained for larger molecules.[13] Zebrafish claudin-5a was identified as the closest homolog of mammalian claudin-5. In addition to its role during brain ventricular expansion at early developmental stages,[10] claudin-5 is expressed in the endothelial tight junction-based BBB and blood–retina barrier of zebrafish.[19,20] Zebrafish claudin-5 proteins are expressed in hyaloid-retinal vessel endothelial cells at around 7 hpf.[21] In comparison, the zebrafish ortholog, which is genetically closest to mammalian claudin-3, has not previously been described. Here, we discuss the expression of several zebrafish orthologs of mammalian BBB-expressed claudins and describe their functional roles during brain morphogenesis.

Zebrafish claudin orthologs of mammalian claudins of the BBB

Genomic analyses have revealed an unexpectedly large family of claudins in teleosts that are the result of a genome duplication that arose in the teleost lineage. In *Takifugu rubripes*, some 35 claudin genes have been annotated as orthologs of 17 mammalian genes and some additional 21 genes as being teleost specific.[22] Based on expression studies, most claudin

genes are differentially expressed in a tissue- or temporal-specific manner, which points at subfunctionalizations of these proteins in teleosts. Another surprising finding is the complete absence of introns among that family of genes in teleosts. Even more surprisingly, comparative analyses among different species revealed only a small overlap among single exon genes to be evolutionarily conserved. Such differences instead point at the evolvement of single exon genes after the evolutionary divergence of species.[23] One likely explanation for the diversification of claudins among teleosts is the involvement of retrotransposition events followed by tandem duplications. Experimentally, subfunctionalizations of these genes have opened the possibility of addressing the role of particular claudins in a tissue- or temporal-specific manner in teleosts.

Within mammalian brain microvessels, claudin-5 is most strongly expressed, while claudins-1, claudins-3, claudins-12, and possibly claudins-11 are expressed at lower levels.[16,18,24,25] Zebrafish claudin-5a and claudin-5b are the orthologs of mammalian claudin-5,[10,19,20] which functions in the BBB.[13] To isolate the zebrafish orthologs of claudin-3, which is prominently expressed at the mammalian BBB, we used mouse and human claudin-3 sequences ([*Mus musculus*] NP˙034032; [*Homo sapiens*] NP˙001297) to search the NCBI database for homologous sequences in zebrafish. Sequence analysis using the BLAST program on ExPASy (http://www.expasy.ch) revealed that zebrafish claudin-3 (NP˙571842, *gi*: 20373151) is the closest homolog of mammalian claudin-3. Zebrafish *claudin-3* (1880 bp; *claudin-h, ZFIN ID: ZDB-GENE-010328-8*) encodes a 214 amino acids protein with a predicted molecular weight of 23 kDa. The amino acid similarity/identity of zebrafish claudin-3 to mouse and human claudin-3 proteins is 84%/62% and 85%/63%, respectively. The identity of zebrafish claudin-3 to zebrafish claudin-5a and claudin-5b is only 53% and 56%, respectively. The second closest zebrafish claudin-3 ortholog, claudin-i, also shows high similarity to mouse claudin-3 (72%). The evolutionary relationships among claudin-3, claudin-i, and claudin-5 from zebrafish, human, and mouse, based on phylogenetical distances calculated according to the DNAMAN multiple alignment method, confirmed that zebrafish claudin-3 is closest to mammalian claudin-3, whereas zebrafish claudin-5a and claudin-5b are genetically closest to mammalian

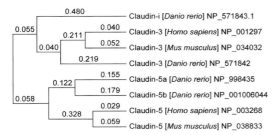

Figure 1. Phylogenetic analysis of claudin-5 and claudin-3 family members. Rooted phylogenetic tree shows the evolutionary relationships between the claudin-3 and claudin-5 proteins of humans, mice, and zebrafish. Zebrafish claudin-5a/b are phylogenetically closest to mammalian claudin-5. Zebrafish claudin-3 is phylogenetically closest to mammalian claudin-3. Phylogenetical distances were calculated according to the DNA-MAN multiple alignment method.

claudin-5 (Fig. 1). Amino acid sequence alignments of zebrafish claudin-3 and claudin-5 were performed by discovery studio gene 1.5 (DS Gene 1.5). The second extracellular loop of claudins is involved in claudin–claudin *trans*-interactions[26] and is conserved among species. It is also the binding site of *Clostridium perfringens* enterotoxin.[27] Within the predicted extracellular loop 2 domain (residues 144 to 158 of claudin-3 according to Swiss-Prot pre-

dictions), residues phenylalanine 146, tyrosine 147, proline 149, glutamine 155, and glutamic acid 158 are highly conserved. Among them, phenylalanine 146, tyrosine 147, glutamine 155, and glutamic acid 158 are critical for the *trans*-interaction and strand formation, whereas Proline 149 is predicted to be important for the loop structure[26] (Fig. 2).

The high protein sequence similarity and close genetic distance indicated that zebrafish claudin-3 may have a role in cerebral barrier function related to that of claudin-5. Because mammalian claudin-5 and claudin-3 are both expressed in brain capillary vascular endothelial cells of the BBB,[17] they may have common and partially overlapping endothelial tight junction barrier functions. In contrast, claudin-i, is mainly expressed within epidermis and periderm at these developmental stages (http://zfin.org). In comparison with zebrafish *claudin-5a*, which is expressed within the entire central nervous system and most strongly within ventral neuroepithelial cells lining the brain ventricles of the hindbrain during early stages of development before 30 hpf,[10] *claudin-3* is strongly expressed in the olfactory placodes, pronephric duct, sensory placodes of the lateral line organ, and hatching gland[28] (http://zfin.org). These differences in

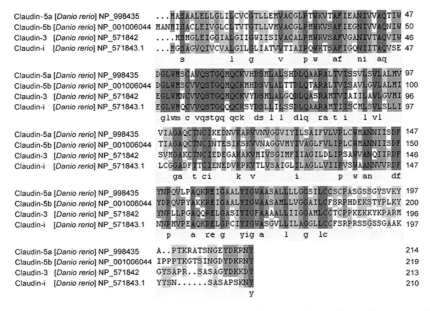

Figure 2. Sequence alignment of different zebrafish claudins. Sequence alignment of zebrafish claudin-5a, claudin-5b, claudin-3, and claudin-i shows that the extracellular loop domain 2 (residues 144 to 158 of claudin-3) residues phenylalanine 146, tyrosine 147, proline 149, glutamine 155, and glutamic acid 158 are highly conserved.

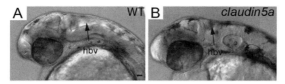

Figure 3. Brain ventricle expansion defects in *claudin-5a* morphants at 30 hpf. (A, B) Injection of *claudin-5a* MO significantly reduces the expansion of midbrain and hindbrain ventricles (hbv) as assessed by differential interference contrast images in comparison to the WT control (black arrow) at 30 hpf (scale bar: 50 μm.)

gene expression between mammalian and zebrafish *claudin-3* genes raised questions about the developmental functions of claudin-3.

Whole-mount *in situ* hybridization of *claudin-3* in *claudin-5a* morphant embryos showed no change in *claudin-3* expression levels at early stages (unpublished results). This finding indicated that the loss of claudin-5a is not compensated by increased expression of *claudin-3*. As evident from the *claudin-5a* morphant phenotype, claudin-3 does not completely rescue the loss of claudin-5a during early stages of zebrafish development. Whether other functional redundancies exist between zebrafish claudin-3 and claudin-5 proteins needed to be addressed.

Brain ventricle expansion defects caused by loss of claudin-5a

As discussed earlier, loss of zebrafish claudin-5a resulted in a failure of brain ventricle opening due to the defects in tightening the paracellular spaces within the neuroepithelial–ventricular barrier prior to 20 hpf[10] (Fig. 3). One explanation for the failure of ventricle expansion is that the neuroepithelium may not maintain the osmotic gradients that are generated by various ion pumps and exchangers that are located on membranes of neuroepithelial cells and that are essential for the water turgor pressure to build up within the ventricular cavity.[10] In comparison to the severe *claudin-5a* morphant phenotypes and consistent with the lack of zebrafish *claudin-3* expression within the neuroepithelium, the specific MO-mediated knockdown of *claudin-3* did not cause any obvious defects in hindbrain ventricle sizes at 30 hpf (unpublished results). Strikingly, the double knockdown of *claudin-5a* and *claudin-3* was more severe with respect to ventricular lumen expansion defects than

that of *claudin-5a* single morphants (unpublished results).

In a previous study, injection of the 0.5 kDA electron dense tracer molecule lanthanum nitrate into the hindbrain ventricle of 20 hpf *claudin-5a* morphant zebrafish embryos and analysis of electron micrographs of hindbrain sections had revealed a breakdown of tight junctions between neuroepithelial cells[10] (Fig. 4). To test whether the enhanced brain morphogenesis phenotype in *claudin-5a/claudin-3* double morphants was due to enhanced tight junctions barrier defects, we injected lanthanum nitrate into hindbrain ventricles of 30 hpf zebrafish embryos and compared electron micrographs of *claudin-5a*, *claudin-3*, or *claudin-5a/claudin-3* double morphants for tight junctions barrier tightness. This analysis revealed that, unlike in *claudin-5a* morphants, the neuroepithelial-ventricular barrier of 30 hpf *claudin-3* morphants is intact and that the 0.5 kDa electron-dense tracer cannot pass through the tight junction that is located between adjacent neuroepithelial cells (unpublished results). Apparently, claudin-3 is not involved in the establishment of neuroepithelial–ventricular barrier properties. Double knockdown of *claudin-3* and *claudin-5a* caused a severe diffusion of tracer particles into the paracellular space, which was comparable but not more severe than the *claudin-5a* morphant phenotype. This finding suggested that claudin-3 and claudin-5a interact by some

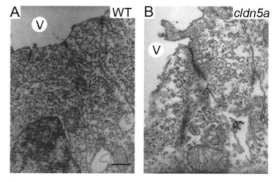

Figure 4. Electron micrographs of neuroepithelial cells lining the cerebral ventricles (V) of 30 hpf zebrafish embryos after intraventricular injection of the electron-dense molecule lanthanum nitrate. (A) In WT, paracellular clefts are tight for the tracer, which accumulates in a dot-like pattern at the tight junction (arrow). (B) Knockdown of *claudin-5a* results in the diffusion of lanthanum nitrate into the paracellular spaces between neuroepithelial cells (arrow). (Scale bars: 2 μm.)

mechanism other than by tightening the neuroepithelial–ventricular barrier during brain ventricular expansion (unpublished results).

Circulation is not required for the initial brain ventricle opening prior to 21 hpf when circulation sets in.[6] However, that the later expansion process requires circulation is evident from the analysis of mutants in *troponin T2a* that lack heart contractility and have smaller brain ventricles.[6,29,30] Because claudin-5a is also expressed within microvessel endothelial cells[19,20] and both *claudin-5a* and *claudin-3* are co-expressed within endocardium, which is the interior lining of the heart tube (unpublished results), these claudins may potentially be required for normal heart function and maintenance of circulation at early stages. Together, tightening of the neuroepithelial–ventricular barrier by claudin-5a during brain ventricle opening and the combined activities of claudin-5a and claudin-3 during the circulation-dependent brain ventricle expansion may contribute to brain morphogenesis (Fig. 5).

Concluding remarks

Mammalian claudin-5 and claudin-3 are important components of the BBB. As the closest homolog to mammalian claudin-5, zebrafish claudin-5a establishes a neuroepithelial–ventricular barrier during early embryonic development. Claudin-5a has been suggested to contribute to the brain ventricle expansion process together with the Atp1α1 ion pump, which generates the fluid pressure within the ventricular cavity (Fig. 5). Zebrafish claudin-3, which is the closest ortholog of mouse and human claudin-3, is detectable with endocardium/endothelium instead of the cerebral barrier systems. Together with claudin-5a, claudin-3 may be involved in maintaining circulation, which is required for the brain ventricular expansion process. However, there are still many unanswered questions concerning claudins and zebrafish brain ventricle development. What are the effects of loss of claudin-3 and claudin-5a for endocardial/endothelial function? Furthermore, the potential roles of some other claudin family members, including claudin-i, claudin-1, and claudin-12, in zebrafish cerebral barrier or brain development remains to be elucidated.

Conflicts of interest

The authors declare no conflicts of interest.

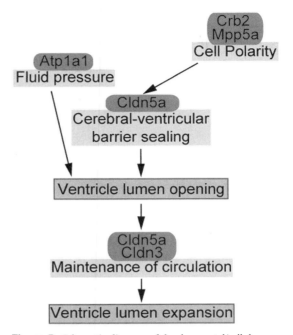

Figure 5. Schematic diagram of developmental/cellular processes contributing to brain ventricle opening and expansion. During the early brain ventricle opening stage between 17 and 21 hpf, the cell polarity regulators Crumbs2 (Crb2) and Membrane protein palmitoylated 5a (Mpp5a) are essential for neuroepithelial integrity and maintenance of the tight junction. Tightness of the tight junction is regulated by claudin-5a, which seals the neuroepithelial layer to maintain the fluid pressure, which may depend on the ion pump activity of Atp1α1. Ventricular fluid accumulation drives the opening and initial expansion of brain ventricles. Between 21 and 30 hpf, claudin-5a and claudin-3 may affect the circulation-dependent phase of brain ventricle lumen expansion potentially via affecting the maintenance of circulation.

References

1. Cushing, H. 1914. Studies on the cerebrospinal fluid. I. Introduction. *J. Med. Res.* **26:** 1–19.
2. Milhorat, T.H., M.K. Hammock, J.D. Fenstermacher & V.A. Levin. 1971. Cerebrospinal fluid production by the choroid plexus and brain. *Science* **173:** 330–332.
3. Pollay, M. & F. Curl. 1967. Secretion of cerebrospinal fluid by the ventricular ependyma of the rabbit. *Am. J. Physiol.* **213:** 1031–1038.
4. Davson, H. & M.B. Segal. 1996. *Physiology of the CSF and Blood-Brain Barriers.* CRC Press. Boca Raton.
5. Lowery, L.A. & H. Sive. 2004. Strategies of vertebrate neurulation and a re-evaluation of teleost neural tube formation. *Mech. Dev.* **121:** 1189–1197.
6. Lowery, L.A. & H. Sive. 2005. Initial formation of zebrafish brain ventricles occurs independently of circulation and requires the nagie oko and snakehead/Atp1α1a.1 gene products. *Development* **132:** 2057–2067.

7. Götz, M. & W.B. Huttner. 2005. The cell biology of neurogenesis. *Nat. Rev. Mol. Cell Biol.* **6:** 777–788.

8. Lowery, L.A. & H. Sive. 2009. Totally tubular: the mystery behind function and origin of the brain ventricular system. *Bioessays.* **31:** 446–458.

9. Ek, C.J., K.M. Dziegielewska, M.D. Habgood & N.R. Saunders. 2011. Barriers in the developing brain and Neurotoxicology. *Neurotoxicology.* [Epub ahead of print].

10. Zhang, J., J. Piontek, H. Wolburg, *et al.* 2010. Establishment of a neuroepithelial barrier by Claudin-5a is essential for zebrafish brain ventricular lumen expansion. *Proc. Natl. Acad. Sci. USA* **107:** 1425–1430.

11. Abdelilah-Seyfried, S. 2010. Claudin-5a in developing zebrafish brain barriers: another brick in the wall. *Bioessays* **32:** 768–776.

12. Furuse, M., M. Hata, K. Furuse, *et al.* 2002. Claudin-based tight junctions are crucial for the mammalian epidermal barrier: a lesson from claudin-1-deficient mice. *J. Cell Biol.* **156:** 1099–1111.

13. Nitta, T., M. Hata, S. Gotoh, *et al.* 2003. Size-selective loosening of the blood-brain barrier in Claudin-5-deficient mice. *J. Cell Biol.* **161:** 653–660.

14. Furuse, M. 2009. Knockout animals and natural mutations as experimental and diagnostic tool for studying tight junction functions in vivo. *Biochim. Biophys. Acta.* **1788:** 813–819.

15. Gupta, I.R. & A.K. Ryan. 2010. Claudins: unlocking the code to tight junction function during embryogenesis and in disease. *Clin. Genet.* **77:** 314–325.

16. Morita, K., H. Sasaki, M. Furuse & S. Tsukita. 1999. Endothelial claudin: claudin-5/TMVCF constitutes tight junction strands in endothelial cells. *J. Cell Biol.* **147:** 185–194.

17. Wolburg, H., K. Wolburg-Buchholz, J. Kraus, *et al.* 2003. Localization of Claudin-3 in tight junctions of the blood-brain barrier is selectively lost during experimental autoimmune encephalomyelitis and human glioblastoma multiforme. *Acta Neuropathol. (Berl.)* **105:** 586–592.

18. Ohtsuki, S., H. Yamaguchi, Y. Katsukura, *et al.* 2008. mRNA expression levels of tight junction protein genes in mouse brain capillary endothelial cells highly purified by magnetic cell sorting. *J. Neurochem.* **104:** 147–154.

19. Jeong, J.Y., H.B. Kwon, J.C. Ahn, *et al.* 2008. Functional and developmental analysis of the blood-brain barrier in zebrafish. *Brain Res. Bull.* **75:** 619–628.

20. Xie, J., E. Farage, M. Sugimoto & B. Anand-Apte. 2010. A novel transgenic zebrafish model for blood-brain and blood-retinal barrier development. *BMC Dev. Biol.* **10:** 76.

21. Hyoung Kim, J., Y. Suk Yu, K.W. Kim & J. Hun Kim. 2011. Investigation of barrier characteristics in the hyaloid-retinal vessel of zebrafish. *J. Neurosci. Res.* **89:** 921–928.

22. Loh, Y.H., A. Christoffels, S. Brenner, *et al.* 2004. Extensive expansion of the claudin gene family in the teleost fish, *Fugu rubripes. Genome Res.* **7:** 1248–1257.

23. Tine, M., H. Kuhl, A. Beck, L. Bargelloni & R. Reinhardt. 2011. Comparative analysis of intronless genes in teleost fish genomes: Insights into their evolution and molecular function. *Mar. Genomics.* **4:** 109–119.

24. Enerson, B.E. & L.R. Drewes. 2006. The rat blood-brain barrier transcriptome. *J. Cereb. Blood Flow Metab.* **26:** 959–973.

25. Wessells, H., C.J. Sullivan, Y. Tsubota, *et al.* 2009. Transcriptional profiling of human cavernosal endothelial cells reveals distinctive cell adhesion phenotype and role for claudin 11 in vascular barrier function. *Physiol. Genomics.* **39:** 100–108.

26. Piontek, J., L. Winkler, H. Wolburg, *et al.* 2008. Formation of tight junction: determinants of homophilic interaction between classic claudins. *FASEB J.* **22:** 146–158.

27. Fujita, K., J. Katahira, Y. Horiguchi, *et al.* 2000. *Clostridium perfringens enterotoxin* binds to the second extracellular loop of Claudin-3, a tight junction integral membrane protein. *FEBS Lett.* **476:** 258–261.

28. Clelland, E.S. & S.P. Kelly. 2010. Tight junction proteins in zebrafish ovarian follicles: stage specific mRNA abundance and response to 17beta-estradiol, human chorionic gonadotropin, and maturation inducing hormone. *Gen. Comp. Endocrinol.* **168:** 388–400.

29. Schier, A.F., S.C. Neuhauss, M. Harvey, *et al.* 1996. Mutations affecting the development of the embryonic zebrafish brain. *Development* **123:** 165–178.

30. Sehnert, A.J., A. Huq, B.M. Weinstein, *et al.* 2002. Cardiac troponin T is essential in sarcomere assembly and cardiac contractility. *Nat. Genet.* **31:** 106–110.

Ann. N.Y. Acad. Sci. ISSN 0077-8923

Modulation of tight junction proteins in the perineurium for regional pain control

D. Hackel,[1,2] A. Brack,[1] M. Fromm,[3] and H.L. Rittner[1]

[1]Klinik und Poliklinik für Anaesthesiologie, Universitätsklinikum Würzburg, Würzburg, Germany. [2]Klinik für Anaesthesiologie m. S. operative Intensivmedizin, Campus Benjamin Franklin, Charité – Universitätsmedizin Berlin, Berlin, Germany. [3]Institute of Clinical Physiology, Campus Benjamin Franklin, Charité – Universitätsmedizin Berlin, Berlin, Germany

Address for correspondence: Heike Rittner, Klinik und Poliklinik für Anaesthesiologie, Universitätsklinikum Würzburg, Oberdürrbacher Strasse 6, D-97080 Würzburg, Germany. rittner_h@klinik.uni-wuerzburg.de

Peripheral neurons are surrounded by the perineurium that forms the blood–nerve barrier and protects the nerve. Although the barrier serves as protection, it also hampers drug delivery of analgesic drugs to the peripheral nerve. We previously showed that opening of the barrier using hypertonic solutions facilitates drug delivery, for example, of hydrophilic opioids, which selectively target nociceptors. The perineurial barrier is formed by tight junction proteins, including claudin-1, claudin-5, and occludin. Under pathophysiological conditions such as nerve crush injury, the perineurial barrier is opened and tight junction proteins are no longer present. After several days, tight junction proteins reappear and the barrier reseals. Similarly, perineurial injection of hypertonic saline transiently opens the barrier, claudin-1 disappears, and hydrophilic analgesic drugs are effective. In the future, these findings could be used to reseal the barrier breakdown and could be applied to other barriers like the blood–brain or the intestinal mucosal barrier.

Keywords: perineurium; tight junction proteins; claudin; metalloproteinase; analgesia; opioids; opioid receptors

Introduction

Peripheral nerves are especially protected from their environment. They are separated from the external space by the blood–nerve barrier (BNB) consisting of the endoneurial microvessels and the perineurium. A detailed anatomical review of these structures has recently been published.[1] The passage of substances from the outside is restricted and regulated by this barrier, which consists of a multi-layered sheath of the perineurium.[1] Besides protection, the barrier prevents the access of hydrophilic (e.g., analgesic) drugs to peripheral neurons.

Analgesic drugs applied near the nerve are limited to lipophilic local anesthetics such as lidocaine, which block all nerve fibers including motor neurons. Ideally, pharmaceutical agents for regional analgesia should only block nociceptors and spare motor and sensory neurons.[2] Therefore, the question arises whether it is possible to open the perineurium transiently to allow for penetration of hydrophilic pharmaceuticals with lower side effects.

Opioids for analgesia via peripheral nociceptive neurons

During pain sensation primary sensory neurons transduce mechanical, chemical, or heat stimuli into action potentials.[3] Opioids are potential candidates for an effective control of pain both in the central nervous system and peripherally at the site of inflammation.[3,4] Sensory neurons have nociceptors that belong to the myelinated (Aδ-fibers) or small-diameter unmyelinated axons (C-fibers). Pain is often associated with inflammation as a result of tissue destruction, nerve injury, or a reaction of the immune system after an infection.

Peripheral application of opioids is only effective under inflammatory conditions.[5,6] At present, three different opioid receptor subtypes are known, the μ-, δ-, and κ-receptors (MOR etc.).[7] Opioid receptors are seven-transmembrane G protein–coupled receptors and are expressed by central and peripheral neurons, and by neuroendocrine, immune, and ectodermal cells.[7,8] Apart from opioid

doi: 10.1111/j.1749-6632.2012.06499.x

receptor upregulation and enhanced G protein coupling, during inflammation a disruption of the perineurial barrier occurs and leads to an enhanced analgesic efficacy of peripherally active opioids as well as of exogenous applied opioids.[9] Analgesic drugs of interest for nociceptive neurons could be opioids, as the μ-opioid receptor agonist (D-Ala2, NMePhe4, Gly5-ol)-enkephalin (DAMGO),[5,6] or voltage-gated sodium channel blockers, such as tetrodotoxin (TTX).[10] In inflamed tissue, opioids bind to opioid receptors on peripheral sensory nerve terminals and induce potent analgesia (antinociception) both in humans and in experimental models of inflammation.[3,4]

In noninflamed tissue, hydrophilic opioids are particularly ineffective, as shown in behavioral, clinical, and electrophysiological studies.[5,11–13] However, if the perineurial barrier is made permeable by hypertonic solutions (i.e., 0.5 M mannitol or 10% NaCl), injection of opioids into the paw increases nociceptive thresholds in noninflamed tissue.[5,13] It has long been postulated that this is due to an opening of the perineurial barrier surrounding peripheral nerves. Peripheral opioid antinociception and opening of the perineurial barrier occur simultaneously at a very early stage (within 12 hours) of the inflammatory reaction. In earlier studies, local hypertonicity induced prolonged opening of the perineurial barrier to permit access of larger hydrophilic molecules such as horseradish peroxidase.[13] Injections of hypertonic solutions applied under anesthesia at the site of the peripheral nerve or its endings do not elicit hyperalgesia. In contrast, injection into the muscle evokes pain and is used for muscle pain studies.[14]

We confirmed these previous findings also for endogenous opioid peptides like Met-enkephalin and β-endorphin using hypertonic saline (10%) and normal saline (0.9%) as a control (Fig. 1).[13] Local hypertonicity induced an immediate and prolonged opening of the perineurial barrier to permit access of opioid peptides to opioid receptors on sensory neurons to elicit significant increases in mechanical nociceptive thresholds (i.e., increased paw pressure threshold; Fig. 1A and C). This effect was inhibited by coinjection of a peripherally restricted dose of the opioid receptor agonist naloxone.[13] Hypertonic saline did not induce cell death and caused no change in the nociceptive thresholds when injected alone.[13] In one of our recent studies, we charac-

terized the requirements and mechanisms responsible for exogenous opioid antinociception in noninflamed tissue in more detail.[15] Local treatment with hypertonic saline in the rat hind paw facilitates μ opioid receptor agonist (DAMGO)-induced increase in nociceptive thresholds up to 4 hours. In electrophysiological recordings of an *in vitro* skin nerve preparation, C- but not Aδ-fiber mechanonociceptors responded to DAMGO after local pretreatment with hypertonic saline only.[15]

Perineurium

The peripheral nerve is surrounded by three different compartments: the endoneurium, the perineurium, and the epineurium. The epineurium, consists of dense irregular fibrous and adipose tissue that bundles nerve fascicles into a single nerve trunk.[16] A nerve fascicle consists of myelinated and unmyelinated fibers and endoneurial vessels embedded in the loose connective tissue of the endoneurium surrounded by the perineurium (Fig. 2) If several fascicles travel together, they are ensheathed by the epineurium. The epineurium, however, has no barrier function. The axonal environment in the endoneurium is isolated from the general extracellular space of the body by a diffusion barrier called the BNB, consisting of endoneurial vessels and the perineurium, and guarantees the constant endoneurial milieu.[16–18] Cells shaping the endoneurial vessels are called pericytes, which completely surround capillary endothelial cells. They are the critical components of the BNB and are supposedly equivalent to those of the blood–brain barrier (BBB). Pericytes play important roles in mediating the development, maintenance, and regulation of the barrier.[19]

The perineurium together with the endoneurium forms the BNB. Thereby, peripheral nerves with Schwann cell–axon units in the endoneurium are isolated from the adjacent tissues by this barrier.[17] Currently, it is not known where exactly the perineurium ends in the periphery.[1] The highly organized structure of the perineurium is formed by concentric layers of polygonal perineurial cells alternating with extracellular matrix.[18] The perineurium, which is the connective tissue ensheathment of nerve fascicles, separates general extracellular space from intrafascicular endoneurial space.[1] It surrounds the endoneurium in concentric

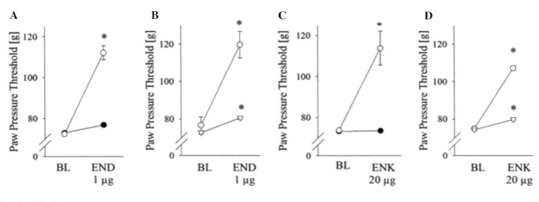

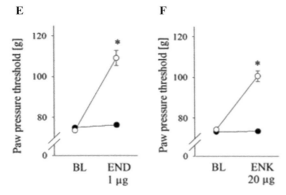

Figure 1. Prolonged opening of the perineurial barrier facilitates antinociception by exogenous opioid peptides. (A–C) To determine the analgesic effects of exogenous opioids in the presence of hypertonicity, rats were treated i.pl. with 100 μL saline (NaCl 0.9% [black circles] and 10% [white circles] in all experiments) followed by an intraplantar injection of the opioid peptides β-endorphin and Met-enkephalin (END, A; ENK, C). (B, D) Rats received opioid peptides in the presence of 10% NaCl (white circles) or in combination with 0.28 ng naloxone intraplantarly (white diamonds). Mechanical nociceptive thresholds were quantified before (baseline, BL) and after injection of opioids ($n = 6$–8 per group, $^*P < 0.05$ ANOVA on ranks, Dunn's method). (E, F) To quantify exogenous opioid-induced antinociception following return to normotonicity, paw pressure threshold was measured in response to intraplantar injection of ENK (E) and END (F) one hour after inoculation of hypertonic or isotonic saline (white and black circles, respectively, $n = 5$–6 per group, $^*P < 0.05$, ANOVA on ranks, Dunn's method). Reproduced with permission from Ref. 13.

cellular layers, each one cell layer thick. In larger nerve fascicles, the perineurium consists of up to 15 cellular layers, fewer layers in smaller fascicles and only one cell layer in the vicinity of motor and some sensory end organs.[1] The isolation is believed to be important for maintaining the appropriate physicochemical environment for the axons. Failure of the barrier causes edema in the endoneurium and makes the peripheral nervous system vulnerable to harmful agents.[17] Previous studies have shown that the perineurium creates

a diffusion barrier, which functions to maintain endoneurial homeostasis.[18]

The perineurial barrier function and its relation to peripheral opioid analgesia have been examined in the rat paw before. Antonijevic *et al.* showed that after intraplantar injection of horseradish peroxidase, the endoneurium is stained only in inflamed paws after intraplantar injection of mannitol or hypertonic saline, but not in noninflamed paws.[5] The staining pattern was similar as seen in complete Freunds adjuvant-induced hind paw inflammation.

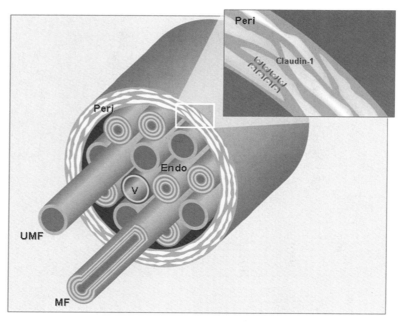

Figure 2. Diagram of the normal perineurium: a small nerve fascicle surrounded by perineurium (Peri) with myelinated (MF) and unmyelinated (UMF) fibers and a single endoneurial vessel (V) embedded in the loose connective tissue of the endoneurium (Endo). The perineurium represents an important element of the BNB. This function is mainly accomplished by the presence of tight junctions. Modified from Refs. 1 and 16.

However, hypertonic solutions did not induce inflammatory cell infiltration.[5] We confirmed these previous findings using hypertonic saline (10%) versus normal saline (0.9%) as a control to evaluate whether, over time, hypertonicity in the tissue and opening are correlated. Hypertonic saline induces a long-lasting opening of the perineurial barrier, as demonstrated by endoneurial horseradish peroxidase staining (Fig. 3A–D).[13] In contrast, sodium concentration normalized after 15 min in the tissue. We did not observe shrinking or bursting of cells, rather, the cells stayed intact (Fig. 3E).[13] Therefore, opening of the perineurium is not solely dependent on the hypertonic surrounding of the tissue, but seems to be dependent on other structural changes.

Tight junctions

Tight junction proteins include the family of claudins and TAMPs (tight junction–associated MARVEL proteins), such as occludin, tricellulin, and marvel-D3. The intracellular backbone of the tight junction is formed by scaffolding proteins such as zonula occludens-1 (ZO-1), which provide an-

choring of claudins and TAMPs and are implicated in a variety of signaling mechanisms.[20,21] Among the 27 members of the claudin family found in mammals, claudin-1, claudin-5, and claudin-12 predominate in tight junctions of the BBB.[22–24] Osmotic stress was previously shown to induce an opening of the BBB, accompanied by a loss of claudin-5, ZO-1, and occludin.[25]

The BNB with the perineurium

An essential part of the BNB in both perineurial and endothelial cells (pericytes) is constituted by the tight junction. (Fig. 2) Pummi *et al.* showed that human adult perineurium contains a high density of tight junction proteins, especially claudin-1 and claudin-3. Endoneurial labeling for claudin-1, claudin-3, and occludin was faint compared with their labeling in the perineurium. The expression of claudin-5, characteristic for endothelial cells of the microendoneurial vessels, was not associated with perineurial cells.[18] Recently, it has been reported that glucocorticoids upregulate the expression of claudin-5 and increase the barrier properties of the BNB. Glucocorticoids have been shown to

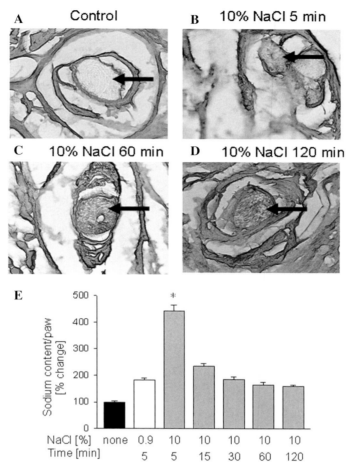

Figure 3. Timecourse of perineurial barrier permeability of sodium content in the paw after hypertonic or isotonic saline. (A–D) Male Wistar rats were intraplantarly injected with 100 μL 10% NaCl (control 0.9% NaCl). Intraneural staining of horseradish peroxidase in the paw tissue was performed 5, 60, and 120 min after injection (magnification 40×; arrows are pointing at the nerve). A representative example of $n = 3$ is shown. (E) Total sodium content was measured in homogenized subcutaneous paw tissue after a single intraplantar injection of 0.9% NaCl or 10% NaCl (white and gray bars, respectively; $n = 6$ per group, $^{*}P < 0.05$, ANOVA vs. 0.9% saline; black bar = noninjected paw). Data are presented as percentage of sodium content in noninjected paws and are means ± SEM. Reproduced with permission from Ref. 13.

effectively restore the BBB in some inflammatory central nervous system diseases.[26]

The main tight junction proteins of the perineurial barrier, claudin-1, claudin-3, and occludin are "sealing" proteins and thus limit paracellular permeability. Claudin-4, claudin-5, and the functionally important channel-forming protein claudin-2 are not expressed in the perineurium of adult human peripheral nerves.[18] Claudin-1 is a major sealing tight junction protein responsible for epithelia barrier function, and claudin-1 knockout mice die during the first day of life due to loss of fluid through the skin. In these mice, claudin-1 is missing in the stratum granulosum leaving occludin-positive (and also claudin-4–positive) tight junction.[27] Typically, the perineurium is positive for epithelial membrane antigens like claudin-1 and is used as a marker.[16] However, the perineurium is thought to originate from fibrous tissue.[1] Claudin-1 and claudin-3, which are mostly located in the innermost sleeves of the perineurium, provide a barrier to the diffusion of various large molecular weight molecules and also restrict diffusion of low molecular weight molecules.[1] Occludin is implicated in the regulation of paracellular trespassing of small hydrophilic molecules and the migration of neutrophils.[18] Perineurial tight junctions can contain occludin and different claudin dimers, claudin-1/-1, claudin-1/-3, and claudin-3/-3. Claudin dimers are considered to be essential for the barrier function of tight junctions. The combination and mixing ratios of claudins within individual paired strands is considered to determine their tightness and selectivity.[28]

The human perineurium matures quite late during fetal life. The tight junctions follow similar spatial developmental expression patterns but in different time frames in achieving their respective adult distributions.[18] Tight junction proteins display different immunohistochemical patterns that differ in various stages of human development. For instance, claudin-1 and claudin-3 have been detected in the perineurium early and remain constant throughout adult life stages of human development.[16] Whereas claudin-1 is already largely restricted to perineurium-derived structures at early stages of fetal development, claudin-3 and occludin are weakly expressed and appear at a later stage, specifically between fetal weeks 22 and 35. ZO-1 appears to acquire its mature profile even later during the third trimester.[18] Interestingly, the tightness of the barrier of the perineurium matures accordingly. In the perineurium of adult rat peripheral nerves, ZO-1, occludin, claudin-1, and claudin-5 have been identified as well.[17,29]

Breakdown of perineurial permeability

Breakdown and recovery of the BNB after injury seems to be associated with changes in the expression of several membrane proteins, among them tight junctions such as claudins and occludins.[17] Early studies have already demonstrated that hypertonic saline increases permeability of the perineurium for (^{14}C) sucrose.[30] Weerasuriya *et al.* speculated that conformational changes in tight junctions induced by cell shrinkage could account for this longer lasting effect (50 min).[30]

The breakdown of the BNB has been studied in a mouse model of crush injury of the sciatic nerve.[17] Both, the permeability of the perineurium and the permeability of the endoneurial vessels, increase and return to normal after seven days. In the same timeframe, tight junction proteins such as claudin-1, claudin-5, and occludin decrease or

disappear. Claudin-1 immunoreactivity reappears after two days, similar to occludin and claudin-5.[17] Our own studies in rats have demonstrated that an intraplantar injection of hypertonic saline opens the perineurial barrier and decreases expression of claudin-1 within the first minutes after treatment, whereas expression of occludin and claudin-5 was unchanged.[15]

In humans, the breakdown of the BNB has been considered an initial key step in many autoimmune disorders of the peripheral nervous system, including Guillain-Barré syndrome, chronic inflammatory demyelinating polyradiculoneuropathy, and paraprotemic neuropathy.[26] These events, which occur during and after a nerve injury or trauma and include breakdown and recovery of the BNB, are associated with changes in expression of intercellular junction proteins, such as claudins, occludin, VE-cadherin, and connexin 43.[17] Tight junctions have also been recognized as dynamic bidirectional signaling complexes.[1] Signals are transmitted from the cell interior to tight junctions to regulate its assembly and function. Conversely, tight junctions receive and transmit information back to the cell interior to regulate gene expression and subsequent cellular responses, such as proliferation and differentiation.[31]

Matrix metalloproteinases on the integrity of the perineurial barrier

Previous studies suggest that metalloproteinases might degrade tight junction proteins, suggesting a role of these enzymes in maintaining the integrity of the barrier.[32–34] Matrix metalloproteinases (MMPs) are believed to be physiologically relevant mediators of the degradation of components of the extracellular matrix such as laminin and type IV collagen. The MMP family includes interstitial collagenases, gelatinases (type IV collagenases), and stromelysins. These enzymes are secreted as proenzymes that become activated by removal of their NH 2 terminal domain.[35] MMPs can open the BBB by degrading tight junction proteins, and an MMP inhibitor prevents degradation of the tight junction proteins by MMPs.[36] An inhibition of MMPs with a broad-spectrum inhibitor provides both acute and long-term neuroprotection in the developing brain by reducing tight junction protein degradation, preserving BBB integrity, and ameliorating brain edema after neonatal injury.[37] Therefore, it seems conceivable that these mechanisms play a role in the opening of the perineurium as well.

Injury to a peripheral nerve is followed by a re-modeling process consisting of axonal degeneration and regeneration. It is not known how the Schwann cell–derived basement membrane is preserved after injury or what role MMPs and their inhibitors play in axonal degeneration and regeneration. La Fleur *et al.* showed that the MMPs gelatinase B (MMP-9), stromelysin-1 (MMP-3), and the tissue inhibitor of MMPs (TIMP)-1 were induced in crush or injury of distal segments of mouse sciatic nerve.[35]

The mechanisms that regulate the opening of the perineurial barrier—as well as the opening of the barrier under pathophysiological conditions like injury or neuropathy—are currently unknown and need to be explored in future studies.[13] An intraplantar injection of hypertonic saline decreases expression of claudin-1 within the first minutes after treatment and facilitates the effect of peripheral opioids. This effect is dependent on metalloproteinases.[15]

Conclusion

When preservation of motor responses and non-painful sensation is desirable, the strategy of nociceptor-restricted blockade of pain could be advantageous for generating local anesthesia. Experiments with hypertonic solutions are a prerequisite for exogenous peripheral opioid action in noninflamed tissue both in behavior and electrophysiological experiments. Co-injection of hypertonic solution and opioid receptor agonists produce analgesia in noninflamed tissue as shown in behavioral experiments in the rat. Furthermore, it was shown that hypertonic saline decreases the expression of claudin-1 in the perineurial barrier dependent on metalloproteinases. Future studies should examine the mechanism of perineurial opening in more detail using hypertonic saline as a tool. Modulating the tight junction protein, claudin-1 could also be considered for treatment of nociceptor-driven chronic pain, such as mononeuropathies or postherpetic neuralgia, with intact barriers or for the transfer of genetic material during nerve degeneration.

Conflicts of interest

The authors declare no conflicts of interest.

References

1. Mizisin, A.P. & A. Weerasuriya. 2011. Homeostatic regulation of the endoneurial microenvironment during development, aging and in response to trauma, disease and toxic insult. *Acta Neuropathol.* **121:** 291–312.
2. Binshtok, A.M., B.P. Bean & C.J. Woolf. 2007. Inhibition of nociceptors by TRPV1-mediated entry of impermant sodium channel blockers. *Nature* **449:** 607–610.
3. Stein, C., M. Schäfer & H. Machelska. 2003. Attacking pain at its source: new perspectives on opioids. *Nat. Med.* **9:** 1003–1008.
4. Rittner, H.L. & A. Brack. 2007. Leukocytes as mediators of pain and analgesia. *Curr. Rheumatol. Rep.* **9:** 503–510.
5. Antonijevic, I., S.A. Mousa, M. Schäfer & C. Stein. 1995. Perineurial defect and peripheral opioid analgesia in inflammation. *J. Neurosci.* **15:** 165–172.
6. Labuz, D. *et al.* 2009. Immune cell-derived opioids protect against neuropathic pain in mice. *J. Clin. Invest.* **119:** 278–286.
7. Kieffer, B.L. & C.J. Evans. 2009. Opioid receptors: from binding sites to visible molecules in vivo. *Neuropharmacology* **56**(Suppl 1): 205–212.
8. Zöllner, C. & C. Stein. 2007. Opioids. In *Handbook of Experimental Pharmacology.* Vol. 177. C. Stein, Ed.: 31–63. Springer-Verlag GmbH. Heidelberg.
9. Stein, C. & C. Zöllner. 2009. Opioids and sensory nerves. In *Handbook of Experimental Pharmacology*. Vol. 194. B.J. Canning, D. Spina, Eds.: 495–518. Springer-Verlag GmbH. Heidelberg.
10. Simons, E.J., E. Bellas, M.W. Lawlor & D.S. Kohane. 2009. Effect of chemical permeation enhancers on nerve blockade. *Mol. Pharm.* **6:** 265–273.
11. Wenk, H.N., J.-D. Brederson & C.N. Honda. 2006. Morphine directly inhibits nociceptors in inflamed skin. *J. Neurophysiol.* **95:** 2083–2097.
12. Likar, R. *et al.* 2001. Efficacy of peripheral morphine analgesia in inflamed, noninflamed and perineural tissue of dental surgery patients. *J. Pain Symptom Manage.* **21:** 330–337.
13. Rittner, H.L. *et al.* 2009. Antinociception by neutrophil-derived opioid peptides in noninflamed tissue–role of hypertonicity and the perineurium. *Brain Behav. Immun.* **23:** 548–557.
14. Capra, N.F. & J.Y. Ro. 2004. Human and animal experimental models of acute and chronic muscle pain: intramuscular algesic injection. *Pain* **110:** 3–7.
15. Rittner, H.L., S. Amasheh, R. Moshourab, *et al.* 2012. Modulation of tight junction proteins in the perineurium to facilitate peripheral opioid analgesia. *Anesthesiology.* In press.
16. Piña-Oviedo, S. & C. Ortiz-Hidalgo. 2008. The normal and neoplastic perineurium: a review. *Adv. Anat. Pathol.* **15:** 147–164.
17. Hirakawa, H., S. Okajima, T. Nagaoka, *et al.* 2003. Loss and recovery of the blood–nerve barrier in the rat sciatic nerve after crush injury are associated with expression of intercellular junctional proteins. *Exp. Cell Res.* **284:** 194–208.
18. Pummi, K.P., A.M. Heape, R.A. Grénman, *et al.* 2004. Tight junction proteins ZO-1, occludin, and claudins in developing and adult human perineurium. *J. Histochem. Cytochem.* **52:** 1037–1046.

19. Shimizu, F. *et al.* 2011. Peripheral nerve pericytes modify the blood-nerve barrier function and tight junctional molecules through the secretion of various soluble factors. *J. Cell. Physiol.* **226:** 255–266.

20. Guillemot, L., S. Paschoud, P. Pulimeno, *et al.* 2008. The cytoplasmic plaque of tight junctions: a scaffolding and signalling center. *Biochimica et biophysica acta* **1778:** 601–613.

21. Shen, L., C.R. Weber, D.R. Raleigh, *et al.* 2011. Tight junction pore and leak pathways: a dynamic duo. *Ann. Rev. Physiol.* **73:** 283–309.

22. Nitta, T. *et al.* 2003. Size-selective loosening of the blood-brain barrier in claudin-5-deficient mice. *J. Cell Biol.* **161:** 653–660.

23. Vorbrodt, A.W. & D.H. Dobrogowska. 2004. Molecular anatomy of interendothelial junctions in human blood-brain barrier microvessels. *Folia histochemica et cytobiologica / Polish Academy of Sciences, Polish Histochemical and Cytochemical Society* **42:** 67–75.

24. Ohtsuki, S. *et al.* 2007. Exogenous expression of claudin-5 induces barrier properties in cultured rat brain capillary endothelial cells. *J. Cell. Physiol.* **210:** 81–86.

25. Dobrogowska, D.H. & A.W. Vorbrodt. 2004. Immunogold localization of tight junctional proteins in normal and osmotically-affected rat blood-brain barrier. *J. Mol. Histol.* **35:** 529–539.

26. Kashiwamura, Y. *et al.* 2011. Hydrocortisone enhances the function of the blood-nerve barrier through the up-regulation of claudin-5. *Neurochem. Res.* **36:** 849–855.

27. Furuse, M. *et al.* 2002. Claudin-based tight junctions are crucial for the mammalian epidermal barrier: a lesson from claudin-1-deficient mice. *J. Cell Biol.* **156:** 1099–1111.

28. Tsukita, S. & M. Furuse. 2002. Claudin-based barrier in simple and stratified cellular sheets. *Curr. Opin. Cell Biol.* **14:** 531–536.

29. Tserentsoodol, N., B.C. Shin, H. Koyama, *et al.* 1999. Immunolocalization of tight junction proteins, occludin and ZO-1, and glucose transporter GLUT1 in the cells of the blood-nerve barrier. *Arch. Histol. Cytol.* **62:** 459–469.

30. Weerasuriya, A., S.I. Rapoport & R.E. Taylor. 1979. Modification of permeability of frog perineurium to [14C]-sucrose by stretch and hypertonicity. *Brain Res.* **173:** 503–512.

31. Terry, S., M. Nie, K. Matter & M.S. Balda. 2010. Rho signaling and tight junction functions. *Physiology (Bethesda)* **25:** 16–26.

32. Bojarski, C. *et al.* 2004. The specific fates of tight junction proteins in apoptotic epithelial cells. *J. Cell Sci.* **117:** 2097–2107.

33. Gorodeski, G.I. 2007. Estrogen decrease in tight junctional resistance involves matrix-metalloproteinase-7-mediated remodeling of occludin. *Endocrinology* **148:** 218–231.

34. Schubert-unkmeir, A. *et al.* 2010. Neisseria meningitidis induces brain microvascular endothelial cell detachment from the matrix and cleavage of occludin: a role for MMP-8. *PLoS Pathog.* **6:** e1000874.

35. La Fleur, M., J.L. Underwood, D.A. Rappolee & Z. Werb. 1996. Basement membrane and repair of injury to peripheral nerve: defining a potential role for macrophages, matrix metalloproteinases, and tissue inhibitor of metalloproteinases-1. *J. Exp. Med.* **184:** 2311–2326.

36. Chen, W. *et al.* 2009. Matrix metalloproteinases inhibition provides neuroprotection against hypoxia-ischemia in the developing brain. *J. Neurochem.* **111:** 726–736.

37. Yang, Y., E.Y. Estrada, J.F. Thompson, *et al.* 2007. Matrix metalloproteinase-mediated disruption of tight junction proteins in cerebral vessels is reversed by synthetic matrix metalloproteinase inhibitor in focal ischemia in rat. *J. Cereb. Blood Flow Metab.* **27:** 697–709.